GEOGRAPHICAL THOUGHT

ABOUT THE AUTHOR

Dr. Lalita Rana (*b.* 1960) is a Gold Medallist and the recipient of UGC Fellowship. She is Associate Professor in the Department of Geography, Shivaji College, University of Delhi. Committed to the teaching and research, she has a special interest in the study of history and philosophy of geography. Her previous book 'Geographical Thought: A Systematic Record of Evolution' (published in 2008) has been widely accepted and well read by the students of geography both at the under and post-graduate levels.

GEOGRAPHICAL THOUGHT
Classical to Contemporary

[REVISED AND ENLARGED EDITION]

Lalita Rana

CONCEPT PUBLISHING COMPANY PVT. LTD.
NEW DELHI-110059

ISBN-13: 978-93-5125-093-7 (PB)

First Published 2008
Revised and Enlarged Edition 2014

Published and Printed by

Concept Since 1974

Concept Publishing Company Pvt. Ltd.
A/15-16, Commercial Block, Mohan Garden
New Delhi-110059 (India)
T: 25351460, 25351794, *F*: 091-11-25357109
E: publishing@conceptpub.com, W: www.conceptpub.com

Editorial Office: H-13, Bali Nagar, New Delhi-110 015, India.

Cataloging in Publication Data--*Courtesy:* D.K. Agencies (P) Ltd. <docinfo@dkagencies.com>

Rana, Lalita, 1960-
Geographical thought : classical to contemporary / Lalita Rana. -- Rev. and enl. ed.
p. cm.
Includes bibliographical references (p.) and index.
ISBN 9789351250920 (HB)
ISBN 9789351250937 (PB)

1. Geography--Philosophy--Textbooks. 2. Geographical perception--History--Textbooks. I. Title.

DDC 910.01 23

Preface

Geography is the study of location and arrangement of phenomena on the earth's surface and the processes that generate these distributions. Geographical thought is the theory and philosophy shaping the discipline. Geographical thought at any point in time is a manifestation of the prevailing viewpoints and methodological approaches in vogue. The development of geography has a history of evolution of over 2000 years. The discipline has been crafted over time to incorporate ideas from various schools of thought and several other disciplines from ancient to modern times. The existing methodology and the subject matter of Geography is a reflection of this assemblage and evolution. The nature of development remained more philosophical up to modern period. The First edition of this book, published in 2008, was an attempt towards compilation of relevant study material on evolution of Geographic thought in a systematic way up to the modern period. The book has been revised with the addition of an elaborated section on the contemporary tradition in geography. Major thrust of the present book is on nature of geographic developments in its contemporary period. More recently, the anthropocentricism and a concern about methodology are more evident in the development of discipline.

The book is divided into five sections. The First section is introductory focusing on the nature and evolution of geography. The Second section deals with (philosophical and methodological) contribution of the Classical Schools of Thought, namely, Greek, Roman, Indian, Chinese, Arab, German, French, American, British and Soviet, that have evolved since ancient to modern times and provided geography with its standard literature. The Third section highlights the paradigmatic shifts in the development and also identifies major paradigms of geographic development. The emphasis of Fourth section is exclusively on the contemporary tradition in geography, i.e. since mid-20th century, focusing on variegated trends, like quantitative revolution; laws, models and theory building; systems analysis; behavioural geography; and different humanistic approaches. The future of geography as a spatial science discipline is outlined in the Fifth section. Manuscript of the book is self-typed.

Lalita Rana

Contents

SECTION-V
FUTURE OF GEOGRAPHY

List of Tables

List of Figures

SECTION-I
INTRODUCTION

CHAPTER-1

Nature and Development of Geography

Like a thread through beads, Geography runs through every that aspect of knowledge which is concerned with the study of earth's surface in part or whole. Geography as a field of learning is related to the study of location and spatial relations of things and events on the surface of earth. Geography is fundamentally a holistic discipline, studying things and events in their total context. This implies study of systems of a really inter-connected and inter-dependent parts of diverse origin. The geographers are concerned with 'territorial space'. Geographical queries begin with asking a simple question, i.e. "where". Asking questions about location is one of the distinguishing characteristics of the field of geography. An answer to this leads to some specified location over terrestrial space. The other questions as 'how' and 'why', etc. follow later. Geography as a discipline is concerned with five sets of questions[1]: *Generic questions*: relating to the kind (class) of phenomena distributed on earth's surface; *Genetic questions*: concerned with the sequences of events and interaction that have gone making of present day landscapes; *Theoretical questions*: relating to formulation of rules, laws and generalizations; *Remedial questions*: concerned with the application of geographic concepts to real life problems; and *Methodological questions*: concerned with the improvement of our scientific skills. Taken together, all these questions take care of distributional and locational aspects of natural and cultural features. The first two questions provide inventoried information of *what* features and *where* located. This remained a popular approach during the ancient and medieval period. However, these two questions did not make geography a scientific discipline till the third question was added. The third question is related to explanation or causal relationships between features and the processes and phenomena. This aspect of geography is related to the question, '*why*'. The fourth question makes geography more and more anthropocentric; whereas attempts to answer fifth question have led to change and development in the methodology of discipline. Today geography is recognized as a spatial science, concerned with the study of the location and arrangement of phenomena (physical, biotic and human) on the surface of earth and the processes that generate these distributions. With the advancement in geography, its nature (meaning, content, purpose, method and scope) has also changed with increased dependence on other disciplines.

Eratosthenes, a 3rd century BC Greek scholar, coined the term 'geography' (from '*geo*' meaning 'the earth' and '*graphe*' meaning 'the description'). The traditional definition of our discipline has been derived from the literal meaning of this term, i.e. '*Geography is the description of Earth*'. Throughout the Classical Phase of its development, geography was considered as a descriptive field of study and the enquiries of geographical nature had begun with the basic question, i.e. "what is where". The emphasis of study of Classical schools of geographic thought was on this very question. The nature of query shifted to "why what is where" after the Age of Discovery. However, a further refinement was seen in the Modern period, when geographers began to seek answer to the questions as "how and why what is where". In an attempt to answer each set of these questions, geography has gained a new and more meaningful definition every time. In its present form Geography may be defined as the study of spatial location and distribution of various phenomena on the surface of earth[2]. Geography is the science of spatial distribution. The 'spatial' here means pertaining to physical space, and not the astronomical, microscopic or abstract space. It is topographical space, i.e. over the surface of earth. The 'distribution' of any element (physical, biotic or human) is spatial in nature, if it occupies a portion of earth's surface, or has a location over it, and which is also a phenomenon of specified magnitude, so that it could be physically identified, measured and represented on a map. Related with the term 'spatial distribution' are the concepts of pattern, process and stage. The various phenomena or activities on the surface of earth are not just localized in any place, anywhere or everywhere in the same manner. They form certain kinds of patterns and there are several processes involved behind the existing distributions, shaping them in various forms, shapes and patterns. These processes help to understand the causal connections among elements of spatial distribution. An understanding of the process helps to answer the question 'how and why what is where'. The interplay of 'time' and 'space' bring change in spatial distribution. The change, whether perceptible or imperceptible, is inherent and is bound to come with time. Thus, every spatial distribution has to be studied in terms of its chronology and chronological perspective.

Geography is not an encyclopaedic description of places. To know by heart a whole gazetteer full of them would not make anyone a geographer. Geography has higher aims than this, i.e. to trace out the great laws of nature and to mark their influence upon man. Geography deals with an enormous range of phenomena, ranging from physical, biotic to human. The four spheres, namely lithosphere, atmosphere, hydrosphere and biosphere constitute the geographic environment. This makes geography highly interdisciplinary, drawing upon and contributing to numerous other fields of knowledge. Geography is an observational science. Both observational and experimental sciences demonstrate how regular occurrences are produced by the predictable interactions in a number of variables. It is possible in experimental sciences, under laboratory conditions, to control and introduce the variables separately in order to observe their effect upon the situation. But in an observational science occurrences must be observed as they naturally occur. Therefore, exploration or fieldwork is a basic research technique in geography. Fundamentally speaking, Geography deals with location, place and regions. Location is studied in terms of its absolute and relative sense. The absolute location is based on coordinate grid system of latitudes and longitudes; the relative location is in relation to other locations. Place is a location having distinctive features which give it meaning and character that is unlike any other location. Then, geography makes sense of the world by synthesizing vast amounts of information into spatial categories of similar traits called 'regions'. The non-overlapping areas of essential uniformity are formal regions, whereas those defined by interaction and connection of a particular activity (function) are functional. Geographers vary considerably in their interests. Some are concerned more with the natural phenomena and some with the human artefacts, while some tend to focus their attention on the man-land interactions and stress the idea of geography as ecology. But implicit in all these branches of the interest is a concern with location and spatial patterns, and a curiosity about the terrestrial space in which the patterns come to be as they are. The physical space is the binding factor. There is no more essentially geographical activity than explaining a spatial distribution. As spatial interrelationships are of vital importance to this synoptic science, the maps are a key tool. The classical cartography has now been joined by great advances in computer technology. With this the analytical and spatial data management tools available to geographers, including Geographic Information Systems and spatial data analysis, are now allowing geographers for more rigorous, quantitative analysis of spatial phenomenon. A geographer's task calls for gathering, representation and analysis (explanation) of the information related to location and distribution of various phenomena on the surface of earth. A geographer is a person who asks question about the significance of location, distance, direction, spread and spatial succession. He deals with problems of accessibility, innovation, diffusion, density and other derivations of relative location[3]. Modern geography is a spatial science, increasingly focusing on the study of spatial aspects of social and economic phenomena, and thus being pre-eminently anthropocentric. The space of geography is that of the earth's surface, the home of man. Thus, the nature of geography is integrative, interdisciplinary, empirical and also anthropocentric.

Geography is a discipline of synthesis. It attempts spatial synthesis. Geography as an integrating discipline has interface with numerous natural and social sciences. Every discipline, concerned with scientific knowledge is linked with geography as many of their elements vary over space. Geography not only takes note of the differences in the phenomena from place to place but also integrates them holistically. A geographer is, therefore, required to have a broad understanding of all the related fields, to be able to logically integrate them. All the disciplines study the earth in one way or the other, but none of them completely. Therefore, none of the disciplines can claim to be self-dependent or independent of others. The claim for geography as a scientific discipline has led to its increased dependence on other disciplines. Today, it cannot survive on its own; in fact, no discipline can do that. Disciplinary interdependence is universal, and is true for every discipline. Moreover, the invention of electronic computer and the revolutionary advances in data collection by satellites orbiting around the earth have made geography more and more inter-disciplinary in its nature. Henceforth, the question regarding what geography does, and the attempt to answer it by drawing boundaries around it with a view to distinguish it from other sciences becomes irrelevant. Thus, various reasons responsible for having geography to adopt an inter-disciplinary approach are three-fold: (i) the claim of geography to be scientific in nature: the methodology of any discipline that claims to be scientific is common for all scientific disciplines, (ii) the accommodative nature of discipline, i.e. any aspect dealing with the distribution over the surface of the earth qualifies to become geographical in study and in fact, the suffix of 'geography' may be added to any branch of learning; and (iii) heavy dependence of geography on the quantitative techniques, statistical methods and now computers, remote sensing and GIS. As far as the interdisciplinary nature of geography is concerned, this has been a characteristic feature since very beginning. The scholars from outside the field (non-geographers) have made valuable contributions to our subject matter. Notable have been astronomers, astrologers, mathematicians, geologists, theocrats, philosophers and travellers. This interdependence has increased with time.

Particularly since modern period, we see geographical studies being based on a number of allied disciplines. Various disciplines geography relies on, may be categorized as: earth sciences, pure sciences, life (biological) sciences, social sciences, behavioural sciences and the sciences dealing with data collection, representation, computation and analysis. The elements covered under the study of most of these disciplines, as history, demography, sociology, economics, botany, zoology, geology, political science, etc. have spatial expressions, and thus they constitute relevant subject matter of geographical study. On the other hand, to some other disciplines, like mathematics, statistics, cartography, computer science, etc. geography looks for assistance in analytical and explanatory procedures of study. The disciplines like geomorphology, geology, meteorology, pedology, hydrology, oceanography, etc. fall under earth sciences. The category of pure sciences includes the basic sciences like physics, chemistry and astronomy. The life sciences encompass a broad spectrum of academic fields as botany, zoology, ecology, anthropology or astrobiology which together address phenomena related to living organisms. The subjects like history, demography, economics, political science, sociology, philosophy, etc. which deal with the study of human society constitute the part of social sciences. The behavioural sciences lie at the crossroads of natural and the social sciences, linking broad areas of scientific exploration. This group of sciences investigates the decision processes and communication strategies within and between organisms in a social system and includes the disciplines like anthropology, psychology, archaeology, etc. Besides, there are a number of other sciences that aid in data collection, representation and analysis in geography and they include as computer science, cartography, surveying, mathematics and statistics, remote sensing and GIS. Thus, geography has strong interface with all the natural and social sciences. It has osmotic relationship with these disciplines. While all the disciplines have their own individuals scope, this individuality does not obstruct the flow of information as in the case of cells in the body that have individual identity separated by membranes but the flow of blood is not obstructed. Geographers use data obtained from sister disciplines and attempt synthesis over space. This disciplinary interdependence of geography also reflects in its various specialized fields of study which constitute its subject matter.

The subject matter of geography is a clear manifestation of its interdisciplinary nature and reflects both the aspects of study, viz. sciences and humanities. Therefore, it covers all the elements of study: physical, human and biotic. Besides, the representation of all these elements with the help of maps and diagrams also falls within the ambit of geography. Physical geography studies the spatial characteristics of various natural phenomena that exist in earth's lithosphere, hydrosphere and atmosphere; its various branches include geomorphology, climatology and hydrology and oceanography. Human geography focuses on the study of patterns and processes that shape human interaction with various environments; it encompasses a vast field of study covering, for instance, historical, demographic, political, cultural, social, economic and developmental aspects. While the major focus of human geography is not the physical landscape of the Earth, it is hardly possible to discuss human geography without referring to the physical landscape on which human activities are being played out, and environmental geography is emerging as a link between the two. The interface between physical geography and human geography has led to the development of biogeography and it is the science which deals with patterns of distribution of species (plants, animals and also humans in their natural setting) and the processes that result in such patterns. Soil geography and environmental geography also fall in this category. Still another branch of knowledge is concerned with the practical work in geography, i.e. related to data collection, representation and analysis. The practical work in geography constitutes both the laboratory and field-work.[4] The data collected in the field, or from the statistical reports, are translated into maps and diagrams in the laboratory. This practical work is based on cartography, surveying, various quantitative techniques, and more recently on remote sensing, computers and GIS. The subject matter of geography as outlined here is enormous and extremely varied in nature and it is under constant change.

EVOLUTION OF GEOGRAPHICAL THOUGHT

Geography was born to satisfy the curiosity of man to acquire knowledge about the lands lying within and beyond his familiar world. Various forms of primitive queries have served the foundation of geography. Some of the earliest queries formulated by man were related to his natural surroundings. These queries were in the form of the questions, like "where to live"; "what to wear" or "what essential means of sustenance are and how can man obtain them". The answers to all these queries have led to accumulation of vast knowledge about the surface of earth. The records of man's interest suggest that his preliminary observations and reactions were of geographical nature. Having a recognizable existence of over 2000 years, Geography is as old as the man himself. However, its development has never been smooth; it is a story of upheavals and downfalls. With time, the definition, nature and methodology of geography have changed. Geography has witnessed the phases of growth and also those of decline alternatively. The chronology of geographic development is presented in Table 1.1. As such no other discipline can claim a longer genealogy than geography. Although, roots of the discipline lie much deeper in antiquity, the scientific geography as we know today is outcome of the developments in past two millennia, beginning from the ancient period and passing to successive phases of development. The nature of geography as we know today is the contribution of various Classical Schools of Thought, namely, European (Greek and Roman), Indian and Chinese in ancient period, Arab in medieval period, and German, French, American, British and Soviet in modern period. These schools have enriched geography from various concepts, terminologies and techniques refining its methodology from time to time. Geography has evolved as an integrating field of study assimilating knowledge from these schools of thought.

TABLE 1.1 Chronology of Geographical Development

Phase	*Time-Period*	
Pre-Historic		**Before 800 B.C. (Appr.)**
Ancient		**800 B.C.–500 A.D.**
	Western (Greek and Roman)	
	Indian	
	Chinese	
Medieval		**500 A.D.–1750 A.D.**
	Dark Age in Europe	
	Rise of Arab School	
	Age of Discovery	
Modern		**1750–World War II**
	(German, French, American, British and Soviet Schools)	
Contemporary		**Post-World War II**
	Quantitative Revolution	
	Behavioural Geography	
	Humanistic Geography	
	Methodological Revolution	

Pre-Historic

Scholarly observations of the Eastern and Western civilizations had begun in

Pre-historic times, approximately about 800 BC and earlier, and the field of geography highly gained by these in its later stage of development. Some of the first truly geographical studies occurred more than four thousand years ago in the region often referred to as "the cradle of civilization", i.e. Mesopotamia. Parallel developments also occurred in the realms of Chinese, Indus and Nile Valley civilizations. These civilizations were first to explore the space within and outside their homelands. The earliest evidence of such explorations comes from archaeological discovery of a clay tablet map. This world map, the first one, is attributed to the Babylonians in 6th century B.C. On this tablet the Earth is shown as a flat circular disk surrounded by ocean and several mythical islands.[5]

Ancient

During ancient period, there was greater and greater concentration of scholarly research in the West and Orient, particularly Europe and Asia. This was the period of rise of first Classical Schools of thought, viz. Greco-Roman, Indian and Chinese. The term 'classical antiquity' is often used to refer to ancient history since the beginning of recorded Greek history in about 776 BC (First Olympiad). This coincides roughly with the traditional date of founding of Rome and the beginning of history of ancient Rome. The span of recorded history is roughly 5,000 years, with Sumerian cuneiform being the oldest form of writing discovered so far. For the most part ancient Greek geography was the academic field. It was the Romans, the immediate successors of Greeks in the West, who made far the more extensive practical use of geography and maps. The Romans, in this manner, carried forward the Greek mathematical and literary traditions. The Indian and Chinese (Oriental World) were the other major regions of geographical learning in ancient period. The geographical studies in India began with the dawn of Indian civilization and these works cover the study of the Universe, the Earth and the Indian Subcontinent. Besides, the ancient Chinese writings of geographical nature cover the fields of human, regional and physical studies, cartographic works and the accounts of travels and explorations.

Medieval

The Medieval period in geography marked its beginning with the onset of Dark Age in Europe, followed by the rise of Arab School and Age of Discovery. In this manner, the medieval phase extends from the fall of the Roman Empire to the beginning of Modern period. It is defined partly by the process of rediscovering ancient culture developed in Greece and Rome in the classical period, and partly by the need to address theological problems. Whereas, Christianity began as a religion of small groups, Islam developed as the religion of an expanding empire. Within a hundred years of Mohammed's death in 632 AD, military conquest extended the Islamic world to India, North Africa and Southern Spain. As a result, a variety of different communities came under Muslim rule, and Islam came into contact with the theological systems of Judaism, Christianity, and Zoroastrianism, and the philosophy of India and Greece.

This led Islamic theologians to use philosophical ideas and principles to interpret Koranic doctrines. The first stage of this process was the translation of Greek philosophical and scientific works into Arabic that had been preserved by Eastern Christians in Mesopotamia, Syria and Egypt.

As far as the West is concerned, little academic progress occurred in geography after the Roman period and middle age was largely a time of intellectual stagnation. This Early Medieval phase (500-1100 AD) is termed as the Dark Age in Europe. This proved to be a long period of decline during which the geographic horizons closed in. The wide-ranging Greek geographical concepts were largely forgotten except among Arab scholars. Observations piled up in Christian monasteries, but the intellectual climate was not favourable for formulation of new interpretation. During this period of regression the simple 'T-in-O' map became the standard depiction of world in Europe. When the medieval Europe had plunged into darkness, the Arab school rose to mark the end of recession in development of geographic literature and return to more scholarly research. This new enlightenment spreaded from Islamic world through the work of various Arab scholars. They not only maintained the Greek and Roman techniques but also developed new ones. The Islamic empire stretched from Morocco to India, and Arab traders travelled throughout Eurasia, Africa, and the Indian Ocean. The Arabs added a great deal of knowledge to expand and correct the classical sources. During this High Medieval phase (800-1400 AD), Arab Civilization was the most advanced and formed the largest empire that world had so far seen. The establishment of vast empire facilitated coming together of intellectual and scientific traditions of different civilizations. The Arabs made all knowledge their own and developed it further. The Arabs played a great role in the promotion of science. Arab civilization, in fact, became main link between the ancient and modern worlds. An early form of globalization began emerging during the Islamic Golden Age, when the knowledge, trade and economies from many previously isolated regions and civilizations began integrating due to contacts with Muslim explorers, scholars and traders. Arabic currency was in circulation throughout the Afro-Eurasian landmass, as far as sub-Saharan Africa in the south and northern Europe in the north, often in exchange for goods and slaves.[6] These factors helped establish the Arab Empire as the world's leading extensive economic power throughout high medieval phase. The Muslim lands also witnessed a fundamental transformation in agriculture from 8th century to the 13th century, a period known as the Islamic Golden Age. 'Arab Agricultural Revolution' or 'Islamic Green Revolution' are terms coined for this change.[7] It is argued that the economy established by Arab and other Muslim traders across the Old World enabled not only the diffusion of many crops and farming techniques but also the adaptation from and to regions beyond the Islamic world. These introductions, along with an increased mechanization of agriculture, led to major changes in economy, population distribution and urban growth in the Islamic world. During the Muslim Agricultural Revolution, the Muslim scientists demonstrated advanced knowledge in areas such as meteorology, climatology, hydrology, soil occupation, and the economy and management of agricultural enterprises.

The Age of Exploration began in Europe in the late medieval phase or the early modern period (1400-1750) and the geographic horizons were again pushed back. Enormous information of geographical nature was gathered. The arrival of all these new observations in Europe started the sequence of events that continues till present. The Age of Discovery was also a renaissance phase in geography. During this period numerous journeys of geographical exploration were commissioned by a variety of nations in Europe. Most of these voyages were financed because of the potential commercial returns from the resource exploitation. But, the voyages also provided an opportunity for scientific investigation and discovery. These voyages added many significant contributions to geographic knowledge. For instance, the trips of venetian explorer Marco Polo and the voyages of Europeans opened up new horizons and stimulated geographic writings.

Modern

Following the journeys of Marco Polo, the interest in geography spreaded throughout Europe. The reintroduction of Ptolemey's writings, which had been lost in Europe but preserved in Islamic world, provided a systematic framework to tie together and portray geographical information. The great voyages of exploration in 16th and 17th centuries revived a desire for both accurate geographic detail, and more solid theoretical foundations. By 18th century, geography had become recognized as a discrete discipline and a part of university curriculum in Europe. With the rise of German School, geography entered the Modern period of its development. A number of other modern schools of thought emerged thereafter, namely French, American, British and Soviet. Before the beginning of Modern Period, geography had already got bifurcated into two basic branches, viz. physical and human. After that it went on splitting, as continued research became an essential feature of the discipline. This bifurcation was essential for healthy and further growth of geography as a scientific discipline. Maximum impetus in the beginning of modern period came from Bernard Varenius (1622-1650), the eminent scholar of pre-Classical phase of German School. Varenius gave the first modern geographical literature in the form of *Geographia Generalis*. *Geographia Generalis* and Gerardus Mercator's world map became prime examples of the new breed of scientific geography. Through his work, Varenius tried to provide a scientific base to geography, initially splitting it into two, viz. 'systematic' and 'regional' and not traditional 'physical' and 'human'. However, the real beginning of Modern period is marked with the writings of Alexander-von-Humboldt (1769-1859) and Carl Ritter (1779-1859). From the early 19th to mid-20th century, geography was developing in several directions under the influence of distinct schools of thought. In the early part of 19th century, the effect of Humboldt and Ritter was very strong on geography, and there were followers of each. Ritter's followers emphasized on historical and human aspect in regional studies, and those of Humboldt on natural aspects in systematic manner. Thus, the geographical studies bifurcated in two directions: study of earth as a natural body and study of earth as a dwelling place of man. This gave birth to two types of controversies, viz. Systematic *versus* Regional and Physical *versus*

Human. However, in the late 19th century, the more expanding frontiers were those of physical and biological sciences than human. Now, the main impact was that of Charles Darwin (1809-1882), the English natural scientist. The year '1859' has a special significance in the history of geographical thought. The year marks not only the death of both Humboldt and Ritter but also the publication of *"On the Origin of Species"*, the most popular work of Darwin. The philosophy of Darwin had a tremendous influence on geographic thought. This influence came through Darwin's theory of "survival of the fittest". Darwin's ideas on adaptation to the environment and evolution stimulated social scientists to reappraise the old concepts. At the same time a number of dualisms also overshadowed the discipline for about 100 years after the emergence of German School. During the entire 19th century the emphasis of geographical studies kept on shifting between physical and human, with none of the field of study being prominent. However, towards the close of the 19th century, the systematic studies on physical geography superseded everything else. It was the time of the stronghold of the followers of Humboldt and Darwin. Maximum contribution came from the American scholars like Davis, Gilbert, Powell, Agassiz, Peschel, Frobel, Koeppen, and so on. They carried out morphological research and produced studies based on landforms and climate. Further, in the process of the study of influence of physical environment on the non-physical world (including human beings) there emerged the doctrine of 'environmental determinism' in later half of 19th century. Fredrick Ratzel, a German Scholar, formally put forward this viewpoint through his famous work entitled *Anthropogeographie*. As an antithesis to 'determinism', the French School of Thought, led by Vidal de la Blache, gave the doctrine of 'possibilism' in the beginning of 20th century. However, a compromise could be reached only by mid-20th century, when 'ecological viewpoint' entered geographical thought balancing the philosophies of both 'determinism' and 'possibilism'. Long phase of dualistic tendencies that characterized geography in the beginning of Modern period continued up to Second World War. During this period, i.e. from late 19th century until Second World War, geographers all-around the world were seeking to establish the status of geography as an independent discipline, distinct in concepts and procedures, from other disciplines.

Thus, the term 'modern' was coined to describe the beginning of a new era. Modern history can be further broken down into the Early Modern Period (1650-1800) and the Late Modern Period (1800-1950) after the French Revolution and the Industrial Revolution. Early Modern European history is usually seen to span from the turn of the 15th century, through the Age of Reason and Age of Enlightenment[8] in the 17th and 18th centuries. The modern period has been a period of significant development in the fields of science, politics and technology. It has also been an age of globalization. During this time, the European powers and their colonies began political, economic and cultural colonization of the rest of the world. By the 20th century, the modernist culture has come to dominate not only Western Europe and North America, but almost every civilized area on the globe. The modern era is closely associated with the development of capitalism, urbanization and a belief in the possibilities of technological and political progress. The concept of the modern world as distinct from an ancient or medieval world rests on a sense that the modern world is not just another era in history, but rather the result of a new type of change. This is usually conceived of as progress driven by deliberate human efforts to better their situation. Advances in all areas of human activity—politics, industry, society, economics, commerce, transport, communication, mechanization, automation, science, medicine, technology, and culture—appear to have transformed an Old World into the Modern or New World.[9] Portions of the Modern world altered its relationship with the Biblical value system, revalued the monarchical government system, and abolished the feudal economic system, with new democratic and liberal ideas in the areas of politics, science and economics.[10] This combination of epoch events totally changed thinking and thought in the early modern period. Particular facets of early modernity include: the Renaissance; the Age of Discovery and Rise of Capitalism; the Scientific Revolution and the Eastern Colonisation. The most important event in the development of early modernity period includes the arrival of printing press. In the Pre-Modern era, on the other hand, the religious officials were the spiritual intermediaries to the common person. It was only through these intermediaries that the general masses had access to the divine. Tradition was sacred to ancient cultures and the social order in a culture was enforced through religion.[11] In contrast to the pre-Modern era, Western civilization made a gradual transition from premodernity to modernity when scientific methods were developed which led to believe that the use of science would lead to all knowledge, thus throwing back the shroud of myth under which pre-Modern peoples lived. New information about the world was discovered *via* empirical observation *versus* the use of reason and knowledge.[12]

The early modern period is marked by the first European colonies, the rise of strong centralized governments and the beginnings of recognizable nation states that are the direct antecedents of today's states. In Africa and the Ottoman Empire, for instance, the Muslim expansion took place in North and East Africa. In West Africa, various native nations existed. The Indian Empires and civilizations of Southeast Asia were a vital link in the spice trade. On the Indian subcontinent, the Great Mughal Empire existed. The archipelagic empires, the Sultanate of Malacca and later the Sultanate of Johor, controlled the southern areas. Concerning the Asia, various Chinese dynasties and Japanese shogunates controlled the Asian sphere. In the Americas, Native Americans had built a large and varied civilization, including the Aztec Empire and alliance, the Inca civilization, the Mayan Empire and cities, and the Chibcha Confederation. In the West, the European kingdoms were in a movement of reformation and expansion. Russia reached the Pacific coast and consolidated its control over the Russian Far East in the 19th century. Later religious trends of the period saw the end of the expansion of Muslims and the Muslim world. Christians and Christendom saw the end of the Crusades and end of religious unity under the Roman Catholic Church. During the early modern period, an age of discovery and trade was undertaken by the European nations. European powers went on a colonial expansion and took possession throughout the world. There was a conquest of new lands and exploitation of their resources. Various European powers set up colonies in North America and Latin America. Towards the end of the early period, Europe was dominated by the evolving system of mercantile capitalism in its trade and the New Economy. The French power and English revolutions dominated the political scene. The end date of the early modern period is usually associated with the Industrial Revolution, which began in Britain in about 1750. In the Orient, the Chinese merchants explored all of the Indian Ocean, reaching East Africa with the voyages of Zheng He. Foreign trade and other contacts with the outside world, particularly Japan, increased considerably. On the Indian subcontinent, the Mughal Empire ruled most of India in the early 18th century. During this period, the Empire was marked by a highly centralized administration connecting the different regions. As part of the Eastern colonisation, the development of New Imperialism saw the conquest of nearly all eastern hemisphere territories by colonial powers. The commercial colonization of India commenced with the colonial rule. While the colonial powers profited much from the region's vast resources and large market, colonial rule did develop the region to a varying extent.[13] Traditionally, the European intellectual transformation of and after the Renaissance bridged the Middle Ages and the Modern era. The Age of Reason (or Enlightenment) in the Western world is generally regarded as being the start of modern philosophy, and a departure from the medieval approach, especially scholasticism.[14] The 18th century saw the beginning of secularization[15] in Europe. The Age of Enlightenment is a term used to describe a time in Western philosophy and cultural life centred upon the 18th century, in which reason was advocated as the primary source and legitimacy for authority. Developing during the Enlightenment era, Renaissance humanism was an intellectual movement spread across Europe. The basic training of the humanist was to speak well and write.[16] Renaissance humanism took a close study of the Latin and Greek classical texts, and was antagonistic to the values of scholasticism with its emphasis on the accumulated commentaries; and humanists were involved in the sciences, philosophies, arts and poetry of classical antiquity. During Scientific Revolution the European ideas saw a fundamental transformation in scientific ideas across physics, astronomy, and biology, in institutions supporting scientific investigation, and in the more widely held picture of the universe. Individuals started to question all manners of things and it was this questioning that led to the Scientific Revolution, which in turn formed the foundations of contemporary sciences and the establishment of several modern scientific fields. Particular facets of the Late Modern period include: increasing role of science and technology; mass literacy and proliferation of mass media; spread of social movements; institution of representative democracy; individualism; industrialization and urbanization. Other important events in the development of the Late Modern period include: the Russian Revolution and the First and Second World Wars. As a result of the Industrial Revolution and the earlier political revolutions, the Modernist worldview's emerged. The great changes of centuries before the 19th century were more connected with ideas, religion or military conquest, and technological advance had only made small changes in the material wealth of ordinary people. During Industrial Revolution the technological and economic progress gained momentum with the development of steam-powered ships and railways, and later in the 19th century with the internal combustion engine and electric power generation. The Industrial Revolutions brought about major technological, socio-economic and cultural changes in late 18th century and early 19th century that began in Britain and spread throughout the world. The effects spread throughout Western Europe and North America, eventually affecting the majority of world. Following the Enlightenment's ideas, the reformers looked to the Scientific Revolution and Industrial progress to solve the social problems. The world witnessed the European Hegemony in various spheres during 19th century. Following the Napoleonic Wars, the British Empire became the world's leading power, controlling one quarter of the World's population and one-third of the land area. Unchallenged at sea, Britain adopted the role of global policeman, a state of affairs later known as the Pax Britannica,[17] and a foreign policy of "splendid isolation". British imperial strength was underpinned by the steamship and the telegraph, new technologies invented in the second half of the 19th century, allowing it to control and defend the Empire. All the major European powers also laid claim to the areas of Africa where they could exhibit a sphere of influence over the area.

Around the turn of the 20th century, Enlightenment philosophy was challenged in various quarters. Darwinism gained acceptance and exposed adaptation in the theory of natural selection. The fields of geology, astronomy and psychology also made strides and gained new insights. Another philosophical trend was that the Chinese philosophy began to integrate concepts of Western philosophy, as steps towards modernization. At the end of the 19th century, Social Darwinism was promoted and included the various ideologies based on a concept that competition among all individuals, groups, nations or ideas was the framework of social evolution in human societies. In this view, society's advancement was dependent on the "survival of the fittest".

Contemporary

All the existing controversies that overshadowed geography during modern period got resolved by mid-20th century, particularly after the publication and popularization of the views of F.K. Schaefer, an American Scholar, through his article entitled "exceptionalism in geography". With this geography is said to have entered its Contemporary phase of development. The important developments of this phase include: the quantitative revolution; the use of behavioural, humanistic and Marxist approaches; and the application of models, theories and systems analysis in geographical studies. From the mid-20th century onwards, the geographic research experienced a new shift in methodology. The geographers began adopting a more scientific approach that relied on quantitative techniques. Soon, the quantitative approach became even more prevalent due to advances in computer and software technologies. Quantitative revolutionaries, often referred to as "space cadets", declared that the purpose of geography was to test general laws about the spatial arrangement of phenomena. They adopted the philosophy of positivism from the natural sciences and turned to mathematics—especially statistics—as a way of proving hypotheses. The quantitative revolution laid the groundwork for the development of geographic information systems. Well-known geographers from this period are Fred K. Schaefer, Waldo Tobler, William Garrison, Peter Hagget, Richard J. Chorley, William Bunge and Torsten Hägerstrand. Though positivist and post-positivist approaches remain important in geography. Critical Geography arose as a critique of positivism. The three strains of critical geography to emerge were the radical geography, the behavioural geography and the humanistic geography. Drawing on the philosophies of existentialism and phenomenology, humanist geographers (such as Yi-Fu Tuan) focused on people's sense of, and relationship with, places. More influential was Marxist geography, which applied the social theories of Karl Marx and his followers to geographic phenomena. Feminist geography also came into existence using ideas from feminism in geographic contexts. The most recent strain of critical geography is post-modernist geography, which employs the ideas of post-modernist theorists to explore the social construction of spatial relations.

The mid-20th century is distinguished from most of human history in that its most significant changes were directly or indirectly economic and technological in nature. Besides, the Space Age in the 20th century is a period encompassing the activities related to the space race, space exploration, space technology, and the cultural developments influenced by these events. The Space Age began with the development of several technologies that culminated with the launch of Sputnik 1 by the Soviet Union. This was the world's first artificial satellite. The launch of Sputnik 1 ushered a new era of political, scientific and technological achievements that became known as the Space Age. Regarding the 21st century and the late modern world, the Information Age and computers were forefront in use, not completely ubiquitous but often present in daily life. The development of Eastern powers was of note, with China and India becoming more powerful. The terms "modernity"and "modernism"refer to a new way of thinking, distinct from medieval thinking. The term 'Post-modern' is applied to the contemporary or more recent events, i.e. belonging to the current. The contemporary era follows shortly afterward with the explosion of research and increase of knowledge known as the Information Age in the later 20th and the early 21st century. The Post-modern era is seen in widespread digitality and describes the span of historic events that are immediately relevant to the present time.

Thus, geography took about 2000 years to evolve as a scientific discipline and obtain a more or less clear-cut definition, content, purpose and methodology. The development of discipline has been like an organism maturing through its infantile, childhood and adolescent stages. Important development of *Infantile Stage* (Ancient and early Medieval period) included the accumulation of knowledge, particularly from the fields like astronomy, mathematics, climatology, physics, astrology and plant studies. The consolidation of discipline into one single entity like a tree-trunk, more or less an independent recognition, and branching or bifurcation into two major streams, viz. physical and human geography, characterized the *Childhood Stage* (early Medieval to late Medieval period). During *Adolescent Stage* (late Medieval to Second World War), the discipline experienced a marked degree of instability. Important developments included the influence of the theory of Charles Darwin, introduction of a large number of dualistic tendencies and the rise of various modern schools of geographic thought like German, French, American, British and Soviet. The post-Second World War period (Second half of the 20th century and later) represents signs of maturity in the discipline. This is reflected in the developments like increase in integrative and inter-disciplinary approaches and the use of more and more utilitarian philosophies. Three important trends have been observed in the field of geography especially after Second World War: (1) Integration; (2) Interdependence; and (3) Utilitarian approach (making geography more and more useful for mankind). The process of integration (unification) has replaced the process of separation (bifurcation) that was evident in the beginning of modern period. The trend is now for all fields of study to come together around specific human problems. The Contemporary Geography is a spatial science increasingly focusing on the study of spatial aspects of social and economic phenomena, and thus being pre-eminently anthropocentric.

Since the basic nature of geography has been shaped by different schools of thought emerging one after the other from ancient to modern times, providing the discipline with its standard literature, the period up to Second World War may be identified as the Classical Phase of development, and the one thereafter as Contemporary Phase. The general nature of geographic development may be called more 'philosophical' during its Classical Phase and more 'explanatory' in the Contemporary Phase.

NOTES

1. Dikshit, 1994, p. 3.
2. Rana, 2008, p. 16.
3. James, 1972.
4. Singh and Dutt, 1979.
5. The Early History of World Maps, www.*maps-gps-info.com.*
6. Roman K. Kovalev, Alexis C. Kaelin (2007), *History Compass*, 5 (2), pp. 560-80.
7. Watson, Andrew M. (1974), pp. 8-35; Ruggles, D. Fairchild (2003), pp. 15-34.
8. The Age of Enlightenment (or Age of Reason) was an elite cultural movement of intellectuals in 18th century Europe that sought to mobilize the power of reason in order to reform society and advance knowledge. It promoted intellectual interchange and opposed intolerance and abuses in church and state. Originating about 1650-1700, it was sparked by philosophers Baruch Spinoza (1632-1677), John Locke (1632-1704), Pierre Bayle (1647-1706), mathematician Isaac Newton (1643-1727) and Voltaire (1694-1778). The Enlightenment flourished until about 1790-1800.The centre of the Enlightenment was France. The new intellectual forces spread to urban centres across Europe, notably England, Scotland, the German states, the Netherlands, Russia, Italy, Austria, and Spain, then jumped the Atlantic into the European colonies, and played a major role in the American Revolution.
9. Henry Kitchell (2004).
10. Edwin Augustus Grosvenor (1899).
11. Fordyce, J. (1888).
12. Baird, F.E., and Kaufmann, W.A. (2008).
13. Bose and Jalal, 2003, p. 76; Peers, 2006, p. 30.
14. Scholasticism is a method of critical thought which dominated teaching by the academics (*scholastics, or schoolmen*) of medieval universities in Europe from about 1100-1500, and a

programme of employing that method in articulating and defending orthodoxy in an increasingly pluralistic context. Scholasticism places a strong emphasis on dialectical reasoning to extend knowledge by inference, and to resolve contradictions.

15. Secularization (or secularisation) is the transformation of a society from close identification with religious values and institutions towards non-religious (or "irreligious") values and secular institutions. *Secularization thesis* refers to the belief that as societies "progress", particularly through modernization and rationalization, religion loses its authority in all aspects of social life and governance. The term secularization is also used in the context of the lifting of the monastic restrictions from a member of the clergy.
16. Paul Oskar Kristeller, *Humanism*, pp. 113-4, in Charles B. Schmitt, Quentin Skinner (editors), *The Cambridge History of Renaissance Philosophy* (1990).
17. *Pax Britannica* (Latin for "the British Peace", modelled after *Pax Romana*) was the period of relative peace in Europe (1815-1914) when the British Empire controlled most of the key maritime trade routes and enjoyed unchallenged sea power. It refers to a period of British imperialism after the 1815 Battle of Waterloo, which led to a period of overseas British expansionism.

REFERENCES

Baird, F.E., and Kaufmann, W.A. (2008), *Philosophic classics: From Plato to Derrida.* Upper Saddle River, N.J: Pearson/Prentice Hall.

Bose, Sugata and Ayesha Jalal (2003), *Modern South Asia: History, culture, political economy,* Parts 2-3, Routledge.

Dikshit, R.D. (1994), *The Art and Science of Geography: Integrated Readings,* Prentice Hall of India.

Fordyce, J. (1888), *The New Social Order.* London: Kegan Paul, Trench and Co.

Grosvenor, Edwin Augustus (1899), *Contemporary history of the world,* T.Y. Crowell & Co. New York and Boston.

Henry Kitchell (2004), *Early European History,* Hutton Webster.

James, P.E. (1972), *All Possible Worlds: A History of Geographical Ideas.* Odyssey Press, New York.

Paul Oskar Kristeller (1990), "Humanism", pp. 113-4, in Charles B. Schmitt, Quentin Skinner (editors), *The Cambridge History of Renaissance Philosophy.*

Rana, L. (2008), *Geographical Thought: A Systematic Record of Evolution,* Concept Publishing Company, New Delhi.

Roman K. Kovalev, Alexis C. Kaelin (2007), "Circulation of Arab Silver in Medieval Afro-Eurasia: Preliminary Observations", *History Compass* 5 (2), pp. 560-80.

Singh, R.L. and P.K. Dutt (1979), *Elements of Practical Geography,* Students' Friends, Allahabad, India.

Watson, Andrew M. (1974), "The Arab Agricultural Revolution and Its Diffusion, 700-1100", *The Journal of Economic History* 34 (1).

"The Early History of World Maps", www.maps-gps-info.com.

CHAPTER-2

Geography and Post-Modernism

The contemporary tradition is reflected in the late 20th century developments. These were the years of intellectual excitement unique in the history of geography.[1] The period saw the works leading to more socially relevant geography, but also the one tied still to a philosophy of science, a set of theories and a methodology compatible with the existing academic and disciplinary power structure. The period after Second World War, in many ways, was a defining point for the discipline. It ushered in major changes in the direction of geographic research and education. The changes were definitional and methodological. Definitionally, the emphasis shifted from regional study to the study of spatial patterns, linkages and interactions. Methodologically, more geographers began to employ the scientific (quantitative) method as the means of enquiry in spatial analysis. The emphasis on the scientific and model building meant an emphasis on a more deductive (theoretical) form of enquiry. In 1953, the paper "Exceptionalism in Geography" by F.K. Schaefer, an American scholar, was published in the Annals. This paper presented a strong and explicit call for more theory in geography and challenged areal differentiation view of Hartshorne, his own country fellow. Schaefer believed that as a social science, geographic research should explore regularities in spatial patterns. It should be a science of those factors governing the spatial distribution of certain features on the surface of the earth. Schaefer believed the geographic methodology should be the basis for formulating generalizations that can be stated as hypothesis to be tested against a large number of cases and the geographers should search for generalizations. Schaefer's influence led to the Quantitative Revolution in geography. But, towards the close of 20th century, another major change in geography began to emerge. A sharp break with spatial analysis came with the advent of Marxist and humanistically oriented geographers. The Marxist analysis offered an appealing alternative to understand social inequalities and investigate its social causes and possibility of alternative social structures. This was strongly reflected in the behavioral movement and the rise of a number of humanistic trends in geography. While the Marxists were emphasizing social inequality, the humanists were concerned about social meaning. Radical geography emerged as the inadequacies of behaviouralist methods became clear. It sought to counter the positivist quantitative methods with normative techniques drawn from Marxist theory: quantitative methods, it argued, were not useful unless alternatives or solutions were given to problems. Positivism is described as "making law like statements about phenomena that can be empirically observed.[2] This means using the scientific method and deductive reasoning to create an explanation of phenomena. The radicals were new kind of intellectuals, who thought hard and taught well in an effort to transform society. Theoretically sophisticated ideas tended to form in areas of radical geography with clear connection with more heavily theorised streams of thought oustside the discipline. As part of growing radicalisation of social thought, questions of gender and gender relations were raised. In an atmosphere of social and academic ferment typical of the early 1970s, geographers began to inquire into relations between woman and space. During the period of methodological change in late 20th century, initially there were symptoms of a move towards a stronger spatial view in geography; but, later, i.e. towards the close of 20th century geography was found becoming more and more a social science discipline.

Besides Behavioural geography, a number of humanistic and Marxist approaches have been advocated in recent trends of geographic development. They constitute the part of 'radical' geography. The radical geographers show their concern for social values and political actions, and reject the traditional concept of geography. They remain interested in the human and environmental relationships, but at the same time, also accept geography as a legitimate field of study and feel that it has much to offer in finding solution to the world problems. The Humanistic and Marxist approaches that constitute the part of radical Geography have been suggested to find socially appropriate uses for the existing techniques and theories in the discipline. 'Humanistic geography' emerged in the late 1960s as the most sophisticated of a series of critical reactions against the logical positivist forms of knowledge dominant in the established academic discipline.[3] There lies a conceptual difference between 'behavioural geography' and the humanistic approaches. 'Humanistic geography' may be defined as an approach to human geography, which is distinguished by the central and active role it accords to human awareness, human agency and human creativity. Viewed thus, humanist geography is an attempt towards "understanding meaning, value and human significance of life events"[4] as well as "an expensive view of what the human person is and can do"[5]. Behavioural Geography, on the other hand, treats man as a response to stimuli, so that it is concerned with identification of how different individuals respond to particular stimuli (and how the same individual responds to the same stimuli in different situations). By isolating the correlated nature of these varying responses, the behavioural geographers have attempted to build models and theories that can predict the probable impact of a given stimulus. The end product of behavioural research is an input to environmental planning with a view to moulding things in a desired manner. As contrast to this, humanistic geography treats man as an individual ceaselessly interacting with his environment and, in that process, transforming both himself and his milieu. This interaction between an individual and his environment is viewed as a particular case, and not as an example of some scientifically defined model of behaviour. The humanistic perspective was stated in definite opposition to the "dehumanizing" characteristics of the methods of scientific rationality in the social sciences. The recent use of humanistic approach in geography is an attempt to make man, the human being, the very core of geographic investigation; to make geography "people geography", about real people and for the people[6]. In other words, to make geography a discipline 'concerned with the social organization of space' rather 'the spatial organization of society'. Several Humanistic approaches have been advocated and canvassed for. The important ones as discussed here include Marxism, Feminism, Positivism, Pragmatism, Phenomenology, Existentialism, Idealism, Functionalism and Realism.

Post-modernism can be seen to mark an unprecedented crisis of intellectual activity within the contemporary crisis of modernity[7]. Pre-modernism, modernism and post-modernism, each of these can be talked about as periods of time and as philosophical systems. From the beginning of history up through the 1650s, the dominant way of viewing the world was largely consistent with the pre-modern philosophical system. Around the 1650s, pre-modernism was losing its influence as the dominant system and was being replaced by the modernist mind-set. For about 300 years, this was the dominant philosophical system in Western culture. The 1950s is also considered the time when the transition from modernism to post-modernism occurred. The primary epistemology[8] of the pre-modern period was based upon revealed knowledge from authoritative sources. In pre-modern times it was believed that Ultimate Truth could be known and the way to this knowledge is through direct revelation. This direct revelation was generally assumed to come from God. The church, being the holders and interpreters of revealed knowledge, was the primary authority source in pre-modern time. Two new approaches to knowing became dominant in the modern period.[9] The first was empiricism (knowing through the senses) which gradually evolved into scientific empiricism or modern science with the development of modernist methodology. The second epistemological approach of this period was reason or logic. As the shift in power moved away from the church, politics (governments, kings, etc.) and universities (scholars, professors) took over as the primary sources of authority. Oftentimes, a religious perspective was integrated into these modern authority sources, but the church no longer enjoyed the privileged power position. Post-modernism (1950s to current times) brought with it a questioning of the previous approaches to knowing. Instead of relying on one approach to knowing, it advocates for an epistemological pluralism which utilizes multiple ways of knowing. This includes the pre-modern ways (revelation) and modern ways (science and reason), along with many other ways of knowing such as intuition, relational and spiritual. Post-modern approaches seek to deconstruct previous authority sources and power.

It is not possible to understand post-modernism without due reference to the history and nature of modernity.[10] The beginnings of the modern world system are traced back to the 16th century, beginning in Europe, and organised around market exchange at international scale. But the process was greatly intensified through the 19th and 20th centuries through global scale of economic operations, so that modernity became identified as a system of market based political-economic structure and organisations. Thus, by the second half of 20th century, a profound crisis of identity and representation had begun to be felt all-round—in the arts, humanities and the social sciences, owing to the accelerated pace of time-space compression. By the end of 1950s, the modern way of viewing phenomena in terms of Cartesian coordinates[11] had begun to be questioned; separation between place and space began to be drawn.[12] All this called for new approaches beyond the modernist tradition of universalism. This was the beginning of post-modernism in social thought. Modernism was closely intertwined with the processes and objectives of colonialism and repressive imperialism, besides ethnocentricism rooted in Europe and cultural values and perspectives.[13] 'Modernism' has become a useful general useful term to capture the specific and changing meaning of the three most basic and formative dimensions of human existence: space, time and being; the spatial, temporal and social orders of human life.[14] The concept of modernity is intimately related to the multiple reconfigurations of social life that have punctuated the historical geography of capitalism since the 16th century. In this context, modernisation refers to "the many different processes of structural change associated with the ability of capitalism to develop and survive, to reproduce successfully its fundamental social relations of production and distinctive divisions of labour." Modernism is thus the explicitly evaluative, culture-shaping and situated consciousness of modernity and is roughly able to split into periods of conjunction with the historical rhythms of intensified capitalist crisis, restructuring and modernisation.[15]

Post-modernism is a recent movement in social sciences. Its distinguishing characteristic is scepticism[16] towards the grand claims and grand theory of the modern era, and the claim to intellectual superiority. Post-modernism (as contrasted to modernism) stresses openness to a range of opinions in social enquiry. Post-modernism represented a response to modernism as a homogenizing force and to its totalizing theoreticm. It lays stress on discontinuities and disjunctures characteristic of everyday life in the real world. The homogeneity in the built-up landscape of the era of 'organised capitalism' (or the Fordist era[17]) is contrasted with the heterogeneity of economic, social and political life in the current phase of 'disorganised capitalism' (or the flexible accumulation of the post-Fordist era).[18] The post-modernist emphasis on heterogeneity, particularity and uniqueness of phenomena in differing contexts of time and space, would appear hard back to the regionalist tradition in geography of the Vidalian era. Post-modern thought provides a theoretical context for the study of spatial/areal diversity in the life-world. Viewed thus, post-modernism in geography is closely linked to the rise of 'new' regional geography. The difference between the old style regional differentiation in geography of the Hettner-Hartshorne era and the present-day post-modernist regional geography lies in that, while the former was indifferent to everyday experience of societal relationships, there is now a declared commitment to the understanding of the condition of man in particular places, and the ways that spaces are socially constructed.[19] Post-modernism does not aspire to generate any grand theory of universal application; its essentially heterogeneous ethos runs counter to the emphasis on consensus on theoretical perspectives (that characterised the discourse in the spatial science era. Instead of universal theory, the emphasis now is on the contemporaneity of social process over time and space, so that geography, like history, is pre-eminently a contextual discipline, engaged in the time-space reconstruction of social life.[20]

The movement toward post-modernism in geography started in 1970s with the introduction of political economy perspective in human geography—a perspective that reoriented human geography to the condition of man in society and gave rise to 'radical' revolution. In the words of Gregory (1989), the cutting edge of the political economy perspective was the logic of capital scrawling its signs on the surface of the landscape—as against the logic of space (i.e. geometry) that dominated spatial science geography until the 1960s. Thus, the focus of post-modern geography became a focus on uneven development in the spatial context. The political economy derived its perspective from the Marxist thought.[21] The post-Marxist theory of Socio-spatial Structure emphasizes that people not only make history but also geography— which is to say that time-space relations are intrinsic to the constitution of societies.[22] Such a view cuts into the sociological concept of societies as totalities with clear-cut boundaries, and renders them extremely unstable, since under the new concept, society becomes the product of the varying spans of time and space. This ends the earlier view of dualism between society and space, between agency and structure. Such a (supposedly) Kantian view of division between the social and the spatial was made all the more distinct in the spatial science episode in human geography. This duality (born essentially out of the urge to apply the methodology of the natural sciences to the study of sociological phenomena) had long been attacked by scholars of hermeneutics (the study and interpretation of meaning) from the time of Schleiemarcher (1768-1834)[23] and Wilhelm Dilthey (1833-1911).[24] Dilthey had argued that the human sciences, owing to the peculiarity of their subject matter, required a special methodology very different from the empirical methodology of natural sciences. This special methodology was hermeneutics that tried to search for meaning in all kinds of activities and objects, including tools and landscapes as well as individual biographies. Hermeneutics went beyond mere physical appearances—physical and physiological processes, and surface patterns—to seek deeper meanings and causes by paying attention to the meaning and motivation behind actions. Hermeneutics was introduced in human geography in the mid-1970s through the writings of Buttimore (1974) and Tuan (1976), and was in late 1970s visible under two different titles: humanistic geography[25] and critical theory.[26] The latter was proposed by Gregory (1978) with a view to linking the hermeneutical approach to a critique of traditional, historical and regional geography. Thus, by around 1980, the supposed dualism between space and society was finally broken, since with such a linguistic/hermeneutic turn, the problem of geographical description became much more than a mere matter of abstract logic. The linguistic turn to the study of human geography through the introduction of hermeneutics now made geography open to experimentation with different modes of representation, so that the discipline reopened its doors to cultural studies through greater interaction across the frontier with other branches of cultural studies. Thus, was completed the post-modernist journey in human geography, beginning in the 1970s, through political economy of the disorganisation of capital (and thereby highlighting the relationship between history and geography, structure and agency), followed by the dissolution of the barrier between the social and spatial, arrived at through revival of the frontier between geography and sociology by giving greater attention to social theory. This was followed by opening of the frontier with other disciplines focused on the study of cultures.

The crisis of 'modern' geography until the mid-1970s lay in that geography had cut itself loose from mainstream philosophy, humanities and the social sciences so that the post-modernist response lay in a realignment with these other streams of knowledge for intellectual leads, particularly with social theory. Indeed, post-modernism focuses on the creative tension among the different theoretical formulations, so that post-modernism is essentially polyphonic: it combines a number of individual but harmonious mselodies, i.e. points of view or perspectives, with a view to arriving at more comprehensive explanation of social reality. Gregory (1989) has identified three distinguishing features of post-modernist thought in geography. First, space-time specificity in social explanation: this implies that post-modernist thought in social sciences insists on the understanding of 'a world which is meaningful for the people who live within it'. Secondly, post-modernism insists on distancing itself from the totalization (i.e. the concept of the society as a totality following a universal process of historical change irrespective of space context). Post-modernist thought holds that: "The ebb and flow of human history is not reduced to the marionette movements of a single structural principle, whatever its location and the differences which make up human geographies are not explained by some central generating mechanism." Thirdly, post-modern human geography represents a critique of spatial science—it is not a continuation of it—but at the same time, it does not represent a break with the past. Post-modern geography focuses on the essential spatiality of social life (rather than the geometry of social behaviour).

Anthropocentricism

The discipline of geography as practised over the past two centuries is essentially a "European science".[27] Its transformation had begun after the Age of Discovery when the scientific methods of observation, classification and comparison were extended to the domain of peoples and societies that made geography.[28] Some of the most seminal cross-fertilisations during this period were those between anthropology and geography, what later became anthropogeography. This thread of relationship is indelibly present in the writings of Humboldt, Ritter, Blache, Ratzel and Sauer.[29] Towards the end of 19th century, a second thread of relationship between sociology and geography emerged: the interactions between the two disciplines were, in part, a way of clarifying their different destinations. While, sociology was to subsume the study of the spatial structure of society—*morphology sociale*—the Vidalian School insisted on the independent course of human geography predicted, on the one hand, on horizontal (spatial) relations and, on the other, on vertical relations between society and nature.[30] By mid-20th century, geography had developed close links with economics alongside the frontiers with anthropology and sociology. Until late 1940s, like the rest of human geography, the human geography-economics frontier was pursued in a resolutely empirical manner; but the emergence of the 'new' geography in the 1950s and the 1960s gave it a pronouncedly theoretical orientation so that human geography now assumed the characteristics of the abstract sciences, in the sense that like sociology and economics, it also became a nomothetic science but it continued to eschew the study of the social element. If the three strands in geography since the beginning of 19th century to mid-20th century are joined together, then human geography as spatial science would appear as a natural growth, rather than representing any break from the past. Geography as a spatial science was "a continuation and culmination of modern geography" in a sense that it helped complete the search for scientific identity of the discipline as a nomothetic science.

NOTES

1. Peet, 67.
2. Johnston, *et al.*, 1994.
3. Peet, 34.
4. Gregory, 1989, p. 348.
5. Buttimore, 1974.
6. Tuan, 1976.
7. Smith, 1977.
8. The theory of knowledge, especially with regard to its methods and validation.
9. "*The Three Major Philosophical Epochs*", Wikipedia, *The Free Encyclopaedia.*
10. Dikshit, 1997, p. 270.
11. Cartesian coordinates are the foundation of analytic geometry, and provide enlightening geometric interpretations for many other branches of mathematics, such as linear algebra, complex analysis, differential geometry, multivariate calculus, group theory, and more. A familiar example is the concept of the graph of a function. Cartesian coordinates are also essential tools for most applied disciplines that deal with geometry, including astronomy, physics, engineering, and many more. They are the most common coordinate system used in computer graphics, computer-aided geometric design, and other geometry-related data processing.
12. Clifford, 1988, p. 2.
13. Gregory, *op. cit.*
14. Soja, 1989a, pp. 320-322.
15. Soja, 1989b.
16. Questioning attitude towards knowledge, fact or opinion.
17. Fordism, named after Henry Ford, is a modern economic and social system based on industrial mass production. The concept is used in various social theories about production and related socio-economic phenomena. It has varying but related meanings in different fields, as well as for Marxist and non-Marxist scholars. In a Fordist system the worker is

paid relatively high wages in order to buy in large quantity the products turned out in mass production.

18. Harvey and Scott, 1989.
19. Dikshit, *op. cit.*, p. 277.
20. Dear, 1988.
21. Dikshit, *op. cit.*, p. 278.
22. Gregory, 1978.
23. Friedrich Daniel Ernst Schleiermacher (1768–1834) was a German theologian and philosopher known for his attempt to reconcile the criticisms of the Enlightenment with traditional Protestant orthodoxy. He also became influential in the evolution of Higher Criticism, and his work forms part of the foundation of the modern field of hermeneutics. Because of his profound impact on subsequent Christian thought, he is often called the "Father of Modern Liberal Theology." The Neo-Orthodoxy movement of the 20th century, typically (though not without challenge) seen to be spearheaded by Karl Barth, was in many ways an attempt to challenge his influence.
24. Wilhelm Dilthey (1833-1911) was a German historian, psychologist, sociologist and hermeneutic philosopher, who held Hegel's Chair in Philosophy at the University of Berlin. As a polymathic philosopher, working in a modern research university, Dilthey's research interests revolved around questions of scientific methodology, historical evidence and history's status as a science. He could be considered an empiricist, in contrast to the idealism prevalent in Germany at the time, but his account of what constitutes the empirical and experiential differs from British empiricism and positivism in its central epistemological and ontological assumptions, which are drawn from German literary and philosophical traditions.
25. Yi-Fu Tuan is a Chinese-American geographer famous for pioneering the field of human geography and merging it with philosophy, art, psychology, and religion. This amalgamation has formed what is known as *humanist geography*. Humanist geography as it is sometimes called is a branch of geography that studies how humans interact with space and their physical and social environments. It also looks at the spatial and temporal distribution of population as well as the organization of the world's societies. Most importantly though, humanistic geography stresses people's perceptions, creativity, personal beliefs, and experiences in developing attitudes on their environments. In addition to his work in human geography, Yi-Fu Tuan is famous for his definitions of space and place. Today, place is defined as a particular part of space that can be occupied, unoccupied, real or perceived (as is the case with mental maps). Space is defined as that which is occupied by an object's volume. During the 1960s and 1970s, the idea of place in determining people's behaviour was at the forefront of human geography and replaced any attention previously given to space. In his 1977 article, "Space and Place: The Perspective of Experience," Tuan argued that to define space, one must be able to move from one place to another, but in order for a place to exist, it needs a space. Thus, Tuan concluded that these two ideas are dependent upon one another and began to cement his own place in the history of geography.
26. Critical theory is an examination and critique of society and culture, drawing from knowledge across the social sciences and humanities. The term has two different meanings with different origins and histories: one originating in sociology and the other in literary criticism. This has led to the very literal use of 'critical theory' as an umbrella term to

describe any theory founded upon critique. In the sociological context, critical theory refers to a style of Marxist theory with a tendency to engage with non-Marxist influences (for instance the work of Friedrich Nietzsche and Sigmund Freud). Modern critical theory arose from a trajectory extending from the antipositivist sociology of Max Weber and Georg Simmel, the Marxist theory of Georg Lukács and Antonio Gramsci, towards the milieu associated with Frankfurt Institute of Social Research. Five "Frankfurt School" theorists were chiefly responsible for establishing critical theory as a specific strand of thought: Herbert Marcuse, Theodor Adorno, Max Horkheimer, Walter Benjamin, and Jürgen Habermas. With the latter, critical theory shed its roots in German idealism and moved closer to American pragmatism. Whilst the critical theorists are usually defined as Marxist intellectuals, their tendency to denounce so many Marxian elements has been attacked as 'revisionism' by stricter Marxists.

27. Dikshit, 273.
28. Gregory, *op. cit.*
29. Dikshit, 274.
30. Berdoulay, 1978.

REFERENCES

Berdoulay, V. (1978), "The Vidal-Durkheim Debate", in Ley, D. and M. Samuels (Eds.), *Humanistic Geography: Prospects and Problems,* Maroufa Press, Chicago.

Buttimore, A. (1974), *Values in geography,* Commission on Geography, Resource Paper 24, Association of American Geographers, Washington D.C.

Clifford, J. (1988), *The Predicament of Culture, Nineteenth Century Ethnography,* Literature and Art, Harvard University Press, Cambridge.

Dear, M. (1988), "The post-modern challenge: restructuring human geography", *Transactions, Institute of British Geographers,* NS 13, pp. 262-274.

Dikshit, R.D. (1997), "*Modern* versus Post-Modern Geographies" in *Geographical Thought: A Contextual History of Ideas,* Prentice Hall of India, pp. 270-282.

Gregory, D. (1978), *Ideology, Science and Human Geography,* Hutchinson, London.

Gregory, D. (1989), "The Crisis of Modernity? Human Geography and Critical Social Theory", in Peet, R. and N.J. Thrift (Eds.), *New Models in Geography,* Vol. 2, Unwin Hyman, London, pp. 348-385.

Harvey, D. and A.J. Scott (1989), "The practice of human geography: theory and empirical specificity in the transition from Fordism to flexible accumulation", in Macmillan, W. (Ed.) *Remodelling Geography,* Basil Blackwell, Oxford.

Johnston, R.J. (2004), *Geography and Geographers: Anglo-American Human Geography Since 1945,* 6th Ed., Edward Arnold, London.

Kuhn, T.S. (1996). *The Structure of Scientific Revolutions* (3rd Ed.).

Ley, D. (1994), "Post-modernism", in Johnston, R.J., Gregory, D. and Smith, D.M. (Eds.), *The Dictionary of Human Geography,* 3rd edition, Basil Blackwell, Oxford, pp. 446-447.

Livingstone, David N., (1993). *The Geographical Tradition,* Oxford, Blackwell.

Peet, R. (1998), *Modern Geographical Thought,* Blackwell.

Smith, D.M. (1977), *Human Geography: A Welfare Approach,* Edward Arnold, London.

Soja, E.(1989a), "Modern Geography, Western Marxism, and the Restructuring of Critical Social Theory", in Peet, R. and N.J. Thrift (Eds.), *New Models in Geography,* Vol. 2, Unwin Hyman, London, pp. 317-347.

Soja, E. (1989b), *Post-modern Geographies: The Reassertion of Space in Critical Theory,* Verso, London.

"The Three Major Philosophical Epochs", Wikipedia, the *Free Encyclopedia.*

Tuan, Yi-Fu (1976), "Humanistic Geography", *Annals of Association of American Geographers,* Vol. 66, Issue 2, pp. 266-276.

Tuan, Yi-Fu (1974), *Trophilia: A Study of Environmental Perception, Attitudes and Values,* Englewood Cliffs, Prentice Hall, NJ.

Wikipedia, the *Free Encyclopedia.*

SECTION-II
CLASSICAL TRADITION

CHAPTER-3

Ancient Period

The scholarly research began in the European and Asian realms during ancient period. Geography evolved as a spatial and interdisciplinary science both in the West and Orient. This is exhibited in the philosophical, literary and methodological parallels discovered between the two worlds. This was the period of rise of three major Classical Schools of thought, viz. Greco-Roman, Indian and Chinese. The main subject of ancient philosophy was to understand the fundamental causes and principles of the Universe. Therefore, important development of this period included the accumulation of knowledge, particularly from the fields like cosmology, astronomy, mathematics, meteorology, physics, astrology, geology, chemistry, etc. The roots of geography in antiquity can be traced back to the works of numerous ancient philosophers, scholars, mathematicians, administrators and explorers. The study of written past begins from the recorded human history in the Old World until the Early Middle Ages in Europe and the Qin Dynasty in China.[1] The period following these events includes the Imperial era in China and the period of the Middle Kingdoms in India. The span of recorded history altogether is roughly 5,000 years. The term 'classical antiquity' is used for this long period of cultural history. Such a wide sampling of history and territory covers three major cultures. The Greeks are the first known culture to have actively explored geography as a science. The developments in Europe were lead by Greek School, subsequently followed by Romans. Similar and parallel developments took place in the ancient Indian and Chinese realms as well and geography has been highly benefited by all.

ANCIENT CIVILIZATIONS

Greco-Roman

In fact, geography, as a discipline, has grown out of three simultaneous and related activities, viz. exploration, charting and mapping, and speculation. First two of the activities were common among other civilizations of the time, but the ability to speculate was observed only among Greek scholars. Therefore, it is them that the real beginning of our discipline is marked with. The Greeks are the first known culture to actively explore geography and to have developed the procedures to be described first as 'scientific'. Although many of the conceptions as put forward by ancient Greeks were based on what they had inherited from earlier civilizations, but their own contributions were also far-reaching. Greeks distinguished themselves from others because of their intellectual, philosophical and scientific approach. The developments in Greek School were followed by Romans who, in general, carried forward the Greek tradition. Due to rise of Roman Empire about 1st century A.D., the interest in Greek literature declined. The Greek Science started to lose its originality and little of worth was produced after 200 A.D. Even the Roman Empire collapsed by 500 A.D. In spite of all odds, the Roman scholars were able to extend further the Greek literary and mathematical traditions. For the most part ancient Greek geography was an academic field; it was the Romans who made far more extensive and practical use of geography. The Greek colonies had established and flourished during 8th and 6th century B.C. (the Golden period of Greeks). The Greek-speaking world in ancient times refers not only to the geographical peninsula of modern Greece, but also to areas of Hellenic culture that were settled in ancient times by Greeks. Roman civilization is often grouped into classical antiquity with ancient Greece. During its twelve-century existence, the Roman civilization shifted from a monarchy to an oligarchic republic and to a vast empire. It came to dominate Western Europe and the entire area surrounding Mediterranean Sea through conquest and assimilation. Most of the Greek accomplishments in geography were passed on to the Romans. Roman military commanders and administrators used this information to guide the expansion of their Empire. The Romans also made several important additions to geographical knowledge. Many of the Roman scholars like Ptolemy, Strabo, Polybius, etc., wrote in Greek itself. Strabo, for instance, wrote a 17-volume series called '*Geographia*'. A century after Strabo another Roman scholar, Ptolemy, launched a similar undertaking and commendably contributed to practical geography; whereas Polybius enriched the field of physical geography. By this time the Roman Empire had expanded through much of Europe and previously unknown areas such as the British Isles had been explored. The Silk Road was also in operation, and for the first time knowledge of the Far East began to be known. A number of centres of Greek learning became popular during ancient period as Athens, Miletus, Halicarnassus, Corinth, Argos, Thebes, Sparta, Amphipolis, Methone, Potidaea, etc. The contribution of ancient Greeks and Romans to various fields of study ranging from physical, human, historical, theoretical, political, regional, mathematical and practical covered almost the entire gamut of geographical knowledge. Besides, Geography as a discipline got enriched by a number of concepts, viewpoints and techniques; and various terminologies were added to its vocabulary. Along with coining the term 'geography', a number of other terminologies are also credited to ancient Greeks.

Ancient Greek philosophy focused on the role of reason and inquiry. The West represents four different philosophical trends, namely naturalistic, metaphysical, ethical and religious.[2] It began in ancient Greece with the emergence of Milesian School and culminated in the Neo-Platonic School. In between there emerged numerous other schools of thought. The period of Naturalism was dominated by the Milesian, Pythagorean, Ephesian, Eleatic, Pluralists' and Atomists' School of thought. Thales, Anaximander, and Anaximenes comprised the Milesian School. The teachings of Milesian (Ionian) School focused on cosmological concerns, particularly the underlying substance of the Universe. For the Milesians, to explain the world and its phenomena was just to show how everything came from the original stuff, such as Thales' water, Anaximander's boundless or Anaximenes' air. Pythagoreanism indicates progress over the Ionic School. The First Principle is elevated from a natural element found in the Ionic School to a conceptual one, the number. The Pythagorean School, revolving around the philosophy of Pythagoras, cultivated the mathematical sciences and the study of mathematics led them to the observation that everything could be represented through a number. The realization that all things are numerable, and can be related to each other in a numerical proportion is one factor which led the Pythagoreans to their emphasis on the value of number in explaining the world order. The Ephesian school, next in sequence, refers to the philosophical thought of Heraclitus. Heraclitus' theory developed as a response to the philosophy of his Ionian predecessors; everything that exists, including man himself, exists because it is in a continuous process of passage from one state to another, i.e. in a state of continual flux. The Eleatic School, on the contrary, was based on the doctrine of Permanence, changelessness. Besides the Milesians, Pythagorians, Ephesians and the Eleatics, there were pluralists among Naturalists. The chief philosophers of this group are Empedocles and Anaxagoras. Empedocles tried to reconcile the world of change with unchanging Being by saying that the world as we know it is composed of four basic material particles which are themselves unchanging and indestructible. He called them "the roots of everything"; they later came to be called elements. According to Empedocles, the four elements are earth, air, fire and water. The objects of the physical world, he said, are made up of a haphazard combination of these four elements. Empedocles' theory endured longest and fell into decline only with the advent of modern chemistry. This theory of the four elements became the standard dogma for the next two thousand years.

Another philosophy to try to resolve the paradox of unchanging being and a changing world was that of a school which came to be called the Atomists. This school was the most radical of all. The best known philosophers of this school were Leucippus and Democritus. Their theory that the Universe is composed of an infinite number of elements which are characterized by quantitative differences has undergone many and important modifications, but has still maintained its validity. Democritus believed that "there exist invisible atoms which are impenetrable, invisible spatial entities differing only in form, weight, size." He declared that nature and the organization of matter is the homogeneity of all bodies and that indeterminate matter is divided into an infinite number of molecules (atoms) differing in size and form but endowed with perpetual motion which is derived from their essence. Because atoms are endowed with motion, Democritus admits a second primordial element, the void, that is, infinite space which surrounds the atoms and gives them the possibility of movement. The differences in gravity cause the atoms to whirl into motion, thus giving origin to the formation of things. This ends the period of the Naturalists. The second period of Greek philosophy occupied the entire 4th century before Christ. The problem which claimed the interest of thinkers during this period is no longer the cosmological question, but man in his concreteness, i.e. in his knowledge, his morality and his rights. This was Metaphysical Period which includes the philosophies of Sophists, Socrates, Plato and Aristotle. Sophistic thought can be considered as a transition from the old cosmological concepts to the new ideas about man. The Sophists were the first to show complete indifference to the problem of the world of matter and to centre their efforts upon man. But man can be an object of study in his sense knowledge as well as in that more profound one of reason. The Sophists stopped at the first, at the immediacy of sense impressions. The analysis of reason was reserved to Socrates and his disciples. Socrates established the fact that true knowledge means knowing through concepts. In fact, the classic Greek philosophy mainly revolves round Socrates. Socrates, like the Sophists, was not concerned with cosmological questions, but concentrated his attention on man. But the man taken into consideration by Socrates is not the individual in particular; rather, he is the universal man in his subjective reason. In this universal man, Socrates discovered knowledge whose characteristics are universality and necessity. The doctrine of Socrates can be summed up in two words—concepts and ethics. The greatest of the Socratic schools, however, referred to as the Major Socratic School, was the Academy of Plato, which stayed closer to the original intent of the teachings of Socrates. The thought of Plato was revived in the later Academies, and in particular in the last important movement of Greek thought, Neo-Platonism. Ethical Period represents the emergence of Stoicism, Epicureanism, Skepticism and Eclecticism. This period extends from the death of Aristotle to the beginning of the era of the decay and dissolution of Greek philosophy—that is, duration of about three centuries. During this time metaphysics falls into decadence and disappears completely. This displacement of metaphysics by ethics, of pure speculation by practical thought, was the unavoidable consequence of the sadness of the times. Greece loses its independence, which was never again to be regained in the ancient world. First, the successors of Alexander the Great, and then the Romans, were to become masters of this piece of earth. Furthermore, this period is characterized by endless wars between the successors of Alexander and the Roman Empire, which bathed in blood not only Greece but all the countries of the Mediterranean basin. If one looks at the problem of life in this way, there are only three possible means of solving it—to appeal to reason; to appeal to pleasures; and to deceive oneself in the belief that every appeal is useless. These are the characteristics of the three streams of thought which philosophy followed in its ethical development; i.e. Stoicism (the appeal to reason); Epicureanism (the appeal to pleasure); and Skepticism (the renunciation of every solution). These three streams of thought formed the major schools of thought during this period. The last period of Greek philosophy extends to the 5th century of Christian era. It is called Religious because of its attempt to resolve the problem of human life by recourse to religion.

Classical culture declined, philosophy became religious mysticism, ancient philosophy ended, as it began, in religion. Greek speculation, Egyptian, Chaldean, and Jewish religions tended to mingle. The problem which claimed the attention of philosophers concerned the presence of evil and death. This period is represented by three main Schools: The Judaic-Alexandrian School, whose greatest representative is Philo—this was an attempt to coordinate Oriental religion with Greek speculation; The Neo-Pythagorean teaching, whose representative is Appolonius of Tiana—the attempt to construct a world religion upon Pythagorean doctrines; and The Neo-Platonic School, which was founded by Ammonius Saccas, and which has Plotinus for its leading representative—the attempt to make a religious philosophy of Plato's teaching. The Greek philosophy culminated in Neo-Platonism, and Plato's system became the framework for a religious world view.

Ancient India

Ancient India is usually taken to refer to the "golden age" of classical Hindu culture, beginning around 500 B.C. with several '*Mahajanapadas*' (monarchies and republics), stretched across the Indo-Gangetic plains from modern-day Afghanistan to Bangladesh.[3] India is estimated to have had the largest economy of the world between the 1st and 15th centuries AD, controlling between one-third and one-quarter of the world's wealth up to the time of the Mughals, from whence it rapidly declined during British rule. Occupying a strategic location in Asia, Indian history is at crossroads of cultures from China to Europe. The geographical studies in India began with the dawn of Indian Civilization in ancient times. Valuable geographical information is contained in Hindu mythology, philosophy, epics, history and sacred laws. Apart from religious records, the travellers' accounts give the description of different regions of the world. The accounts of these travellers reveal that India had close links with the neighbouring lands and Indian scholars were familiar with the geographical conditions of China, Southeast Asia, Central Asia and Mesopotamia. The contributions of Indian scholars in ancient period are parallel to that of Greeks and Romans. Major sources of information are *Ramayana, Mahabharata, Vedas, Purana* and the works of Buddhists and Jains. According to Winternitz[4] the Vedas belonged to a period stretching from 2500 B.C. to 7000 B.C. It would, then, mean that the Indian civilization is much older than that of European. The ancient *Rishis* and sages are our first *Gurus* (teachers) who had blessed humanity with the ancient knowledge and wisdom. All current scientific and spiritual teachings are based on the knowledge passed on from these great divine souls. Science and technology in ancient India covered all the major branches of human knowledge including mathematics, astronomy, physics, chemistry, medical science and surgery, fine arts, architecture, sports and games. In fact several inventions and discoveries believed to have originated from the Western world have been made centuries earlier by our ancestors. 'Geography' in Hindi (Indian Language) is called "*Bhugol*'; '*bhu*' meaning 'the Earth' and '*gol*' meaning 'round', i.e. 'the study of the round earth'. Thousands of years before Columbus, our scriptures (*Vedas, Puranas* and *Itihaasaas*) used the term 'Bhoogola' for Earth. Indian astronomers propounded the theory that the earth is a sphere. They also formulated ideas about gravity and gravitation (*gurutvakarshan*). The ancient Indian scholars wrote elaborate treatises on astronomy, mathematics, geometry, medicine, etc. During Aryabhatta's time (around 500A.D.), our ancestors evinced keen interest in motion of bodies. Sir Isaac Newton gained fame since his discovery of the principle of gravity, but Bhaskaracharya II talked about gravity much before Newton. Aryabhatta's treatise, *Aryabhateeyam* contains many theorems of motion. The point is that Aryabhatta developed these equations centuries before Newton did. It is widely believed that Christopher Columbus was the first person to discover that the Earth was round, during his voyages around 1500 A.D. However, in 499 A.D., the famous Indian astronomer Aryabhatta had already defined the shape of Earth in his work, *Aryabhateeyam*, as: *Mrujjalashikhivaayumayo Bhoogola: sarvatho vruttha* (Meaning: Earth which is made of soil, water, fire and air is circular when viewed from all sides i.e. The Earth is spherical). The vast galaxy of seer-scientists, philosopher-poets, savants and sages left behind a rich history of thought that ranged from science, philosophy, religion and the living arts of man.

The contributions to ancient Indian geographical literature came largely from astronomers, astrologers, astro-physicists, mathematicians, scientists and philosophers. The views of these scholars are available on the study of the Universe, the Earth and the Indian Subcontinent. The ancient Indian literature deals with problems pertaining to cosmology, cosmogony and cosmography. The theory of Philosophical origin of the Universe is very close the presently accepted theory of Nebular Origin. In the *Rig-Veda*, there is a description of various heavenly bodies including the Sun, the Moon, different planets and constellations. Observations about the earth are related to its origin, sphericity, eclipses, size and dimensions, latitudes-longitudes and local time, cardinal points, earthquakes and volcanoes, atmosphere and seasons, and physical divisions of the planet. The estimates regarding the earth's volume, circumference and mass, as given in '*Suryasidhanta*', the ancient treatise of Indian astronomy, are nearly correct and reliable. Geographical knowledge about the Indian Subcontinent is associated with its physical identification, people, culture, relief and drainage.

The Indian mystics found answers to the questions related to the understanding of meaning and goal of human life. It is because of this special perception their thoughts are called *darshana* which actually means vision or realization of the Self. The Indian philosophical system can be classified into two broad categories, viz. orthodox (*astika*) and heterodox (*nastika*), conveying a division of thinkers into 'theist' and 'atheist'. Orthodox systems are those which accept the authority of the Vedas, while the heterodox systems are those which reject it. To the latter group belong three systems of Charvaka, Buddhism and Jainism. The '*Shaddarshanas*', or the six systems of Indian philosophy belong to the former group. These systems are called *Nyaya*, *Vaisesika*, *Samkhya*, *Yoga*, *Purva Mimamsa* and *Uttara Mimamsa* or *Vedanta*. They generally deal with four topics: Existence and Nature of *Brahman*, Nature of the *Jiva* or individual soul; Creation of the *Jagat* or the world; and *Moksha* or liberation.

Ancient China

Besides the Christian Europe and Indian Subcontinent, there was another major centre of geographical learning in the world in ancient period. This was China, a region in East Asia, known as the People's Republic of China (PRC). It refers to a long-standing civilization comprising successive cultures dating back more than 5,000 years. Essentially, the European and Chinese worlds remained isolated, each discovering the other step by step. Yet there are fascinating parallels in the concepts and methods of study that seem to require the existence of contacts, however indirect and remote. From about second century before Christ until at least the 15th century after Christ, the people of China enjoyed the highest standard of living of any people on earth. The developments that began here in ancient times continued up to the Age of Discovery and later.[5] With one of the world's longest period of largely uninterrupted civilization and the world's longest continuously used written language system, China's history has been largely characterized by repeated divisions and reunifications amid alternating periods of peace and war, and violent imperial dynastic change (Table 3.1). For centuries, Imperial China

TABLE 3.1 History of China

Ancient
3 Sovereigns and 5 Emperors
Xia Dynasty 2100-1600 BCE
Shang Dynasty 1600-1046 BCE
Zhou Dynasty 1045-256 BCE
Spring and Autumn Period (770-470 B.C.)
Warring States Period (475-221 B.C.)
Imperial
Qin Dynasty 221 BCE–206 BCE
Han Dynasty 206 BCE–220 CE
Three Kingdoms 220-280
Wei, Shu & Wu
Jin Dynasty 265-420
16 Kingdoms 304-439
Southern and Northern Dynasties 420-589
Sui Dynasty 581-618
Tang Dynasty 618-907
(Second Zhou 690-705)
5 Dynasties and 10 Kingdoms 907-960
Liao Dynasty 907-1125
Song Dynasty 960-1279
Yuan Dynasty 1271-1368
Ming Dynasty 1368-1644
Qing Dynasty 1644-1911
Modern
Republic of China 1912-1949
People's Republic of China 1949–present

was also one of the world's most technologically advanced civilizations and East Asia's dominant cultural influence, with an impact lasting to the present day. The country's territorial extent expanded outwards from a core area in the North China Plain, and varied according to its changing fortunes to include multiple regions of East, Northeast, and Central Asia. The term "China Proper" is used to describe this core territory. China Proper is bounded by the Great Wall and the edge of the Tibetan Plateau. As in Egypt, Mesopotamia, and along the Indus River, Chinese civilization began within a major river valley, the Yellow River Valley. Around 4000 B.C. the modern China, with a huge geographical expanse, contained an almost infinite number of ethnic groups and languages. Most Chinese dynasties were based in the historical heartland of China, the China Proper. China was one of the earliest centres of human civilization. Chinese civilization was also one of the few to invent writing independently. The Chinese script is still in use today by the Chinese, Japanese and Koreans. This script is the only logographic script used in the world.

Chinese civilization, as described in mythology, begins with Pangu, the creator of universe, and a succession of legendary sage-emperors and culture heroes who taught the ancient Chinese to communicate and to find sustenance, clothing, and shelter. But, the written history of China as recorded at the traditional historical records extends as far back as the Three Sovereigns and Five Emperors about 5000 years ago. China is one of the world's oldest continuous civilizations. Turtle shells with markings reminiscent of ancient Chinese writing from the Shang Dynasty have been carbon dated to around 1500 B.C. Chinese civilization originated with city-states in the Yellow River (Huang He) Valley. The conventional view of Chinese history is that of a country alternating between periods of political unity and disunity and occasionally becoming dominated by foreign peoples, most of whom were assimilated into the Han Chinese population. Cultural and political influences from many parts of Asia, carried by successive waves of immigration, expansion, and assimilation, merged to create the Chinese culture. The first dynasty according to Chinese sources was the Xia Dynasty, from about 2100 to 1600 B.C. But its reference has been believed to be legendary. However, the earliest discovered written record of China's past dates from the Shang Dynasty and takes the form of inscriptions of divination records on the bones or shells of animals—the so-called *oracle bones.* Its civilization was based on agriculture, augmented by hunting and animal husbandry. Two important events of the period were the development of a writing system and the use of bronze metallurgy. Workmanship on the bronzes attests to a high level of civilization. The Shang were invaded and by the end of the 2nd millennium B.C., the Zhou Dynasty began to emerge in the Yellow River valley. Sharing the language and culture of Shang, the early Zhou rulers, through conquest and colonization, gradually extended Shang culture in much of China Proper north of the Yangtze River. In 771 B.C. the Zhou court was sacked, and its king was killed by invading barbarians who were allied with rebel lords. The capital was moved eastward to Luoyang in present-day Henan Province. Because of this shift, historians divide the Zhou era into Western Zhou (1027-771 B.C.) and Eastern Zhou (770-221 B.C.). With the royal line broken, the power of the Zhou court gradually diminished; the fragmentation of the kingdom accelerated. Eastern Zhou divides into two sub-periods. The first, from 770 to 476 B.C., is called the Spring and Autumn Period, after a famous historical chronicle of the time; the second is known as the Warring States Period (475-221 B.C.). But, the territory was again unified under one emperor in 221 B.C. by Qin Shi Huang, ushering in the Qin Dynasty, the first unified centralized Chinese state. Historians often refer to the period from Qin Dynasty to the end of Qing Dynasty as Imperial China. The Qin Dynasty is well known for beginning the Great Wall of China, which was later augmented and enhanced during the Ming Dynasty. The major contributions of the Qin include the concept of a centralized government, the unification of the legal code, written language, measurement, and currency of China after the tribulations of the Spring and Autumn and Warring States Periods. After the fall of a Qin Dynasty in 206 B.C. emerged the Han Dynasty which lasted until 220 A.D. Han Dynasty was the first dynasty to embrace the philosophy of Confucianism, which became the ideological underpinning of all regimes until the end of imperial China. Under Han Dynasty China made great advances in many areas of arts and sciences. Han power declined again amidst land acquisitions, invasions, and feuding between consort clans and eunuchs. In the ensuing turmoil, three states tried to gain predominance in the Period of the Three Kingdoms. Though the three kingdoms were reunited temporarily in 278 by the Jin Dynasty, the contemporary non-Han Chinese ethnic groups controlled much of the country in the early 4th century and provoked large-scale Han Chinese migrations to south of the Chang Jiang. *Sixteen kingdoms* were a plethora of short-lived non-Chinese dynasties that came to rule the whole or parts of northern China in the 4th and 5th centuries. Many ethnic groups were involved, including ancestors of the Turks, Mongolians and Tibetans. Most of these nomadic people had to some extent been sinicized long before their ascent to power. Signalled by the collapse of East Jin Dynasty in 420, China entered the era of the Southern and Northern Dynasties. The Han people managed to survive the military attacks from the nomadic tribes of the north, and their civilization continued to thrive. An increasing number of nomadic people in Northern China adopted Confucianism as personal life guidance and state ideology while becoming gradually assimilated into the Han Chinese civilization. During the rivalry between Northern and Southern China, Buddhism propagated throughout China for the first time. In 589, Sui annexed the last Southern Dynasty, Chen, through military force, and put an end to the era of Southern and Northern Dynasties. A period of disunion followed again. In the late 6th century, China was reunited under the Sui. Under the succeeding Tang and Song dynasties, China reached its golden age. For a long period of time, especially between the 7th and 14th centuries, China was one of the most advanced civilizations in the world. The Sui brought China together again and set up many institutions that were to be adopted by their successors, the Tang. Like the Qin, however, the Sui overused their resources and collapsed. In 618 A.D. the Tang Dynasty was established, opening a new age of prosperity and innovations in arts and technology. Buddhism, which had gradually been established in China from the first century, became the predominant religion and was adopted by the imperial family and many of the common people. The Tang and the Han are often referred to as the most prosperous periods of Chinese history. The Tang, like the Han, kept the trade routes open to the west and south and there was extensive trade with distant foreign countries and many foreign merchants settled in China. From about 860 the Tang Dynasty began to decline due to a series of rebellions within China itself. The period of political disunity between the Tang and the Song, known as the *Five Dynasties* and *Ten Kingdoms Period*, lasted little more than half a century, from 907 to 960 A.D. During this brief era, when China was in all respects a multi-state system, five regimes succeeded one another rapidly in control of the old Imperial heartland in northern China. During this same time, 10 more stable regimes occupied sections of southern and western China, so the period is also referred to as that of the Ten Kingdoms. In 960 A.D., the Song Dynasty (960-1279 A.D.) gained power over most of China. The Jurchen Jin Dynasty (1115-1234 A.D.) emerged to prominence, annihilating the Liao Dynasty. Meanwhile, in what are now the northwestern Chinese provinces of Gansu, Shaanxi, and Ningxia, there emerged a Western Xia Dynasty from 1032 up to 1227 A.D., established by Tangut tribes. In the ensuing years China was divided between the Song Dynasty, the Jin Dynasty and the Tangut Western Xia. The Song Dynasty is considered by many to be classical China's high point in science and technology, with innovative scholar-officials such as Su Song (1020-1101 A.D.) and Shen Kuo (1031-1095 AD). The Song Dynasty fell to the invading Mongols in 1279 A.D. The Mongols, under Kublai Khan, established the Yuan Dynasty. A peasant named Zhu Yuanzhang overthrew the Mongols in 1368 A.D. and founded the Ming Dynasty, which lasted until 1644 A.D.. After the Ming dynasty, came the Qing (Manchu) dynasty, which lasted until the overthrow of Puyi in 1911 A.D. During the era after the war, later called the *Pax Mongolica*, adventurous Westerners such as Marco Polo travelled all the way to China and brought the first reports of its wonders to Europe. Despite the xenophobia and intellectual introspection characteristic of the increasingly popular new school of neo-Confucianism, China under the early Ming Dynasty was not isolated. Foreign trade and other contacts with the outside world, particularly Japan, increased considerably. Chinese merchants explored all of the Indian Ocean, reaching East Africa with the voyages of Zheng He. During the Ming Dynasty the last construction on the Great Wall was undertaken to protect China from foreign invasions. The Qing Dynasty (1644-1911 A.D.) was founded after the defeat of the Ming, the last Han Chinese Dynasty, by the Manchus. Over the next half-century, the Manchus consolidated control of some areas originally under the Ming, including Yunnan. They also stretched their sphere of influence over Xinjiang, Tibet and Mongolia. But during the 19th century, Qing control weakened. Frustrated by the Qing court's resistance to reform and by China's weakness, young officials, military officers, and students began to advocate the overthrow of the Qing Dynasty and creation of a republic. By the 19th century, however, China's political, economic, and military influence declined relative to the growing regional power of Japan and the influence of Western powers. The imperial system in China ended with the establishment of the Republic of China (RoC) in 1912. After its victory in the Chinese Civil War, the Communist Party of China established the People's Republic of China (PRC) in 1949. The region is known by this name till date.

The ancient Chinese scholars and travellers possessed a rich heritage of geographical knowledge and this has passed on to the succeeding generations in spite of physical isolation from the West and the communication (language) barrier. Although, the European and Chinese worlds remained isolated, there are fascinating parallels in the concepts and methods of study. Written records of China's past date from the Shang Dynasty in 13th century B.C., and take the form of inscriptions on the bones or shells of animals—the oracle bones. The earliest Chinese writing system was developed at around 6000 B.C.[6] From the 3rd century onwards, the Chinese methods of documenting geography became more systematic. The earliest known reference to the application of a geometric grid and mathematically graduated scale to a map was contained in the writings of the cartographer Pei Xiu (224-271).[7] The ancient Chinese historian Ban Gu (32-92) started the trend of the Gazetteer in China, which became prominent in the Southern and Northern Dynasties period and Sui Dynasty. Local Gazetteers would feature a wealth of geographic information, although its cartographic aspects were not as highly professional as the maps created by professional cartographers[8]. From 5th century B.C. onward, Chinese geographical writing provided more concrete information and less legendary element. In later periods of the Song Dynasty (960–1279 A.D.) and Ming Dynasty (1368–1644 A.D.) there were much more systematic and professional approaches to geographic literature. The Song Dynasty scholar and government official, Fan Chengda (1126–1193), wrote the geographical treatise known as the 'Gui Hai Yu Heng Chi'.[9] It focused primarily on the topography of the land, along with the agricultural, economic and commercial products of Chinese provinces. The polymath Chinese scientist Shen Kuo (1031–1095) devoted a significant amount of his written work to geography, particularly geomorphology. The Ming Dynasty geographer Xu Xiake (1587–1641) travelled throughout the provinces of China to write his enormous geographical and topographical treatise, documenting various details of his travels, such as the locations of gorges, mineral beds, etc.[10] The Chinese were also concerned with documenting geographical information of foreign regions far outside. The most important contributions of Chinese scholars go to the fields of cosmology, human and regional geography, physical geography, mathematical and practical geography, economic geography and accounts of travels and explorations. The Chinese mathematics had developed the Decimal System, which was introduced to the Arab world in about A.D. 800 from the Indians. The Chinese were expert in map-making on the latitudinal-longitudinal grid, and this art is believed to have developed much earlier than in Greece or Rome. Also commendable are their observations on weather and climate, since they are known to have prepared a number of weather reports dating back several centuries before Christ. These reports are now preserved in the National Archives. The discovery of the rest of world by Chinese travellers is an aspect of the history of geography that is often overlooked in western writings. The ancient Chinese scholars and travellers possessed a rich heritage of geographical knowledge and this heritage has passed on to the succeeding generations in spite of physical isolation and the communication (language) barrier. In fact, the record of geographical work in ancient China is highly impressive. In fact, the developments that began here in ancient times continued up to the Age of Discovery and later.

However, with time the West became centre of continuously increasing scholarly research. Towards West the earliest developments took place in Western Asia (present

Turkey, Lebanon, Syria and Israel), Egypt and Europe, and various cultural groups, namely Sumerians, Assyrians, Babylonians, Phoenicians, Egyptians, Greeks and Romans were involved. Although, the credit the beginning of geographical writings in the West is given to Homer, a Greek scholar, but the earliest known map was made by the Pre-Greek civilizations (particularly Sumerians) about 2700 B.C. The scholars of Babylon had already collected a large number of data on the motions of stars and planets, and they had developed the concept that the position of the celestial bodies had a fundamental effect on human activities. The roots of geography lie in antiquity and can be traced back to the works of numerous ancient schools of thought, philosophers, scholars, mathematicians, administrators and explorers. But, most of the literature dealing with historical geography has a western bias, contributions of schools other than Greco-Roman being overlooked. The literary and other evidences suggest that that there were schools even before and also parallel to Greeks that have been instrumental in the process of evolution of Geography as a discipline of scholarship during the ancient period. Geography is still indebted to all these developments that took place in classical antiquity.

Pre-Greek Civilizations

The ancient civilization, particularly the Western, had a deep sharing and inheritance of culture, knowledge, science and technology from many of its preceding civilizations that emerged in Mesopotamian region, commonly known as the "cradle of civilization". Various cities here served as capitals of the Sumerian, Akkadian, Babylonian, Assyrian, Phoenician and Carthaginian empires. The region was famous as one of the riverine civilizations as that of Indus Valley in Indian subcontinent and the Yellow River Valley in China. Mesopotamian history extends from the emergence of urban societies in Southern Iraq in the 5th millennium B.C. to the arrival of Alexander the Great in 4th century B.C. which is seen as the hallmark of the Hellenization of the Near East. The earliest language written in Mesopotamia was Sumerian, an agglutinative language isolate and later the cuneiform script was invented. Mesopotamians used a sexagesimal (base 60) numeral system. This is the source of the current 60-minute hours and 24-hour days, as well as the 360 degree circle. This mathematical knowledge was used in mapmaking. The Babylonians were familiar with the general rules for measuring the areas and distances. This measurement for distances eventually was converted to a time-mile used for measuring the travel of the Sun, therefore, representing time. Mesopotamian people invented many technologies, notably the wheel, the most important mechanical invention in history; other inventions being flood control, water storage and irrigation techniques. They were also one of the first Bronze Age people in the world. Early on they used copper, bronze and gold, and later they used iron. The Sumerians and Babylonians developed the earliest system of economies. Food supply in Mesopotamia was quite rich due to the location of the two rivers, Tigris and Euphrates. The Tigris and Euphrates River valleys formed the north-eastern portion of the Fertile Crescent. Although land nearer to the rivers was fertile and good for crops, portions of land further from the water were dry and largely uninhabitable. This is why the development of irrigation was very important for settlers of Mesopotamia. The earliest known people of Fertile Crescent were the Sumerians, who lived in southern Mesopotamia. The challenge for any population attempting to dwell in Iraq's arid southern floodplain was to master the river waters for year-round agriculture and drinking water. The Sumerians were farmers, who had developed a social organization and a technology that enabled them, through their control of the water, to survive and prosper in a difficult environment.

Besides the cuneiform script, the Sumerians also developed astronomy, basic algebra, geometry and the number system based on the unit of 60. Using this number system, they developed mathematics and invented the clock and 12-month calendar still used today. An extremely large body of texts in Sumerian language has survived on clay tablets. With the invention of writing, Sumerians were able to record their knowledge and pass it on, particularly, to the Greeks and Romans. Sumerian written literature achieved a high degree of excellence in the Akkadian period. The name Akkad became synonymous with the Sumerians. Southern Mesopotamia became known as the "land of Sumer and Akkad". Combining the territories of Sumer and Akkad there emerged a new state, Babylonia, in southern Mesopotamia. Babylon was one of the most fertile areas despite its location in the desert. But, the country was completely destitute of mineral wealth, and possessed no stone or metal. The Babylonian civilization was like the Sumerian that preceded it, urban in character, although based on agriculture rather than industry. The Babylonians were an inventive and intelligent people who made large advances in the techniques of agriculture and science. Babylonians adopted the Sumerian style of writing. Among the sciences, astronomy and astrology occupied a conspicuous place in Babylonian society. Astronomy was of old standing here. Observatories were attached to the temples, and reports were regularly sent by astronomers to the king. The stars had been numbered and named at an early date. Great attention was naturally paid to the calendar. Babylonian astrology was based on the belief that the entire universe was created in relation to the earth. Thus the ancients saw it as no accident that the stars and planets were set in a certain divine order at the time of creation. The Babylonian texts contain catalogues of stars and constellations as well as schemes for predicting heliacal risings and the settings of the planets, lengths of daylight measured by a water-clock, gnomon, shadows, and intercalations. During the 8th and 7th centuries B.C., Babylonian astronomers developed a new approach to astronomy. They began studying philosophy dealing with the ideal nature of the early universe and began employing an internal logic within their predictive planetary systems. This was an important contribution to astronomy and the philosophy of science and the scholars have thus referred to this new approach as the first scientific revolution. This new approach to astronomy was adopted and further developed in Greek and Hellenistic astronomy.

The Babylonian development of methods for predicting the motions of the planets is considered to be a major episode in the history of astronomy. Since the Babylonian civilisation replaced the Sumerian and Akkadian civilizations, the Babylonians inherited their sexagesimal number system. The city of Assur, a colony of Babylonia, grew to turn out to be a great empire, Assyria, in the northern reaches of Mesopotamia, the southern half being Babylonia with Nineveh as its capital. Assyria spanned four countries: Syria, Turkey, Iran and Iraq. Two large areas comprised the Assyrian breadbasket: the Arbel plain and the Nineveh plain. The Assyrian culture showed a dramatic growth in science and mathematics in the pre-historic times. Among the great mathematical inventions of the Assyrians was the division of circle into 360 degrees and Assyrians were the first to invent longitude and latitude in navigation. Among others, there were phonecians, an enterprising maritime culture that spreaded across the Mediterranean from 1200 to 800 B.C. The Phoenicians owed a great deal of their prosperity to trade. Fernand Braudel, a French historian, remarked in 'The Perspective of the World' that Phoenicia was an early example of a 'world-economy' surrounded by empires. This league of independent city-state ports was ideally suited for trade between the Levant area[11], rich in natural resources, and the rest of the ancient world. The most important of their early cities were Tyre, Sidon, Berytus (modern Beirut), and Byblos. The most important Phoenician colony was at Carthage, others being in Sicily, Corsica, Sardinia, and Spain (modern Cadiz and Cartagena). The early Phoenician economy was built on timber sales, wood working, and cloth dyeing. The Phoenicians' initial trading partners were the Greeks. In the centuries following 1200 B.C., the Phoenicians formed the major naval and trading power of the region. The Phoenicians established commercial outposts throughout the Mediterranean, the most strategically important being Carthage in North Africa. The Phoenicians were the first state level society to make extensive use of the alphabet. The Phoenician alphabet arose around 1400 B.C. from a need to communicate with the diverse languages of their trading partners encircling the Mediterranean Sea. The alphabet served as the origin of the Hebrew, Aramaic, Greek and Arabic alphabets.

This system of writing consonants and leaving out vowels is still in use today. Through their maritime trade, the Phoenicians spread the use of the alphabet to North Africa and Europe where it was adopted by Greeks, who later passed it on to the Romans. Carthage, the civilization founded by Phoenician colonists, became a large and major power of the Mediterranean. Carthage was built on a promontory with inlets to the sea to the north and south. The city's location made it master of the Mediterranean's maritime trade. The Carthaginian Empire was one of the longest living and largest empires in ancient Mediterranean. The empire was in a constant struggle with the Roman Republic, which led to a series of conflicts known as the Punic Wars. After the third and final Punic War, Carthage was destroyed and fell into Roman hands from then on. The empire of Carthage depended heavily on its trade. Its trade relations with the Iberians and the naval might that enforced Carthage's monopoly on trade with tin-rich Britain and the Canary Islands allowed it to be the sole significant broker of tin and maker of bronze. Maintaining this monopoly was one of the major sources of power and prosperity for Carthage.[12] In addition to manufacturing, Carthage practiced highly advanced and productive agriculture, using iron plows, irrigation and crop rotation. Carthage's merchant ships, which surpassed even those of the cities of the Levant, visited every major port of the Mediterranean, Britain, the coast of Africa, and the Canary Islands. The commercial fleet of Carthage was comparable in size and tonnage to the fleets of major European powers in the 18th century. Carthage traded in almost every commodity wanted by the ancient world. When the Mesopotamian civilizations stagnated and declined, there arose in the Mediterranean an aggressive and inquisitive sea faring group of people—the Greeks. The Greeks assimilated technology and knowledge from almost all cultures they encountered—picking and choosing the best. The main feature of Greek thought represents the knowledge of these pre-Greek civilizations. These civilizations bequeathed later civilizations the sexagesimal system of numerals, the earliest known form of writing (cuneiform) and a system of alphabet, besides a number of valuable mathematical and astronomical contributions.

CONTRIBUTORS

The contributions to ancient geographical knowledge came largely from astronomers, astrologers, astro-physicists, mathematicians, scientists, philosophers and explorers belonging to Greco-Roman, Indian and Chinese realms. These all are listed in Table 3.2.

Greco-Roman

There is a long list of scholars and travellers, whose efforts laid the foundation of geographic knowledge in the West in ancient period. They included, for instance, Homer, Thales, Anaximander, Pherecydes, Anaximenes, Pythagoras, Xenophanes, Hecateus, Heraclitus, Parmenides, Leucippus, Anaxagoras, Protagoras, Empedocles, Zeno of Elea, Gorgias, Herodotus, Philolaus, Socrates, Melissus, Prodicus, Hippias, Democritus, Xenophon, Plato, Archytas, Eudoxus, Aristotle, Theophratus, Eudemus, Pyrrho, Epicurus, Zeno of Citium, Euclid, Archimedes, Chrysippus, Eratosthenes, Apollonius, Polybius, Crate, Hipparchus, Posidonius, Marcus Tullius Cicero, Strabo, Philo, Pliny the Elder, Ptolemy, Claudius Galen, Ammonius Saccas, Porphyry and Proclus. In addition, there were explorers like Hanno The Navigator, Pytheas and Alexander The Great.

The first European poet, Homer, was the author of two literary works, the 'Iliad' and 'Odyssey', which contain a great deal of geographical information and give an insight into early Greek civilization. Moving beyond traditional mythology, Thales was much involved in the problems of astronomy and provided explanations for cosmological events that traditionally involved supernatural entities. His successor Anaximander was the first person to have attempted to create a scale map of existing world and to have introduced the instrument functioning like a Sun-dial (*Gnomon*). Pherecydes is the first ancient author to introduce the everlastingness of time (*chronos*). Anaximenes, a younger contemporary of Anaximander and Thales, continued their philosophical inquiries into the first material principle, but with different results. He believed that air was divine and common substance, and that it is responsible for the creation, destruction, change or evolution of everything. Known as "the father of numbers", Pythagoras, a 6th century B.C. mathematician, believed that everything was related to mathematics and that numbers were the ultimate reality

TABLE 3.2 The Earliest Scholars and Explorers

- Acharya Kapil (3000 B.C.)
- Acharya Bharadwaj (800 B.C.)
- Baudhāyana (800 B.C.)
- Homer (8th Century B.C.)
- Thales (639-544 B.C.)
- Anaximander (610-547 B.C.)
- Pherecydes (600-550 B.C.)
- Acharya Charak (600 B.C./300 B.C.)
- Acharya Kanad (600 B.C.)
- Acharya Sushrut (600 B.C.)
- Laozi (6th Century B.C.)
- Anaximenes (585-525 B.C.)
- Pythagoras (582-507 B.C.)
- Xenophanes (570-480 B.C.)
- Gautama Buddha (563 to 483 B.C.)
- Confucius (551-479 B.C.)
- Hecateus (550-476 B.C.)
- Heraclitus (535-475 B.C.)
- Hanno the Navigator (500 B.C.)
- Parmenides (510 B.C.)
- Leucippus (500-450 B.C.)
- Anaxagoras (500-428 B.C.)
- Protagoras (490-420 B.C.)
- Empedocles (490-430 B.C.)
- Zeno of Elea (490-430 B.C.)
- Gorgias (487-376 B.C.)
- Herodotus (485-425 B.C.)
- Philolaus (480-385 B.C.)
- Socrates (470-399 B.C.)
- Melissus of Samos (470 B.C.)
- Prodicus (465-415 B.C.)
- Hippias (460-399 B.C.)
- Democritus (460-370 B.C.)
- Xenophon (430-354 B.C.)
- Plato (428-348 B.C.)
- Archytas (428-347 B.C.)
- Eudoxus (408-347 B.C.)
- *Pāṇini* (4th century B.C.).
- Aristotle (384-322 B.C.)
- Pytheas (380-310 B.C)
- Theophratus (372-7 B.C.)
- Eudemus of Rhodes (370-300 B.C.)
- Pyrrho (360-0 B.C.)
- Alexander the Great (356-323 B.C.)
- Epicurus (342-270 B.C.)
- Zeno of Citium (334-262 B.C.)
- Euclid (325-265 B.C.)
- Archimedes of Syracuse (287-212 B.C.)
- Chrysippus (280-207 B.C.)
- Eratosthenes (276-194 B.C.)
- Apollonius (262-190 B.C.)
- Polybius (203-122 B.C.)
- Crate (2nd Century B.C.)
- Hipparchus (190-120 B.C.)
- Chang-Chien (172-114 B.C.)
- Sima Qian (145-90 B.C.)
- Posidonius (135-51 B.C.)
- Marcus Tullius Cicero (106-43 B.C.)
- Strabo (64 B.C.-20 A.D.)
- Philo (20 B.C.–50 A.D.)
- Ban Gu (32-92)
- Chang Heng (78-139)
- Nagarjuna (100 A.D.)
- Ptolemy (90-168 A.D.)
- Claudius Galen (129-200 A.D.)
- Ammonius Saccas (176-242 A.D.)
- Pei Xiu (224-271 A.D.)
- Porphyry (234-305 A.D.)
- Proclus (411-485 A.D.)
- Āryabhatta I (476-550 A.D.)
- Varahamihir (499-587 A.D.)
- Brahmagupta (598-668 A.D.)
- Bhāskara I (600-680 A.D.)
- Huan-Tsang (602-664 A.D.)
- I Ching (635-713 A.D.)
- Jia Dan (730-805 A.D.)
- Madhav (8th century)
- Adi Shankara (788-820 A.D.)
- Aryabhata II (about 920 A.D.)
- Sridharacharya (991 A.D.)
- Su Song (1020-1101 A.D.)
- Shen Kuo (1031-1095 A.D.)
- Brahmadeva (1060-1130 A.D.)
- Bhāskara II or Bhaskaracharya (1114-1183 A.D.)
- Fan Chengda (1126-1193)
- Jayadeva (1200 A.D.)
- Zhou Daguan (13th century)
- Zheng He (1371-1433)
- Xu Xiake (1587-1641 A.D.)

for everything that could be predicted and measured. He was the first to propose a spherical world. Plato embraced this idea and Aristotle presented evidence to verify this. Turning his back on anthropomorphic polytheism of Homer and others, Xenophanes, the founder of Eleatic school, maintained that there was only one god—namely the world, the one incorporeal eternal being, and spherical in form. Because of his development of the concept of a "one god" that is abstract, universal, unchanging, immobile and always present, Xenophanes is often seen as one of the first monotheists in Western philosophy. Hecateus, a historian, initiated the prose form of geography. To him goes credit of originating the 'literary tradition' in geography. The Greeks credit Hecateus with two major works entitled *Ges Periodos* ('Travels round the Earth' or 'World Survey') and *Genealogiai* (an account of the traditions and mythology of Greeks), a break with the epic myth-making tradition. His works came to be known as 'new geography' of his times. Heraclitus, a pre-Socratic philosopher, shifted his focus from the cosmic to the human realm for the first time. He propounded a distinctive theory in which he expressed that things are constantly changing (universal flux), that opposites coincide (unity of opposites), and that fire is the basic material of the world. The Stoics adopted Heraclitus's physical principles as the basis for their theories.

Promoting the Eleatic philosophy as its chief representative, Parmenides, a disciple of Xenophanes, authored the work, *On Nature*, a poem. The work has two divisions describing two views of reality, viz. the *Way of Truth* and the *Way of Opinion*. The first discusses the truth and the second the world of illusion—i.e. the world of the senses and the erroneous opinions of mankind founded upon them. Parmenides attempted to distinguish between the unity of nature and its variety, insisting in the *Way of Truth* upon the reality of its unity, which is therefore the object of knowledge, and upon the unreality of its variety, which is therefore the object, not of knowledge, but of opinion. In the *Way of Opinion* he propounded a theory of the world of seeming and its development, pointing out however that, in accordance with the principles already laid down, these cosmological speculations do not pretend to anything more than mere appearance. Under the *Way of Opinion*, Parmenides set out a contrasting but more conventional view of the world, thereby becoming an early exponent of the duality of appearance and reality. For him, the phenomena of movement and change are simply appearances of a static, eternal reality. These thoughts strongly influenced Plato, and through him, the whole history of Western philosophy. Leucippus' theory originated from Parmenides' denial of the void, from which the impossibility of multiplicity and motion had been deduced. Leucippus was the first philosopher to affirm the existence of empty space. The Pythagorean void had been more or less identified with 'air', but the void of Leucippus was really a vacuum. Leucippus was among the earliest philosophers of Atomism. His pupil, Anaxagoras, a contemporary of Leucippus, is considered to be both the geographical and theoretical successor to the earliest Ionian philosophers. He was, in fact, influenced by two strains in early Greek thought. First, there is the tradition of inquiry into nature founded by the Milesians and carried on by Xenophanes and by Heraclitus. Second, there is the influence of Eleatic arguments due to Parmenides, concerning the metaphysical requirements for a basic explanatory entity within this Milesian framework. While Anaxagoras proposed theories on a variety of subjects, he is most noted for two theories. First, he speculated that in the physical world everything contains a portion of everything else. The second theory is Anaxagoras' postulation of Mind (*Nous*) as the initiating and governing principle of the cosmos. Anaxagoras propounded a physical theory of "everything-in-everything," and claimed that *nous* (mind or intellect) was the motive cause of the cosmos. Belonging to the group of Older Sophists, Protagoras, founded a new method of discussion, known as 'antilogies', i.e. finding a better argument by discarding the less viable one. Along with Protagoras, Gorgias forms the first generation of Sophists. He used to go to great lengths to exhibit his ability of making an absurd, argumentative position appear stronger. The last Greek philosopher to write in verse was Empedocles. In his two works, *On Nature* and *Purifications,* lies the origin of cosmogenic theory of the four classical elements—fire, air, earth, and water; these elements are moved by two opposing forces, Love and Strife. Empedocles four-element theory of matter was one of the earliest theories of particle physics, put forward to rescue the phenomenal world from the static monism of Parmenides. Empedocles' world-view is of a cosmic cycle of eternal change, growth and decay, in which two personified cosmic forces, Love and Strife, engage in an eternal battle for supremacy. Parmenides had taught that the world of sense is an illusion because it consists of motion (or change) and plurality (or multiplicity or the many). True Being is absolutely one; there is in it no plurality; and it is absolutely static and unchangeable. Commonsense says there is both motion and plurality. This is the Pythagorean notion of reality against which Zeno directed his arguments. Zeno of Elea, a member of the Eleatic School, laid the foundations of modern logic. Popularly known as the 'Father of History', Herodotus' narrative style served as a model for historical writing. He is known for writing the *Histories,* a work similar to that of Hecateus, his predecessor. Following the ancient Pythagorean tradition, Philolaus was the contemporary of the mathematician, Hippocrates. Like most philosophers of 5th century his work is a response to Parmenides. He argued that the universe is determined by numbers. He is credited with originating the theory that the earth was not the centre of the universe. Philolaus argues that the cosmos and all matter are composed of two basic things, limiteds and unlimiteds. Unlimited includes the traditional Pre-Socratic material elements such as earth, air, fire and water but also space and time; they are undefined by any structure or quantity. Limited set limits in such unlimiteds and include shapes and other structural principles. The cosmos comes to be when the unlimited fire is fitted together with the centre of the cosmic sphere (a limited) to become the central fire. It is, in fact, Socrates who is widely credited for laying the foundations for Western Philosophy. Socrates' contribution to Western thought is his *dialectic* method of inquiry, known as the *Socratic Method*. For this, Socrates is customarily regarded as the father of political philosophy and ethics, and as a fountainhead of all the main themes in Western philosophy.

Another pupil of Parmenides, whose philosophical ideas he extended, was Melissus of Samos. His works consist of long series of arguments. Being, he says, is infinite, having no beginning or end. He goes on to show that eternal being must also be unlimited in magnitude and, therefore, one and unchangeable. Following this logic Melissus also posits that movement is impossible as it is always movement into nothing or void, which according to monism cannot exist. Belonging to the first generation of Sophists was Prodicus, a rhetorician and a disciple of Protagoras. While Protagoras' chief instruments of education were rhetoric and style, Prodicus made linguistics prominent in his curriculum. Prodicus' outlook, like those of his fellow Sophists, was humanistic, and he interpreted religion through the framework of naturalism. Hippias, a younger contemporary of Protagoras and Socrates, is also claimed to be regarded as an authority on poetry, grammar, history, politics, archaeology, mathematics and astronomy. He invented a system of mnemonics[13]. Democritus, a pioneer of mathematics and geometry, was a student of Leucippus and co-originator of atomist theory. Democritus was the first philosopher we know to realize that the celestial body we perceive as the Milky Way is formed from the light of distant stars. Democritus was among the first to propose that the universe contains many worlds, some of them inhabited. Xenophon is known for his writings on the history of his own times, preserving the sayings of Socrates, and the life of ancient Greece. Plato is, by any reckoning, one of the most dazzling writers in Western literary tradition and one of the most penetrating, wide-ranging, and influential authors in the history of philosophy. He was a student of Socrates and the teacher of Aristotle. Plato's dialogues are philosophical discussions. It is from Plato that we get *the theory of Forms*. Plato extended his theory beyond the realm of mathematics. Plato's influence throughout the history of philosophy has been monumental. Neoplatonism, founded by the 3rd century philosopher Plotinus, was an important later development of Platonism. Platonic ideas have had a crucial role in the development of Christian theology and also in medieval Islamic thought. The tradition identifies Archytas as a Pythagorean. He was first to identify the group of four canonical sciences (logistic/arithmetic, geometry, astronomy and music), which would become known as the *quadrivium* in the middle ages. Trained in the Platonic method, Eudoxus of Cnidus continued to work out his own philosophy as well. His fame is due to the introduction of the astronomical globe and his early contributions to understanding the movement of planets. His work on proportions shows tremendous insight into numbers. Eudoxus used the idea of a sphere to explain how the sun created differing climatic zones based on latitude. This led the Greeks to believe in a division of world into five regions. The work of Eudoxus and Archimedes as precursors of calculus was only exceeded in mathematical sophistication and rigour by Indian Mathematician Bhaskara II (13th century) and by Isaac Newton (17th century). Aristotle, the student of Plato and teacher of Alexander the Great, founded a school of his own, called the "peripatetic school". Aristotle is one of the few figures in history who studied almost every subject possible at the time. As a mathematician, Aristotle made important contribution by systematizing deductive logic. He placed great value on knowledge gained from senses and would correspondingly be better classed among modern empiricists. His combined works practically constitute an encyclopaedia of Greek knowledge. He was the author of a philosophical and scientific system that through centuries became the support and vehicle for both medieval Christian and Islamic scholastic thought. Until the end of 17th century, Western culture was Aristotelian and, even after the intellectual revolutions of centuries to follow, Aristotelian concepts and ideas remained embedded in Western thinking. His student, Theophratus, the immediate successor of Aristotle, held Aristotle's writings and also presided over the Peripatetic School for 35 years. The most important of his works are two large botanical treatises, *On the History of Plants* and *On the Causes of Plants*. Although Theophrastus generally followed Aristotle's lead in philosophy, he continued important empirical and philosophical investigations of his own as well. Important among other students of Aristotle, was Eudemus, the first historian of science. Although Eudemus too conducted original research, his *forte* lay in systematizing Aristotle's philosophical legacy, and in a clever didactical presentation of his teacher's ideas. At the insistence of Aristotle, Eudemus wrote histories of Greek mathematics and astronomy. The philosophy of Pyrrho belonged to the post-Aristotelian systems and is known as Pyrrhonism. Pyrrho is credited as being the first Skeptic philosopher. Epicurus, a key figure in the Axial Age[14], was the founder of the school of Epicureanism. Epicurus taught that the basic constituents of the world are atoms, undivisible bits of matter, flying through empty space, and he tried to explain all natural phenomena in atomic terms. Zeno of Citium was the founder of the Stoic school of philosophy, which along with its rival Epicureanism, came to dominate the thinking of Hellenistic world, and later, the Roman Empire. Based on the moral ideas of the Cynics, Stoicism laid great emphasis on the goodness and peace of mind which would arise from living a life of virtue in accordance with nature. Euclid was a mathematician and is often referred to as the Father of Geometry. He is best known for his treatise on mathematics *The Elements*. The geometrical system described in the *Elements* is often referred to as *Euclidean Geometry* to distinguish it from other so-called *Non-Euclidean Geometries* that mathematicians discovered in the 19th century. Often regarded as a designer of mechanical devices, Archimedes also made contributions to the field of mathematics. Archimedes used the method of exhaustion to approximate the value of P. Archimedes was able to use infinitesimals in a way that is similar to modern integral calculus. A prolific writer, Chrysippus expanded the fundamental doctrines of Zeno of Citium, which earned him the title of Second Founder of Stoicism. Chrysippus was one of the first to organise propositional logic as an intellectual discipline. This allowed Stoics to make major advances in mathematics and science. As Homer was master among poets, Chrysippus came to be renowned as one of the foremost logicians of ancient Greece. Among all, Eratosthenes is often identified as the 'Father of Geography' because, he was the first to coin the term "geography". The size and shape of Earth were important questions to ancient Greeks. Eratosthenes attempted to calculate its circumference and Posidonius improved upon these calculations. A Greek geometer and astronomer noted for his writings on conic sections was Apollonius of Perga. His innovative methodology and terminology, especially in the field of conics, influenced many later scholars including Ptolemy, Isaac Newton and others. It was Apollonius who gave the ellipse, the parabola, and the hyperbola the names by which we know them. The hypothesis of eccentric orbits and epicycles to explain the apparent motion of planets are attributed to him. Polybius, a Greek historian and statesman of the Mediterranean world, wrote on the rise of

Rome to world prominence. Polybius provided a unified view of history rather than a chronology. His views are contained in his *The Histories*, the 40-volume book, describing the Constitution of Rome and the sequence of Roman conquests. It has strong political slant and represents an early attempt at a universal history. Crates, a 2nd Century B.C. Stoic philosopher and grammarian, is said to have made one of the earliest globes representing the Earth. Crates gave the greatest attention to the works of Homer, and was also surnamed *Homerikos*. One of the most influential astronomers of antiquity was Hipparchus, whom Ptolemy considered as his most important predecessor and made extensive use of his work. Hipparchus was the first whose quantitative models for the motion of the Sun and Moon survive. For this he made use of the observations and mathematical techniques accumulated over centuries by Babylonians. He developed trigonometry and constructed trigonometric tables, and solved several problems of spherical trigonometry. With his solar and lunar theories and trigonometry, he was the first to develop a reliable method to predict solar eclipses.

Hipparchus is also credited with the invention or improvement of several astronomical instruments, such as the gnomon, the astrolabe, and the armillary sphere. His famous star catalog was incorporated into the one by Ptolemy. Hipparchus seems to have been the first to exploit Babylonian astronomical knowledge and techniques systematically. He was the first Greek known to divide the circle in 360 degrees of 60 arc minutes. Hipparchus was the first to show that the stereographic projection is conformal. With an astrolabe Hipparchus was the first to measure the geographical latitude and time by observing stars. Noted for his learning, Posidonius gave a new life to *stoicism* by fortifying it with contemporary learning. He made contribution to stoic, physics and ethics—notably the theory that a vital force emanating from the sun permeated the world, through which man and all things in the universe are united. Marcus Tullius Cicero introduced Romans to Greek philosophy. An impressive orator and successful lawyer, Cicero is appreciated primarily for his humanism and philosophical and political writings. While Cicero the humanist deeply influenced the culture of Renaissance, Cicero the republican inspired the Founding Fathers of the United States and the revolutionaries of the French Revolution.

Strabo, another philosopher of the Roman Empire, was a historian and geographer. Strabo's work is one of the most important sources of information on classical geography. In his *Geography*, he provides us with a fascinating description of the 'Inhabited World', as envisaged at that time. The work is in fact a sort of encyclopaedia of the ancient world. A century after Strabo, Ptolemy launched a similar undertaking. By this time the Roman Empire had expanded through much of Europe, and previously unknown areas such as the British Isles had been explored. The Silk Road was also in operation, and for the first time knowledge of the Far East began to be known. Philo of Alexandria, a Hellenized Jew, is a figure that spans two cultures, the Greek and the Hebrew. Philo produced a synthesis of both traditions. In the process, he laid the foundations for the development of Christianity. Philo's philosophy represented contemporary Platonism.

Gaius Plinius Secundus, better known as Pliny the Elder, wrote an encyclopaedic work, *Naturalis Historia*, which became a model for all such works written subsequently.

The *Naturalis Historia* is one of the largest single works to have survived from the Roman Empire to modern day and purports to cover the entire field of ancient knowledge. The work became a model for all later encyclopaedias in terms of the breadth of subject matter examined, the need to reference original authors, and a comprehensive index list of the contents. Pliny is still remembered in volcanology where the term *Plinian* (or *Plinean*) refers to a very violent eruption of a volcano marked by columns of smoke and ash extending high into the stratosphere. Ptolemy, the ancient scholar, attempted to advance the sciences of astronomy and geography. He codified the Greek geocentric view of the planets, and rationalized the apparent motions of the planets, as they were known in his time. He was the author of several scientific treatises, three of which have been of continuity importance to later Islamic and European science. They are: the astronomical treatise, *Almagest;* a description of the Greco-Roman world, the *Geography;* and the astrological treatise, the *Tetrabiblos.*

Claudius Galen was the most accomplished medical researcher of Roman period. His theories dominated and influenced Western medical science for well over a millennium. He also wrote about philosophy and believed that Nature expressed a divine purpose, a belief that became increasingly popular with the rise of Christianity. His writings pay homage to, amongst others, Plato, Aristotle and the Stoics, but above all to Hippocrates, whom he refers to as "divine". Hippocrates and Galen form important landmarks of 600 years of Greek medicine. Ammonius Saccas is often referred to as one of the founders of Neoplatonism. He is mainly known as the teacher of Plotinus and was the biggest influence on Plotinus in his development of Neoplatonism. Porphyry was a key figure in promulgation of Neoplatonic thought. His writings on Aristotle's logical works contain attempts to harmonize Aristotle's logical writings with Platonism. He applied Neoplatonic doctrines to traditional pagan religion. Proclus Lycaeus was another Neoplatonist and one of the last major Classical philosophers. He set forth one of the most elaborate and fully developed systems of Neoplatonism. Standing near the end of the classical development of philosophy, he was very influential on Western Medieval Philosophy (Greek and Latin) as well as Islamic thought.

The theories of ancient Greek scholars clashed with the evidence of explorers. Hanno the Navigator, for instance, travelled as far as Sierra Leone and circumnavigated Africa in 5th century B.C. In 4th century B.C. the Greek explorer Pytheas travelled through northwest Europe, and encircled the British Isles. He found that the region was considerably more habitable than expected, but his discoveries were largely dismissed as fanciful by his contemporaries. Conquerors also carried out exploration, for example, Caesar's invasions of Britain and Germany, expeditions sent by Augustus to Arabia, and the military expeditions of Alexander the Great, the Macedonian king. Hanno the Navigator was a Carthaginian explorer. Carthage dispatched Hanno at the head of a fleet of sixty ships to explore and colonize the northwestern coast of Africa, he sailed through the straits of Gibraltar, founded several colonies along the African coast of Morocco, and explored further along the Atlantic coast of the continent. At the terminus of his voyage Hanno found an island heavily populated with what were described as hirsute and savage people. The interpreters called them *gorillae*. Pytheas made a voyage to explore northwestern Europe around 325 B.C. He travelled over a considerable part of Great Britain, circumnavigating in between 330 and 320 B.C. Pytheas was first to describe the midnight sun, the aurora and polar ice, and the first to mention Germanic tribes. Pytheas described his travels in a document entitled *On the Ocean*. Pytheas estimated the circumference of Great Britain within 2.5 per cent of modern estimates. He used the Pole Star to fix latitude and understood the relationships between tides and phases of the Moon. In northern Spain, he studied the tides and discovered that they are caused by the Moon. Pytheas obtained latitudes, which he expressed in proportions of the gnomon, or trigonometric tangents of angles of elevation to celestial bodies. Pytheas sailed northward with the intent of locating the Arctic Circle and exploring the 'frigid zone' at the extreme of earth. Pytheas visited an island six days sailing north of Britain, called Thule and noted that the night in Thule was only two to three hours. The greatest explorer of the Ancient Greeks was Alexander the Great, who was the king of Macedonia and one of the greatest Generals in history. Alexander deliberately set out to learn more about the east through his military expeditions. With his army he brought a large number of geographers and writers who recorded their observations as they moved east. Alexander conquered much of what was then the civilized world and brought Greek culture to all the countries he conquered. He conquered much of Persia, Egypt and Asia. Alexander founded a city on a strip of land between Lake Mareotis and the Mediterranean Sea.

This city, Alexandria, became a world centre of commerce and learning.[15] His conquests included Anatolia, the Levant, Egypt, Bactria and Mesopotamia, and he extended the boundaries of his own empire as far as Punjab, India. He planned to make Asia and Europe one country and combine the best of the East with the West. He chose Babylon as his capital city. He encouraged the spread of Greek ideas, customs, and laws into Asia. Alexander had vast plans, including his governmental reorganization and an expedition to Arabia, but he died in 323 B.C. Alexander died after twelve years of constant military campaigning. By the time of his death, he had conquered the Achaemenid Persian Empire, adding it to Macedon's European territories; this was most of the world as known to the ancient Greeks. His legacy and conquests lived on long after him and ushered in centuries of Greek settlement and cultural influence over distant areas. This period is known as the Hellenistic period, which featured a combination of Greek, Middle Eastern and Indian culture. Alexander's conquests also had long-term cultural effects, with the flourishing of Hellenistic civilization throughout the Middle East and Central Asia. At the time of Alexander there existed the concept of *oikouménē* (Ecumene), meaning literally "the inhabitated world". Alexander, with his conquers, changed it to mean "the civilized world", the group of civilized nations which has common interests, as opposed to the barbarian nations. Throughout the Roman world, the one language spoken everywhere was Alexander's Greek. Alexander is remembered as a legendary hero in Europe and much of both Southwest and Central Asia.

Indian

The ancient Indian *Rishis* were greatest scientists ever present who were adepts in all fields known and unknown to humanity. Some of these scholars, as discussed below, include Acharya Kapil, Acharya Bharadwaj, Baudhāyana, Acharya Charak, Acharya Kanad, Acharya Sushrut, Gautama Buddha, Pānini, Nagarjuna, **Āryabhatta I,** Varahamihir, Brahamagupta, Bhāskara I, Adi Shankara, Aryabhata II, Sridharacharya, Brahmadeva and Bhaskaracharya. India's priceless heritage is unveiled by Acharya Kapil as the founder of Sankhya philosophy. Acharya Kapil's concept of transformation of energy and profound commentaries on the soul (*atman*), nature, matter and the subtle elements of cosmos places him in a class by himself, incomparable to the discoveries of other cosmologists. He contended that 'Prakriti' and 'Purusha' are the mother of cosmic creation and all energies. With this he contributed a new chapter in the science of cosmology. His extraordinary and extrasensory perceptions, observations and revelations on the secrets of creation, brought him the recognition as the Father of Cosmology. Through his innovative and brilliant discoveries, Acharya Bharadwaj has been recognized as the pioneer of space science. He authored the "*Yantra Sarvasva*" which includes astonishing and outstanding discoveries in aviation science, space science and flying machines. Belonging to Yajurveda School Baudhāyana was an Indian mathematician, who was also a priest. He is noted as the author of the earliest *Sulba Sutra*—appendices to the Vedas giving rules for the construction of altars—called the *Baudhāyana Śulbasūtra* (texts dealing with geometry and mathematical principles). Baudhayana wrote the *Śulbasūtra* to provide rules for religious rites. The mathematics in *Śulbasūtras* enabled the accurate construction of altars needed for sacrifices.

Charaka was one of the principal contributors to the ancient art and science of Ayurveda, a system of medicine and lifestyle thought to be developed about 5000 years ago. *Charak Samhita* is his most renowned work, considered as the encyclopaedia of Ayurveda. In *Charak Samhita* he described the functions and the medicinal properties of 100,000 herbal plants. As the founder of "*Vaisheshik Darshan*"—one of six principal philosophies of India—Acharya Kanad was a genius in philosophy. He was the pioneer expounder of realism, law of causation and the atomic theory. He has classified all the objects of creation into nine elements, namely: earth, water, light, wind, ether, time, space, mind and soul. He believed that 'Every object of creation is made of atoms which in turn connect with each other to form molecules.' His statement ushered in the Atomic Theory for the first time ever in the world nearly 2500 years before John Dalton[16]. Kanad also described the dimension and motion of atoms and their chemical reactions with each other. The eminent historian, T.N. Colebrook, has said, 'Compared to the scientists of Europe, Kanad and other Indian scientists were the global masters of this field.' Acharya Sushrut, born to sage Vishwamitra, was recognized as a genius in medical science. The medical treatise *Sushruta Samhita*—compiled in Vedic Sanskrit—is attributed to him. A unique encyclopaedia of surgery, 'Sushruta Samhita' is an unparalleled work of the medical science of ancient India, popular as Ayurveda. Siddhārtha Gautama was the founder of Buddhism. Siddhārtha tried to find enlightenment through near total deprivation of worldly goods, practising self-mortification.

The Buddhist system of insight and meditation practice is not believed to have been revealed divinely, but by the understanding of the true nature of the mind, which must be discovered by personally treading a spiritual path guided by the Buddha's teachings. Pāṇini is known for his Sanskrit grammar, particularly for his formulation of the rules of Sanskrit morphology in the grammar, known as *Ashtadhyayi* (meaning 'eight chapters'). It is the earliest known work on descriptive linguistics, generative linguistics, and stands at the beginning of the history of linguistics. Pāṇini's grammar defines Classical Sanskrit. Nagarjuna was an extraordinary wizard of science. His dedicated research produced maiden discoveries and inventions in the faculties of chemistry and metallurgy. Textual masterpieces like "*Ras Ratnakar*," "*Rashrudaya*" and "*Rasendramangal*" are his renowned contributions to the science of chemistry. Indian scientists have shaped the course of mathematics and astronomy for the world to marvel upon. One of the very early pioneers in astronomy and mathematics was **Aryabhatta-I**, the first in the line of great mathematician-astronomers from the classical period. His most famous works are the *Aryabhatiyam* and *Arya-Siddhanta*. 'Aryabhatiyam.' (498 B.C.), written in verse, was the summary of Hindu mathematics up to the time. Aryabhatiyam presented a number of innovations in mathematics and astronomy which were influential for many centuries. The work was recognized as a masterpiece and was translated into Latin in 13th century. Through this translation, European mathematicians got to know methods for calculating the areas of triangles, volumes of spheres as well as square and cube root.

Aryabhatta was the first one to have put forward the theory that the earth is a sphere. He also presented ideas about eclipses and the sun being the source of moonlight. Considering that Aryabhatta discovered these facts 1,500 years ago, and 1,000 years before Copernicus and Galileo, makes him an early pioneer of this field. He was also the earliest to discover that the orbits of the planets around the Sun are ellipses. He made an accurate approximation of the Earth's circumference and diameter, and also discovered how the lunar and solar eclipses happen for the first time. He is the first known astronomer to have used a continuous system of counting solar days. His working was later adopted by Greeks and Arabs. In his name has been launched the First Indian Experimental Satellite on April 19, 1975. Calendric calculations worked out by Aryabhatta have been in continuous use in India for the practical purposes of fixing the *Panchangam*, the Hindu calendar. These were also transmitted to Islamic world, and formed the basis for the Jalali calendar introduced in 1073, versions of which are the national calendars in use in Iran and Afghanistan today. Varahmihira is known through his unique contributions to the science of astronomy and astrology. His book '*Panchsiddhant*' holds a prominent place in the realm of astronomy and his *Bṛahat Samhita* is an encyclopaedia of astrology. Varahmihira is considered the greatest name among all the authors on *Jyotisa*. He enriched all the major branches of the science, as *Gaṇita* (Vedic mathematics) and *Samhita* (meaning 'joined' or 'collected'). *Samhita* is the basic text of each of the *Vedas*, comprising collections of hymns and ritual texts. What make Varahmihira unique among ancient scientists are his versatility, encyclopaedic knowledge, poetic talent, and his deep grounding in Sanskrit grammar. Brahmagupta, another Indian mathematician and astronomer, belonged to the Ujjain School. He was the head of the astronomical observatory at Ujjain, the foremost mathematical centre of ancient India. During his tenure there, he wrote two texts on mathematics and astronomy: the *Brahmasphutasiddhanta* (The Opening of Universe) and the *Khandakhadyaka*. The *Brahmasphutasiddhanta* is the earliest known text to treat 'zero' as a number in its own right and this work is a compendious volume of astronomy. Bhāskara is commonly called Bhāskara-I to avoid confusion with the 12th century mathematician Bhāskara II. Along with Brahmagupta, he is the most renowned Indian mathematician of 7th century. He was the first to use a circle for zero in the Hindu-Arabic decimal system. He is considered to be a follower of Aryabhatta I and one of the most renowned scholars of Aryabhatta's astronomical school. Bhaskara I wrote two treatises, the *Mahabhaskariya* and the *Laghubhaskariya*. The *Mahabhaskariya* deals with mathematical astronomy. Some of the contributions of Bhaskara I to mathematics are: numbers and symbolism, the categorization of mathematics, the names and solution of the first degree equations, quadratic equations, cubic equations and equations which have more than one unknown value, symbolic algebra and the algorithm method to solve linear indeterminate equations which was later suggested by Euclid.

The teachings of Adi Shankara are based on the unity of the soul and Brahman. Shankara travelled across India and other parts of South Asia to propagate his philosophy through discourses and debates. His works concern themselves with establishing the doctrine of *Advaita* (Nondualism). He travelled widely, preaching to the local populace and debating philosophy with Hindu, Buddhist and other scholars and monks along the way. Adi Shankara founded four Mam̀has (monasteries) to guide Hindu religion. Adi Shankara stressed the importance of the Vedas and his efforts helped Hinduism regain strength and popularity. Adi Shankara, along with Madhava and Ramanuja, was instrumental in the revival of Hinduism. These three teachers formed the doctrines that are followed by their respective sects even today. In their writings and debates, they paved the way for Vedanta to be the dominant and most widely followed tradition among the schools of Hindu philosophy. Āryabhama II was an Indian mathematician and astronomer, and the author of the *Maha-Siddhanta*. He is so called to distinguish him from the earlier and more influential Āryabhama I. Āryabhama-II's most famous work is *Mahasiddhant*, written in Sanskrit, the work discusses topics on astronomy, geometry, geography and algebra. Sridhara is referred to as a distinguished mathematician. Two arithmetical works by him are popular—*Patiganita* and *Trisatika*. For multiplication, he uses a new term *pratyutpanna* (re-produced) and discusses the *kapata-sandhi* (door-junction) method which became very popular among later Hindu writers and was transmitted to the West through Arab works. Brahmadeva is known for his work, *Karanaprakasa*, which is a commentary on the *Aryabhatiya*. Topics covered include the longitudes of the planets, problems relating to their rotation, eclipses of the sun and moon, risings and settings, the lunar crescent, and conjunctions of the planets. The work also contains contributions to trigonometry, motivated by its application to mathematical astronomy. Bhāskara II or Bhaskaracharya was an Indian mathematician and astronomer who extended Brahmagupta's work on number systems. He is more famous as Bhaskara-II and is also known as Bhaskaracharya meaning 'Bhaskara the Teacher'. Bhaskara was the lineal successor of Brahmagupta as the head of astronomical observatory in Ujjain. The outstanding mathematicians such as Varahmihira and Brahmagupta had worked there and built up a strong school of mathematical astronomy. Bhaskara was a genius in algebra. He wrote the first work with a full and systematic use of the decimal number system. Bhaskara is regarded without question as the greatest mathematician of all time. He represents the peak of mathematical knowledge in the 12th century. He reached an understanding of the number systems and solving equations that was not to be achieved in Europe for several centuries. Bhaskaracharya's work in Algebra, Arithmetic and Geometry catapulted him to fame and immortality. He wrote his famous book *Siddhanta Shiromani* in 1150. It is divided into four parts: (arithmetic)[17], *Bijaganita* (algebra), *Goladhyaya* (celestial globe), and *Grahaganita* (mathematics of the planets). Their translations in several languages of world bear testimony to its eminence. In his treatise *'Siddhant Shiromani'* he writes on planetary positions, eclipses, cosmography, mathematical techniques and astronomical equipment. In *Siddhanta Shiromani,* he makes a note on the force of gravity: "Objects fall on earth due to a force of attraction by the earth. Therefore, the earth, planets, constellations, moon, and sun are held in orbit due to this attraction." Bhaskaracharya was the first to discover gravity, 500 years before Sir Isaac Newton. He was the leading mathematician of ancient and medieval India.

Chinese

A number of Chinese philosophers, travellers and scholars contributed to geographical knowledge in ancient period, beginning from Laozi and Confucius and followed by Zhang Qian, Sima Qian, Ban Gu, Chang Heng, Pei Xiu, Huan-Tsang, I Ching, Jia Dan, Sima Guang, Su Song, Shen Kuo, Fan Chengda, Chou Ta-Kuan, Zheng He and Xu Xiake[18]. Of these, Zhang he is known for his scholarly works; Confucius, Loazi and I Ching for their philosophical contribution and others for their travels. The contribution of these scholars goes to the fields of astronomy, astrology, human and regional Geography, physical Geography, mathematical and practical Geography, and accounts of travels and explorations.

Laozi was a central figure in Taoism (or "Daoism") in ancient China. Laozi's magnum opus, the *Daodejing,* is one of the most significant treatises in Chinese cosmogony. The *Daodejing,* often called simply *the Laozi* after its reputed author, describes the Dao (or Tao) as the mystical source and ideal of all existence: it is unseen, but not transcendent, immensely powerful yet supremely humble, being the root of all things. According to the *Daodejing,* humans have no special place within the Dao, being just one of its many manifestations. The influence of the *Laozi* on Chinese culture is both deep and far-reaching. Daoist beliefs and practices have contributed also to the formation of Korean and Japanese culture. The work has been translated into Sanskrit and several Western languages. Confucius was a thinker, political figure, educator, and founder of the *Ru* School of Chinese thought[19]. Confucius' influence in Chinese history is compared with that of Socrates in the West. His philosophy emphasized personal and governmental morality, correctness of social relationships, justice and sincerity. These values gained prominence during Han Dynasty in China. His lifetime almost exactly coincided with that of Buddha. Confucius' teachings were later turned into a very elaborate set of rules and practices by his numerous disciples and followers who organized his teachings into the *Analects*. The *Analects* depicts him spending his last year's teaching disciples and transmitting the old wisdom *via* a set of texts, the *Five Classics*. After more than a thousand years, the scholar Zhu Xi created a different interpretation of Confucianism which is now called Neo-Confucianism, to distinguish it from the ideas expressed in the *Analects*. Chang-Ch'ien or Zhang Qian was a Chinese Discoverer. Also a diplomat, he was an imperial envoy to the world outside of China in 2nd century B.C. Almost 1400 years before Marco Polo, Chang-Chien travelled as far west as Samarkand and Usbekistan. His travel report stimulated Chinese trade relations with the West, and led to the introduction into China of a superior breed of horses and new plants, such as grapes and alfafa. He was the first official diplomat to bring back reliable information about Central Asia to the Chinese imperial court, then under Emperor Wu of Han, and played a pioneering role in the Chinese colonization and conquest of the region now known as Xinjiang. Today Zhang Qian's travels are associated with the major route of transcontinental trade, the Silk Road. In essence, his missions opened up to China the many kingdoms and products of a part of the world then unknown to the Chinese. Zhang Qian visited directly the kingdom of Dayuan in Ferghana, the territories of the Yuezhi in Transoxiana, the Bactrian country of Daxia with it remnants of Greco-Bactrian rule, and Kangju. He also made reports on neighbouring countries that he did not visit, such as Anxi (Arsacid territories), Tiaozhi (Mesopotamia), Shendu (India) and the Wusun. Following Zhang Qian's embassy and report, commercial relations between China and Central as well as Western Asia flourished, initiating the development of Silk Road. Sima Qian laid the foundation for Chinese historiography. He journeyed throughout the country, collecting useful first-hand historical records for his main work, *Shiji*. *Shiji* is respected as a model of biographical literature with high literary value, and still stands as a "textbook" for the study of classical Chinese worldwide. Ban Gu was another prominent historian. He took over the responsibility for writing a history of the former Han Dynasty, a book known as the *Book of Han*. The book became a model for many other works about later dynasties. Chang Heng (or Zhang Heng) was the leading scientist of the Han Dynasty. He is credited with creating the first seismograph, devising a celestial globe, proposing the concept of lunar eclipse, developing graticule and inventing the odometer. A visionary mathematician, Heng computed the value of *pi*. He constructed a sundial to measure the position of sun. Heng perfected the science of latitude and longitude, and applied grid system to maps. His book, *Discourse on New Calculations,* established the basis for the mathematical use of the grid with maps. Heng's mechanical inventions influenced later Chinese inventors such as Yi Xing, Zhang Sixun, Su Song, and Guo Shoujing.

During the Western Jin Dynasty, there was a famous geographer and cartographer named Pei Xiu, whom the historian Joseph Needham (1900-1995) claimed was "the father of China's Scientific Cartography." Pei's major contribution to cartography lies in his establishment of a theoretical foundation for the making of traditional Chinese maps for the first time by summing up the experiences of his predecessors.[20] He put forward six principles in map-making, representing a milestone in the development of cartography. The six principles boil down to three factors: scale, direction and distance. Huan-tsang (or Xuanzang) was a famous Chinese Buddhist monk, scholar, traveller and translator that brought up the interaction between China and India in the early Tang period. Huan-tsang's travel accounts from the Silk Road provide reliable information about distant countries whose terrain and customs had been known, at that time, in only the sketchiest way. During his travels he studied with many famous Buddhist masters, especially at the famous centre of Buddhist learning at Nâlanda University. His book, *Great Tang Records on the Western Regions*, is the most detailed account of the countries of Central and South Asia. I Ching or Yi Jing was another Tang Dynasty Buddhist monk, originally named Zhang Wen Ming. The written records of his travels contributed to the knowledge of the ancient kingdom of Srivijaya (today's Palembang of Sumatra), as well as providing information about the other kingdoms lying *en route* between China and India. Jia Dan was a Chinese scholar-official, general, geographer and cartographer during the Tang Dynasty of China. In the 40-volume work that he completed in 801, Jia wrote of two common sea trade routes in his day: one from the coast of the Bohai Sea towards Silla in Korea and another from Guangzhou through Malacca towards the Nicobar Islands, Sri Lanka and India, the eastern and northern shores of the Arabian Sea to the Euphrates River. Sīmā Guāng was a Chinese historian, scholar, and high chancellor of the Song Dynasty. Sīmā Guāng is best remembered for his *Zizhi Tongjian*, the most famous historical work. Su Song was a renowned Chinese statesman, astronomer, cartographer, horologist, pharmacologist, mineralogist, zoologist, botanist, poet, antiquarian, and ambassador of Song Dynasty (960-1279).[21] Like his brilliant contemporary Shen Kuo, Su Song was a polymath. Su Song had in-depth knowledge of cartography. He made a celestial atlas which was a work of great precision. Shen Kuo or Shen Gua was another polymath of Song Dynasty excelling in many fields of study and statecraft. His famous work, *Meng Xi BiTan* (*Dream Pool Essays*) was a milestone in the history of Chinese science. Shen Kuo was hundreds of years ahead of the Western scientists in the fields of anatomy, astronomy, meteorology, geography, mineralogy, mathematics and the calendar making. Shen Kuo was the first to describe the magnetic needle compass, which would be used for navigation (first described in Europe by Alexander Neckam in 1187). Shen discovered the concept of true north in terms of magnetic declination towards the North Pole.[22] This was the decisive step in human history to make compasses more useful for navigation. He was the first to discover that the compass does not point directly north, but to the magnetic north pole. In the field of mathematics, he developed techniques that laid the foundations for spherical trigonometry and high-order arithmetic progressions. Shen wrote extensively about movable type printing invented by Bi Sheng (990-1051), and because of his written works the legacy of Bi Sheng and the modern understanding of the earliest movable type has been handed down to later generations.[23] The British sinologist, historian, and biochemist Joseph Needham (1900-1995) stated that Shen Kuo was 'one of the greatest scientific minds in Chinese history.'[24] Fan Chengda was one of the best-known Chinese poets, a government official, and an academic authority in geography of Song Dynasty. His written work falls under the category of 'travel record literature' (*youji wenxue*), a narrative and prose style approach to writing about one's travel experiences. He wrote an important geographical treatise known as the *Gui Hai Yu Heng Chi*; the book focused primarily on the topography of the land and commercial products of China's southern provinces. This followed a long geographical literary tradition, continuing from the *Shu Jing* (Historical Classic) of the 5th century B.C., the *Huainan Zi* of the 2nd century B.C., and predating the famous written works by the Ming Dynasty geographer Xu Xiake. Zhou Daguan or Chou Ta-Kuan was a Chinese diplomat under the Temür Khan, Emperor Chengzong of Yuan. He is most well known for his accounts of the customs of Cambodia and the Angkor temple complexes. Zhou's portrayal is today one of the most important sources of understanding of historical Angkor and the Khmer Empire. A series of parallels have been drawn between the voyage of Zhou Daguan and the travels of Marco Polo. Decades before Christopher Columbus sailed the ocean in search of a water route to Asia, the Chinese were exploring the Indian Ocean and Western Pacific with voyages of the "Treasure Fleet" that solidified Chinese control over much of Asia in the 15th century. The Treasure Fleets were commanded by a powerful eunuch admiral named Zheng He (or Cheng Ho). He was a Chinese explorer, voyaging between 1405 and 1433. Zheng He led seven expeditions (Table 3.3) to what the Chinese called 'the Western Ocean' (Indian Ocean).

TABLE 3.3 Seven Voyages of Zheng He

Order	*Time*	*Regions along the way*
1st Voyage	1405-1407	Champa, Java, Palembang, Malacca, Aru, Sumatra, Lambri, Ceylon, Kollam, Cochin, Calicut
2nd Voyage	1407-1409	Champa, Java, Siam, Cochin, Ceylon
3rd Voyage	1409-1411	Champa, Java, Malacca, Sumatra, Ceylon, Quilon, Cochin, Calicut, Siam, Lambri, Kaya, Coimbatore, Puttanpur
4th Voyage	1413-1415	Champa, Java, Palembang, Malacca, Sumatra, Ceylon, Cochin, Calicut, Kayal, Pahang, Kelantan, Aru, Lambri, Hormuz, Maldives, Mogadishu, Barawa, Malindi, Aden, Muscat, Dhufar
5th Voyage	1416-1419	Champa, Pahang, Java, Malacca, Sumatra, Lambri, Ceylon, Sharwayn, Cochin, Calicut, Hormuz, Maldives, Mogadishu, Barawa, Malindi, Aden
6th Voyage	1421-1422	Hormuz, East Africa, countries of the Arabian Peninsula
7th Voyage	1430-1433	Champa, Java, Palembang, Malacca, Sumatra, Ceylon, Calicut, Hormuz... (17 states in total)

Zheng's voyages are collectively referred to as the travels of *"Eunuch Sanbao to the Western Ocean"* or *"Zheng He to the Western Ocean"*. Xu Xiake was a Chinese travel writer and geographer of the Ming Dynasty (1368-1644) known best for his famous geographical treatise. He travelled throughout China for more than 30 years, documenting his travels extensively (which were compiled posthumously into the The Travel Diaries Xu Xiake).[25]

FIELDS OF LEARNING

The spheres of knowledge of these ancient scholars and travellers, through which they contributed to geography in ancient period, are summarized as follows (Table 3.4). The scholars like Aristotle, Posidonius and Hippias were polymaths whereas the others had shown the interest in certain specialized fields of learning.

TABLE 3.4 Ancient Fields of Learning and the Contributors

Philosophy:	Xenophanes, Heraclitus, Leucippus, Anaxagorus, Philolaus, Mellisus, Plato, Pyrrho, Epicurus, Zeno of Citium, Chrysipus, Strabo, Philo, Claudius Galen, Ammonius Saccas, Porphyry, Proclus, Socrates, Cicero, Gautam Buddha, Adi Shankara, Laozi, Confucius and Shen Kuo.
Astronomy:	Thales, Anaximander, Anaxagorus, Democritus, Eudoxus, Appolonius, Hipparchus, Ptolemy, Aryabhatta I and II, Brahamgupta, Bhaskara I, Varahmihir, Brahamdeva, Chang Heng and Shen Kuo.
Astrology:	Ptolemy and Varahmihir.
Mathematics:	Thales, Pythagoras, Democritus, Archytas, Eudoxus, Euclid, Archimedes, Eratosthenes, Hipparchus, Ptolemy, Shen Kuo, Baudhayana, Aryabhatta I and II, Brahamgupta, Bhaskara I, Sridharacharya and Brahamadeva.
Algebra:.	Bhaskaracharya.
Geometry:	Euclid.
Cosmology:	Anaximander, Anaximenes, Heraclitus, Parmenides, Anaxagorus and Acharya Kapil.
Physics:	Democritus, Archimedes and Acharya Kanad.
Meteorology:	Anaxagorus, Chang Heng and Shen Kuo.
Geology:	Chang Heng.
Chemistry:	Nagarjun.
History:	Homer, Pherecydes, Hecateus, Herodotus, Xenophon, Eudemus, Polybius, Strabo, Pliny the Elder, Ban Gu and Sima Guang.
Historiography:	Sima Qian.
Cartography:	Ptolemy, Chang Heng, Pei Xiu, Jia Dan, Su Song.
Botany:	Theophratus.
Political Science:	Empidocles.
Medicine:	Claudius Galen, Acharya Charak and Acharya Sushrut.
Literature and Poetry, etc.:	Valmiki, Homer, Hecateus, Fan Chengda, Thales, Pherecydes, Protogoras, Empidocles, Gorgias, Prodicus, Cicero, Zeno of Elea, Zeno of Elea, Socrates, Crate of Mallus and Panini.
Exploration:	Hanno the Navigator, Pytheas, Alexander the Great, Chang Chien, Huan-Tsang, I-Ching, Zhou Daguan, Zheng He and Xu Xiake.
Engineering:	Acharya Bhardwaj and Chang Heng.

Among all the ancient schools of thought (Greek and Roman, Indian, Chinese), at least, one aspect of development was common and it was related to the interest of scholars in the fields of philosophy, astronomy, mathematics, history and literature. Astrology, cosmology, physics and medicine were especially studied by the ancient Europeans and Indians together, as those of meteorology and cartography by Europeans and Chinese. Certain studies were more or less exclusive to these worlds as geometry, botany, political science to the Europeans; algebra and chemistry to the Indians; and geology and historiography to the Chinese. As far as the explorations are concerned the Europeans and Chinese travellers handled them more efficiently. Altogether, the scholars and explorers of classical antiquity had an inclination towards Philosophy, Astronomy, Astrology, Mathematics, Algebra, Geometry, Cosmology, Physics, Meteorology, Geology, Chemistry, History, Historiography, Cartography, Botany, Political Science, Medicine, Engineering and Literature and Poetry. Geography as a spatial science has been carved out of these fields of knowledge. Besides, travels and explorations have also aided in this process of evolution.

Philosophy

The main areas of philosophical study included Metaphysics, Logic, Epistemology and Ethics. Metaphysics investigates principles of reality transcending those of any particular science. Cosmology and ontology are traditional branches of metaphysics. It is concerned with explaining the fundamental nature of being and the world. The earliest speculators (i.e. Thales, Anaximenes and Anaximander) were philosophers of nature. The Pythagoreans succeeded these with mathematical abstractions. The level of pure thought was reached partly in the Eleatic philosophers (such as Parmenides) and Anaxagoras, but more completely in the work of Socrates. In its universal character, metaphysics resembles dialectics and sophistry. Logic is the art and science of reasoning which seeks to identify and understand the principles of demonstration and inference. As a discipline, logic dates back to Aristotle and remains integral to fields such as mathematics, computer science, and linguistics. Both Aristotle and Plato regarded philosophy as concerned with the universal; Aristotle, however, found the universal in particular things, and called it the essence of things, while Plato stated that the universal exists apart from particular things, and is related to them as their prototype or exemplar. For Aristotle, therefore, philosophic method implies the ascent from the study of particular phenomena to the knowledge of essence, while for Plato philosophic method means the descent from knowledge of universal ideas to a contemplation of particular imitations of those ideas. In a sense, Aristotle's method is inductive, while that of Plato is deductive. Epistemology or theory of knowledge is the branch of philosophy concerned with the nature and scope (limitations) of knowledge. There are two kinds of knowing, the one he calls "legitimate" (genuine) and the other "bastard" (obscure). The "bastard" knowledge is concerned with the perception through the senses; therefore, it is insufficient and subjective. The second sort of knowledge, the "legitimate" one, can be achieved through the intellect, in other words, all the sense-data from the "bastard" must be elaborated through reasoning. Ethics addresses questions about morality Ethics is conceived as an art of caring for the soul analogous to medicine's care for the body).[26]

The ancient Indian philospphy belongs to the *āstika* ("orthodox") and *nāstika* ("heterodox") schools. The *āstika* schools of thought[27] or *darshanas*, accept the Vedas as supreme revealed scriptures and they are: Samkhya, Yoga, Nyaya, Vaisheshika, Mimamsa and Vedanta. The *nāstika* schools, that do not accept the Vedas as supreme, are Buddhism, Jainism and Cārvāka (a skeptical materialist school). Samkhya is the oldest of orthodox philosophical systems in Hinduism. It is a dualistic philosophy and postulates that everything in reality stems from *purusha* (self, *atma* or soul) and *prakriti* (matter, creative agency or energy). In the West, dualism is between the mind and the body, whereas in Samkhya it is between the soul and matter. The *Yoga* philosophical system is closely allied with the Samkhya school[28]. Originally, Samkhya was not theistic, but in confluence with Yoga it developed a theistic variant. *Yoga* treats specifically of the dynamics of the process for disentanglement, and outlines practical techniques for the gaining of release, or 'isolation-integration' (*kaivalya*). The foundational text of the Yoga school is the *Yoga Sutras* of Patanjali, who is regarded as the founder of the formal Yoga philosophy. The Nyaya school is based on the *Nyaya Sutras*. They were written by Aksapada Gautama in the 2nd century B.C. The most important contribution made by this school is its methodology which is based on a system of logic. This is comparable to the relationship between Western science and philosophy, which was derived largely from Aristotelian logic. According to Nyaya, there are exactly four sources of knowledge: perception, inference, comparison, and testimony. Knowledge obtained through each of these is either valid or invalid. In this sense, Nyaya is the closest Indian equivalent to analytic philosophy. The Vaisheshika School, an empiricist school of atomic pluralism, was founded by Kanada. All objects in the physical universe are reducible to certain types of atoms, and Brahman is regarded as the fundamental force that causes consciousness in these atoms. Although the Vaisheshika School developed independently from the *Nyaya*, the two eventually merged because of their closely related metaphysical theories. In its classical form, however, the Vaisheshika School differed from the *Nyaya* in one crucial respect: where *Nyaya* accepted four sources of valid knowledge, the Vaisheshika accepted only two—perception and inference. The main objective of the Mimamsa School was to establish the authority of the Vedas. Consequently, this school's most valuable contribution to Hinduism was its formulation of the rules of Vedic interpretation. Its adherents believe in the power of the *mantras* and *yajñas* to sustain all the activity of universe. The Vedanta, or *later* Mimamsa School, concentrates on the philosophical teachings of the *Upanishads* rather than the ritualistic injunctions of the Brahmanas. Vedantic thought drew on Vedic cosmology, hymns and philosophy. The most significant contribution of Vedantic thought is the idea that self-consciousness is continuous with and indistinguishable from consciousness of Brahman.

In ancient China, on the other hand, the period of Hundred Schools of Thought (around 500 B.C.) is considered the golden age of Chinese philosophy. Of many schools founded at this time, the three most influential ones were Confucianism, Taoism, and Buddhism. Developed from the teachings of Confucius, Confucianism is an ethical system, a moral code. Like Socrates, Confucius taught that vice sprang from ignorance and that knowledge led unfailingly to virtue. The knowledge on which he insisted was not purely scientific learning, but an edifying acquaintance with the sacred texts and the rules of virtue and propriety. Unlike most Western philosophers, Confucius did not rely on deductive reasoning to convince his listeners. Instead, he used figures of rhetoric such as analogy and aphorism to explain his ideas. As a foundation for the life of perfect goodness, Confucius insisted chiefly on the four virtues of sincerity, benevolence, filial piety and propriety. Confucianist doctrine remained a mainstream Chinese orthodoxy for two millennia until 20th century, eventually culminating in its repression during the Cultural Revolution.[29] Confucianism as passed down to the 20th and 21st century derives primarily from the school of the Neo-Confucians, led by Zhu Xi, who gave Confucianism renewed vigour. Neo-Confucianism combined Taoist and Buddhist ideas with existing Confucian ideas to create a more complete metaphysics than had ever existed before. Laozi is traditionally regarded as the founder of Taoism (or Daoism). Through his book, *Daodejing* (*Tao-te ching*), commonly translated as the "Classic of the Way and Virtue", Laozi emphasizes Three Jewels of the Tao: compassion, moderation, and humility. Taoist thought generally focuses on nature, men-cosmos correspondence, health, longevity, *wu wei* (action through inaction), liberty, and spontaneity. Next to the Bible, the *Daodejing* is the most translated work in world literature. It is concerned with the Dao or "Way" and how it finds expression in "virtue" (*de*), especially through what the text calls "naturalness" (*ziran*) and "nonaction" (*wu wei*). Until the arrival of Buddhism in China, Taoism was the chief philosophical system of people. Fundamental to Taoism is the belief that the Tao is the eternal reality and source of all existence. The goal of individual lives is to avoid earthly things and to live in harmony with the Tao and with nature in hopes of achieving immortality.

Despite the spread of Confucianism and Buddhism, Taoism has continued to influence the Chinese thought and culture up to the present day. Although, the character *Tao* (*Dao*) literally means "path" or "way", in Daoism it refers more often to a meta-physical term that describes a force that encompasses the entire universe but which cannot be described nor felt. All major Chinese philosophical schools have investigated the correct *Way* to go about a moral life, but in Taoism it takes on the most abstract meanings, leading this school to be named after it. Although it serves as a rival to Confucianism, a school of active morality, this rivalry is compromised and given perspective by the idiom "Confucianism on the outside, Taoism on the inside." Most of Taoism's focus is on what is perceived to be the undeniable fact that human attempts to make the world better actually make the world worse. Therefore, it is better to strive for harmony. Buddhism was introduced from India into China in 6th Century A.D. Buddhism, just like Confucianism and Taoism, guides people to behave, to be honest and responsible. Buddhism is a religion, a practical philosophy, focusing on the teachings of Gautama Buddha. It promotes harmony and peaceful mind, sharing and compassion. Buddhism emphasizes in "awakening of mind". Through learning, one will develop intellectual capacity to the fullest so as to understand, to love and be kind to other beings. The basic Buddhist concepts base on Eightfold path leading to overcome the suffering. The Eightfold path includes Perfect Understanding, Perfect Thought, Perfect Speech, Perfect Action, Perfect Livelihood, Perfect Effort, Perfect Mindfulness and Perfect Concentration. Buddhist teaching introduces the way and concepts of meditation, rebirth, intellectual development and comparisons. It leads to self-understanding, instead of blind believing.

Cosmology and Cosmogony

The early Milesian philosophers, namely Thales, Anaximander and Anaximenes, and the works of some later scholars as Heraclitus, Anaxagoras, Philolaus and Empedocles sought to explain the cosmos and all its phenomena. The Milesians based their explanations on the observed regular behaviour of the materials that make up the cosmos. There is the influence of Eleatic arguments due to Parmenides, concerning the metaphysical requirements for a basic explanatory entity within this Milesian framework. Parmenides argues that any acceptable cosmological account must be rational. Anaxagoras offers an ambitious scientific theory that attempts to explain the workings of the cosmos. The original state of the cosmos was an unlimited (*apeiron*) mixture of all the ingredients. Up to some point in the past, it was motionless and it was everywhere undifferentiated. At some point *nous* set the mixture in motion and caused it to begin to revolve. The rotary motion caused the ingredients in the mass to shift. This shifting produced "separating off" and the continued spreading rotation produces more and more separation. The Cosmological theory of Anaxagoras suggests that all things have existed from the beginning. But originally they existed in infinitesimally small fragments of themselves, endless in number and inextricably combined. All things existed in this mass, but in a confused and indistinguishable form. These parts had to be eliminated from the complex mass before they could receive a definite name and character. Mind arranged the segregation of like from unlike. Mind causes motion. It rotated the primitive mixture, starting in one corner or point, and gradually extended until it gave distinctness and reality to the aggregates of like parts, working something like a centrifuge, and eventually creating the known cosmos. Anaximenes, like Anaximander, gave an account of how our world came to be out of previously existing matter. According to Anaximenes earth was formed from air by a felting process. From evaporations from the earth fiery bodies arose which came to be the heavenly bodies. According to this theory there is only one substance, in this case air, from which the whole world and everything in it are composed. The several stuffs: wind, cloud, water, etc., are only modifications of the real substance that is always and everywhere present. Anaximenes' notion of successive change of matter by rarefaction and condensation was influential in later theories. It was developed by Heraclitus and criticized by Parmenides. Both Melissus and Plato see Anaximenes' theory as providing a commonsense explanation of change. Philolaus' ideas about the cosmology of the universe were drastically different and he developed one of the first non-geocentric views of the universe. These new ways of thinking quite literally revolved around a hypothetical astronomical object he called the Central Fire. His ideas about the nature of the Earth's place in the cosmos influenced Nicolaus

Copernicus. He supposed the Sun to be a disk of glass which reflects the light of the universe. He made the lunar month consist of 29½ days, the lunar year of 354, and the solar year of 365½ days. Empedocles also posits two cosmic forces which work upon the elements in both creative and destructive ways. These he personifies as Love—a force of attraction and combination and Strife—a force of repulsion and separation. These two forces are engaged in an eternal battle for domination of the cosmos and that they each prevail in turn in an endless cosmic cycle. The most widely accepted interpretation of this cosmic cycle is represented in Figure 3.1.

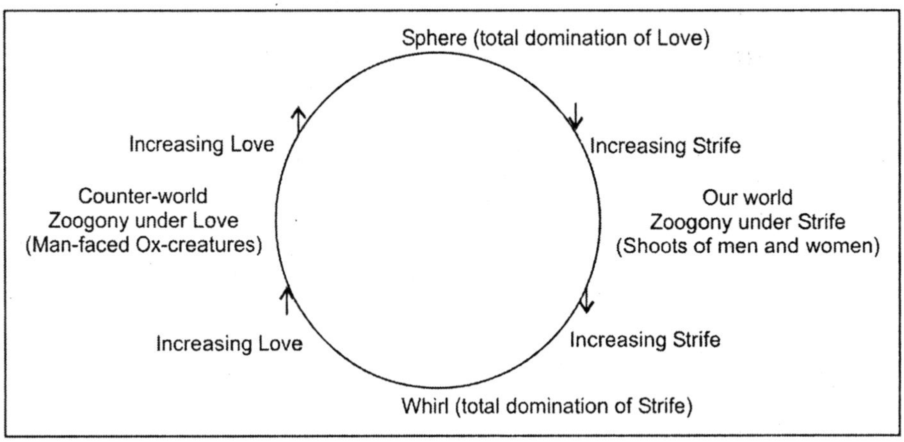

FIGURE 3.1 Empedocles' Cosmic Cycle

Beginning from the top of diagram and proceeding clockwise, when Love is completely dominant she draws all the elements fully together into a Sphere. The Sphere then, is an acosmic state during which no matter can exist, and no life is possible. Then as Love's power gradually weakens and Strife begins to grow in power; he gradually separates out the elements from the Sphere until there is enough separation for matter to come into existence, for the world to be created and for all life to be borne. When Strife has achieved total domination we again get an acosmic state in which the elements are separated completely and the world and all life is destroyed in a Whirl. Then Love begins to increase in power and to draw the elements together again, and as she does so the world is again created and life is again borne. When Love has achieved full dominance we return once more to the sphere. As Empedocles puts it: The cosmos exists in a state of constant flux then, beneath which there is a certain sort of stability in the eternity of the elements. The world is in a constant state of organic evolution. Although Heraclitus used, for the first time the word *kosmos* (order) to mean "world". He identifies the world with fire. Heraclitus described the transformations of elementary bodies: Fire turns into water ("sea") and then half of that quantity turns into earth and half into "fire burst" (a fiery, windy kind of storm phenomenon). The portion that becomes earth turns back into water, in the same quantity it had previously. Here, Heraclitus envisages a law like transformation of stuff from fire to water to earth; the transformation is reversible, and in it the same relative quantities of stuff are preserved. What would make the world to be continuous would be the fact that when one portion of fire turns into water, an equivalent portion of water turns into fire. The overall equilibrium is preserved, even if the water that is now in the sea is not the same water as was in it before. Thus, cosmogony, or cosmogeny, explains the origin of the universe. In other words, it refers to theories of creation of the Solar System. Cosmogony can be distinguished from cosmology, which studies the universe at large and throughout its existence, and which technically does not inquire into the source of its origins. Anaxagoras' theory of the origins of the world is reminiscent of the cosmogonies that had been previously developed in the Ionion tradition, particularly through Anaximenes and Anaximander. The traditional theories generally depict an original unity which begins to become separated off into a series of opposites. Prior to the beginning of world as we know it everything was combined together in such a unified manner that there were no qualities or individual substances that could be discerned. At some point, the unity is spurred into a vortex motion. This motion begins the separation and it is "air and ether" that are the first constituents of matter to become distinct. As the air and ether became separated off, all other elements become manifest in this mixture as well. Therefore, the origin of the world is depicted through this process of motion and separation from the unified mixture. According to Anaxagoras, the agent responsible for the rotation and separation of the primordial mixture is Mind or nous. Anaxagoras' theory of Mind served as an impetus towards the development of cosmological systems that speculated on final causes. The Mind of the cosmos is a dynamic governing principle which is imminent to the entire natural system while still maintaining its transcendental determining power. From Anaxagoras' perspective it appears to be a principle which is both natural and divine. Following Empedocles' cosmology, since the time of the sphere, Strife has gained more sway; and the actual world is full of contrasts and oppositions, due to the combined action of both principles. Empedocles attempted to explain the separation of elements, the formation of earth and sea, of sun and moon, of atmosphere. It is possible (although anachronistic) to see this theory as a crude anticipation of Darwin's theory of natural selection.

The ancient Indian scholars dealt with many problems pertaining to *Cosmology* (the science of Universe), *Cosmogony* (the origin of Universe) and *Cosmography* (the description of Universe). Four types of theories have been advanced about the origin of the Universe[30]. This Universe has been named as *Brahamand* in the ancient Indian literature. It is conceived as very immense and wide; and is believed to be consisting of seven upper and seven internal divisions (*Pataal*). The theories related to the Universe are of Artistic, Mechanical, Instrumental and Philosophical Origin. According to the Theory of Artistic Origin, the universe has been compared to a house and *Rigveda* refers various stages in the construction of this universal house. *Rigveda* mentions a number of Gods who performed various functions during the process of creation. These Gods were the artists who contributed their skills to the construction and completion of the Universe. They wove various materials into a pattern, and shaped the universe by blasting and smelting. The views regarding the mechanical origin of the universe developed in the last phase of Rigvedic period.

The Theory of Mechanical Origin suggests the sacrifice (or disintegration) of a primeval body known as *Adi-Purush*. He was conceived as the soul and the nucleus of the Universe and an embodiment of the supreme spirit. The 'sky', the 'wind', the 'moon' and the 'sun'—all these elements were believed to be the results of dismemberment of *Adi-Purush* as a part of sacrifice ceremony. The scholars having faith in the Theory of Philosophical Origin believed that there was nothing earlier, i.e. no atmosphere, no sky, no days and nights, etc. The space was empty (a vacuum). Within this a unit was born on its own, perhaps due to inherent heat and it hardened with time. This led to the formation of the Universe later. The Theory of Instrumental Origin, on the other hand, talks about the occurrence of twin parent bodies, *Dharti* (Earth) and *Swarga* (Heaven), from which the Universe was created. The union of these two produced the Sun, considered as the most important agent in the creation of the world. It is the soul of all that 'moved' or that 'not moved'. It has been later identified as Rajapati, Vishwakarma and sometimes with the 'Golden Egg' and 'Unborn Being'. This 'Unborn Being' is also named as *Atma* (Soul) who produced the Universe through intermediary body.

Today scientists rely on powerful telescopes and sophisticated computers to formulate cosmological theories. In former times, people got their information from traditional books of wisdom and the ancient schools of philosophy. Followers of India's ancient culture, for example, learned about the cosmos from scriptures like the *Srimad-Bhagavatam*, or *Bhagavata Purana* and the Samkhya School of Thought. Celebrated as the founder of Samkhya philosophy, Acharya Kapil is recognized as the Father of Cosmology. His pioneering work threw light on the nature and principles of the ultimate Soul (*Purusha*), primal matter (*Prakriti*) and creation. On his assertion that *Prakriti*, with the inspiration of *Purusha*, is the mother of cosmic creation and all energies, he contributed a new chapter in the science of cosmology. The *Srimad-Bhagavatam* describes that the universe lies within a series of spherical shells which is divided in two by an earth plane called *Bhu-mandala*. A series of *dvipas*, or 'islands,' and oceans make up *Bhu-mandala*. In the centre of *Bhu-mandala* is the circular 'island' of Jambudvipa, whose most prominent feature is the cone-shaped Mount Meru. The *Srimad-Bhagavatam* presents an earth-centred conception of the cosmos. It tells of innumerable universes, each one is contained in a spherical shell surrounded by layers of elemental matter that mark the boundary between mundane space and the unlimited spiritual world. The region within the shell (Figure 3.2) is called the *Brahmanda*, or "Brahma egg." It contains an earth disk or plane—*Bhu-mandala*—that divides it into an upper, heavenly half and a subterranean half, filled with water. *Bhu-mandala* is divided into a series of geographic features, traditionally called *dvipas*, or "islands," *varshas*, or "regions," and oceans. In the centre of *Bhu-mandala* (Figure 3.3) is the circular "island" of Jambudvipa, with nine *varsha* (subdivisions); *Bharata-varsha* being one of them. In the centre of Jambudvipa stands the cone-shaped Sumeru Mountain, which represents the world axis and is surmounted by the city of Brahma, the universal creator.

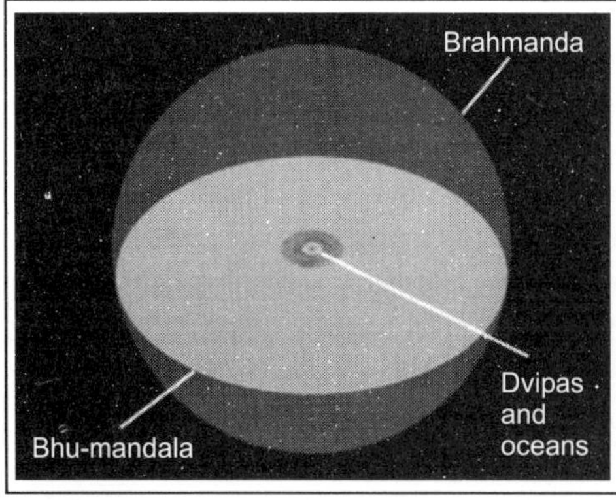

FIGURE 3.2 Brahmanda

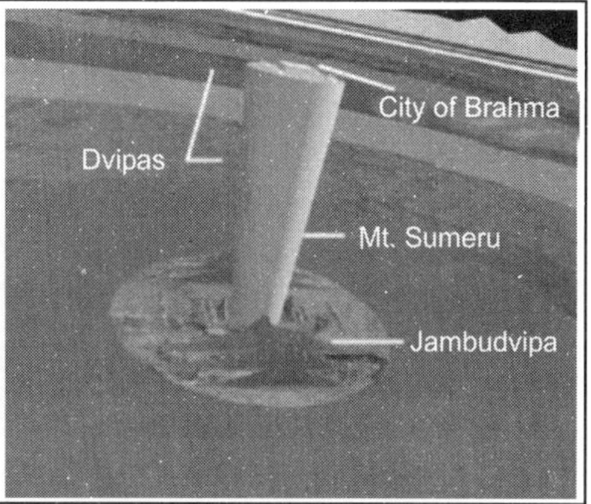

FIGURE 3.3 Bhu-Mandala

We can compare *Bhu-mandala* with an astronomical instrument called an astrolabe, popular in Middle Ages. On the astrolabe, an off-centred circle represents the orbit of the sun—the ecliptic. The Earth is represented in stereographic projection on a flat plate, called the *mater*. The ecliptic circle and important stars are represented on another plate, called the *rete*. Different planetary orbits could likewise be represented by different plates, and these would be seen projected onto the Earth plate when one looks down on the instrument. The *Bhagavatam* similarly presents the orbits of the sun, the moon, planets and important stars on a series of planes parallel to *Bhu-mandala*. Not only does the *Bhagavatam* use the ecliptic model, but it turns out that the disk of *Bhu-mandala* corresponds in some detail to the solar system (Figure 3.4). The solar system is nearly flat. The sun, the moon, and the five traditionally known we compare the rings of *Bhu-mandala* with the orbits of Mercury, Venus, Mars, Jupiter, and planets—Mercury through Saturn—all orbit nearly in the ecliptic plane. The structures

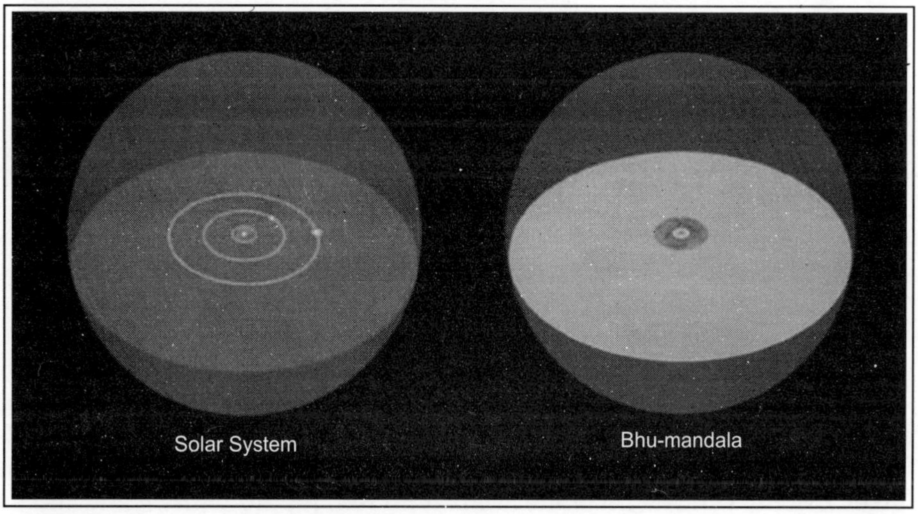

FIGURE 3.4 The Ecliptic Model

of *Bhu-mandala* correspond with the planetary orbits of the solar system. If Saturn, we find several close alignments that give weight to the hypothesis that *Bhu-mandala* was designed as a map of the solar system. It is remarkable that the dimensions of *Bhu-mandala* in the *Bhagavatam* are consistent with modern data on the size of the sun's orbit and the solar system as a whole. Jambudvipa (Figure 3.5), the central hub of *Bhu-mandala*, can be understood as a local topographical map of part of south central Asia. Jambudvipa represents the northern hemisphere of the Earth globe. Six horizontal and two vertical mountain chains divide Jambudvipa into nine regions, or *varshas*. The southernmost region is called *Bharata-varsha*. Careful study shows that this map corresponds to India plus adjoining areas of south-central Asia.

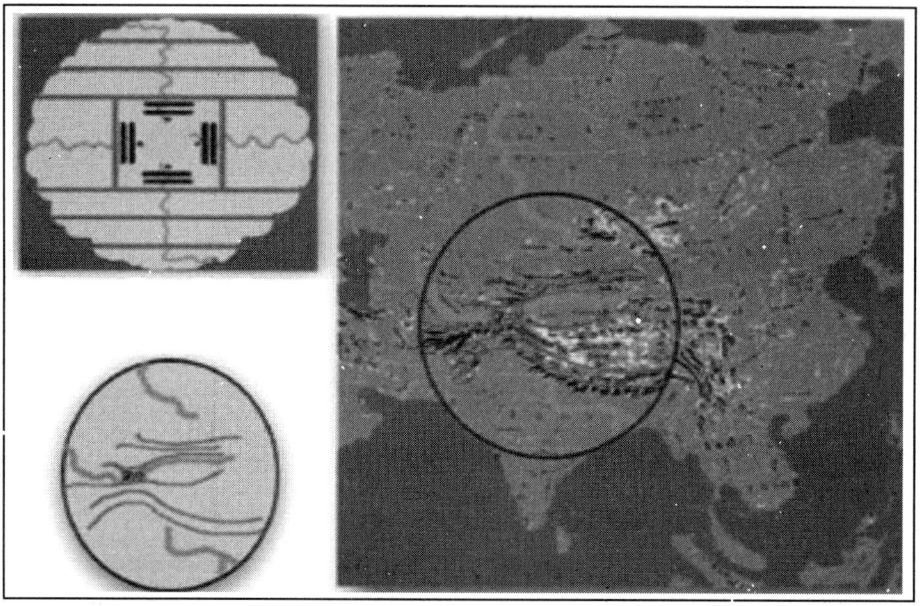

FIGURE 3.5 Jambudvipa: As a Local Map of South Central Asia

The cosmology of the *Bhagavata Purana* is a sophisticated system of thought. It combines practical understanding of astronomy with spiritual conceptions to produce a meaningful picture of the universe and reality. Unlike our modern scientists, the ancient Indian astronomers believed in geo-centric universe, i.e. the Earth at the centre and the Sun being as one of its planets (*Grah*). In the *Rigveda*, there is a description of 34 heavenly bodies including the Sun, the Moon, five *Grah* (Planets) and 27 constellations. The five Planets have been described as the five Gods. But, the astronomers of Puranic period established nine planets, viz. the Sun, the Moon, the Mars (*Mangala*), the Mercury (*Budha*), the Jupiter (*Brahaspati*), the Venus (*Shukra*), the Saturn (*Shani*), *Rahu* and *Ketu*. The astrophysical characteristrics of some of these planets have been described in the classical literature. For instance, Mercury has been shown with green colour, Venus with white, Mars with red, Jupiter with yellow and Saturn with black.

By combining the basic concepts of Taoism and Confucianism, the ancient Chinese created a traditionally consistent and uniquely harmonious view of the universe. Taoism, for instance, is a philosophy of duality and balance. The most obvious example of this focus lies in the concept of *yin* and *yang*. These two terms represent almost all opposable forces in the universe. Yin is cool, wet, and female, while yang tends to be hot, dry, and male. Various other contrasts include: day and night, light and dark, and honest and deceitful. The second primary school of ancient Chinese philosophy followed the teachings of Confucius. The Confucian version of Tao and Te were referred to as Jen and Li. The Chinese did in fact distinguish between stars and planets, and that they acknowledged the erratic behaviour of many celestial bodies. There were primarily three models of celestial orientation (Figure 3.6).

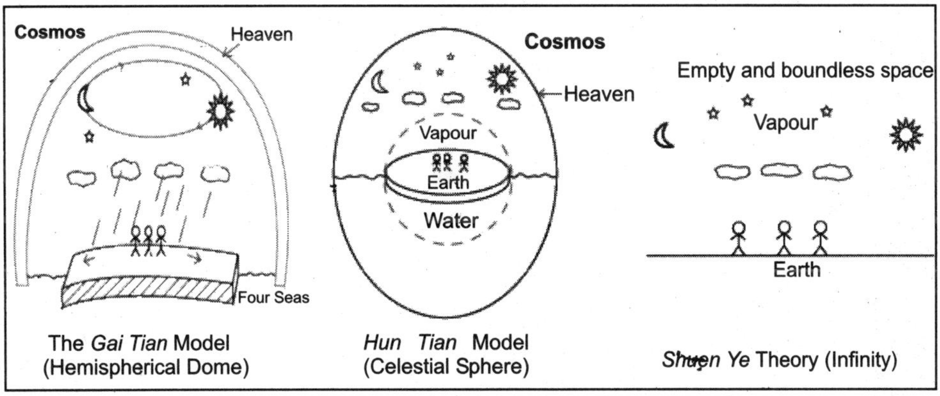

FIGURE 3.6 Three Early Chinese Models of the Universe

The most archaic Chinese cosmological model, *Gai Tian*, consists of a flat earth and umbrella—like heavens, whereas its centuries-long opponent, *Hun Tian* model, was presented through the analogy of "egg". The flat earth was situated in the middle of the egg yolk and surrounded by water, while the heavens were like the egg shell. It should be noted that the heavens, no matter in which form, were always filled with *Chi* (vapour) and accommodated the freely floating heavenly bodies, and, therefore, have nothing to do with the solid celestial shells as presented in Greek astronomy. For the Chinese, the heavens and the earth altogether form an observable and researchable world, beyond which is the infinite cosmos. The third model, *Shuen Ye*, is said to be more a philosophical notion than a scientific theory. It did not discuss the forms of the earth and the heavens, but merely concentrated on the free moving objects in the sky and the empty and boundless space. Of the three dominant cosmological models in early China two of them (Hun Tian and Gai Tian) maintained a flat Earth, either like a disk or a square, and the round heavens, either spherical or hemispherical, whilst the third (Shuen Ye) did not give descriptions of the shapes. Briefly, the main characteristics of the ancient Chinese models of the universe can be listed as: (1) The phrase, "the heavens and the earth" was used to refer to their researchable universe, which is an observable space based in the earth; beyond the heavens there is unknown infinite cosmos. That is, for the Chinese, the universe was in fact infinite, but they confined their models to a space called "the heavens and the earth"; (2) The earth is flat; (3) The heavens are round; the heavenly bodies are floating, unattached to the heavens, and moving freely; and (4) There is a lack of emphasis on reasoning and consequently the structural view of nature. These features interestingly are distinguished from those of the Greek cosmological models: (1) The universe is finite, where "the heavens and the earth" are located; (2) The earth is round (spherical); (3) The heavens are layers of solid spheres; the heavenly bodies are attached to the layers respectively; and (4) Efforts are made to search for reasons and to establish a structural view of nature. In the Greek cosmological models, the universe was finite, Earth was round (spherical) in its centre, and the heavens were layers of solid spheres; the heavenly bodies were attached to the layers respectively. Another common perception in Chinese cosmology was the shape of Heaven and Earth.

The earth was divided into nine continents, each surrounded by ocean, and further divided into nine provinces. The Chinese perceived Heaven to be round. It had nine levels; each of which was separated by a gate and guarded by a particular animal. The highest level, the Palace of Purple Tenuity, was where the Emperor of Heaven lived in the constellation we call Ursa Major. At the centre of Heaven were the North Pole and the polar star. The celestial pole was a critical characteristic of Chinese cosmology. To the Chinese, the centre was the most important geographical point because it was the closest to Heaven. They believed that the heart of civilization lay at the centre of the earth, and as the land spread out, the lands and its inhabitants became more savage. Naturally, this emphasis on the centre point lead to the polar axis as a pivotal aspect of Chinese astronomy. While the Greeks focused on the constellations on the horizon and created a solar calendar, the Chinese observed the circum polar stars, which lead them to devise a lunar calendar instead. The polar axis which ran from the polar star, south to Earth was the pivot of the heavens. The heavenly vault slid up and down this axis while the earth itself oscillated along it to create the seasons. The stars around the pole were also an integral part of Chinese cosmology. At the equator, the Chinese divided the sky arbitrarily into twenty-eight divisions, each corresponding to an equatorial and a circumpolar constellation. Based on which mansion the moon occupied at night, the Chinese created their lunar calendar. The Chinese calendar was particularly notable because the equinoxes and solstices marked the centre of the seasons rather than the beginnings. Like many other cosmologies, Chinese interpretation of the universe was not completely secular. There was an incredibly strong emphasis on harmony and balance that originated from philosophical ideas. In accordance with the Taoist philosophy of *yin* and *yang*, everything in the universe had a counterpart. The fusang tree in the east over which the sun would rise, complimented the ruo tree in the west which watched the sun set every evening. The Yellow Springs that flowed under and supported the earth opposed the sky above. Another example of this balance is the Chinese dependence on numerology and various characteristics to determine and maintain a celestial and earthly balance. Five was a very auspicious number.

There were the five directions: north, south, east, west, and centre. Each had a corresponding colour, animal, element, and flavour. Each of these aspects had to be mixed and balanced in everyday life, be it in the food people ate, the houses they built, or the furniture they chose. Nine was also a very divine number. It was the number of perfect yang and highest completeness. There were nine continents with nine provinces. There were also nine layers to Heaven. The Round Mound at the Temple of heaven has nine rings at its zenith, each comprised stones in multiples of nine. It is overwhelmingly obvious the great extent to which the Chinese took this numerology. To them, following the rules of numerology was more than showing respect to heaven and nature, but more importantly, it was continuing the celestial balance between Heaven and Earth. Ancient Chinese cosmology is more than a cultures interpretation of the universe. In his publication called *The Spiritual Constitution of the Universe* (*Ling Xian*), Zhang Heng theorized that the universe was like an egg "as round as a crossbow pellet" with the stars on the shell and the Earth as the central yolk.[31] This universe theory is congruent with the geocentric model. Zhang supported the "radiating influence" theory to explain solar and lunar eclipses. Some other contemporary writers, as Jing Fang (78-37 BCE), also wrote about eclipses and the sphericity of heavenly bodies. The theory posited by Zhang and Jing was supported by later pre-modern scientists such as Shen Kuo (1031-1095), who expanded on the reasoning of why the Sun and Moon were spherical."[32] Much like what Zhang Heng had said, Shen Kuo likened the Moon to a ball of silver, which does not produce light, but simply reflects light if provided from another source (the Sun). Shen's hypotheses were similar to the concept of the epicycle in the Greco-Roman tradition. Shen likened the relationship of the Moon's path to the ecliptic, the path of the Sun, "to the figure of a rope coiled about a tree".

Earth Studies

A number of queries have frequently made by ancient scholars related to the earth. Thales, for instance, solved problems related to its shape; its size; the cause of earthquakes; the dates of solstices; and the size of the sun and moon. Thales's hypothesis was substantiated by sound observation and reasoned considerations. At the busy city-port of Miletus, Thales had unlimited opportunities to observe the arrival and departure of ships with their heavier-than-water cargoes, and recognized an analogy to floating logs. Aristotle wrote that the earth rests on water. He explained his theory by adding the analogy that the earth is at rest because it is of the nature of wood and similar substances which have the capacity to float on water, although not on air. Thales envisaged the earth as spherical. Aristotle used several arguments to support this view. First is the fact that during a solar eclipse, the shadow caused by the interposition of the earth between the sun and the moon is always convex; therefore, the earth must be spherical. Second, Thales, who is acknowledged as an observer of the heavens, observed that stars which are visible in a certain locality may not be visible further to the north or south, a phenomenon which could be explained within the understanding of a spherical earth. Third, from mere observation the earth has the appearance of being curved. From observation, it appears that the earth is covered by a dome. When observed from an elevated site, the sky seems to surround the earth, like a dome, to meet the apparently curved horizon. If observed over the seasons, the dome would appear to revolve, with many of the heavenly bodies changing their position in varying degrees, but returning annually to a similar place in the heavens. From the shore, a ship can be seen to be descending, gradually, below the horizon, with the hull disappearing from view first, to be followed by masts and sails. Further, Thales's theory about the cause of earthquakes is consistent with his hypothesis that earth floats upon water. It seems that he applied his floating on water simile to the natural phenomena of earthquakes. Thales and Democritus found in water the cause of earthquakes.

The geographical observations of the ancient Indian scholars are about the earth in general and the Indian subcontinent in particular. The earth studies deal with its origin, sphericity, eclipses, size and dimensions, latitudes, longitudes and local time, directions or cardinal points, earthquakes and volcanoes, atmosphere and seasons, and its physical divisions. As far as the origin of the earth is concerned, many of the facts as put forward by the ancient Indian scholars were more or less accurately known. They believed in the solidification of earth from gaseous matter. The earth's crust, according to them, is made of hard rocks (*sila*), clayey material (*bhumih*) and sandy material (*asma*). The *Puranas* mention the earth to be apparently floating on the water like a sailing boat on the river. They were also aware of the fact that there is more land surface in the Northern Hemisphere. The concept of *Prithvi* (Earth) was the most basic in the study of geography. It has been profusely used in the Vedas and *Puranas*. The study of the earth or *Prithvi* has been called *'Bhugol'* or Geography. Thus, the use of the term *'Bhugol'* for the discipline of Geography was the most appropriate and it clearly suggests that the ancient Indians endorsed the earth being a sphere, and not a flat disc as believed by some of their parallel civilizations. This fact was further affirmed by: (1) the periodical occurrence of eclipses and the circular shadow of the earth casted on moon during lunar eclipse; and (2) the rise and setting of Sun, which according to the prevalent belief, never sets but goes on the other side of the sphere earth. The ancient Indian scholars were aware of the causes of occurrences of eclipses (*grahanas*). Some rituals and ceremonies of Indian society were being performed on the days when eclipses occurred. The Aryans, for instance, considered an eclipse as unlucky (inauspicious) and a messenger of disaster. It was also believed that if a solar and lunar eclipse occurs in the same month, it becomes more disastrous. Varahamihira considered the effects of eclipse month-wise and emphasized that eclipse in *Pusa* (December) leads to famine and its occurrence in April and May results in good rainfall, while and eclipse in *Phagun* (March) and *Asadh* (June) is unlucky. It was believed that the looking at eclipses is bad, because seeing the eclipse (particularly Solar) is harmful for the naked eyes. The facts related to the size and dimension of the earth were quite near to accuracy. It was well known to the ancient Indian scholars that the earth is an oblate spheroid slightly flattened at the poles. Also, its equatorial diameter is greater than its polar diameter. The literature of the 5th and 6th centuries A.D. on astronomy gives convincing information on earth's volume, its equatorial circumference, mass or weight. The estimates made by Aryabhatta in this regard were very close to those established in the modern period. *Akshansh* and *Deshantar* are the terms used for 'latitudes' and 'longitudes' respectively in ancient Indian literature. *Puranas* have a reference of three imaginary lines of latitudes passing through Equatorial belt, North Pole and South Pole. Accordingly, three major regions have been identified in the Literature, viz. Equatorial (*Nirakshadesha*), Northern Polar (*Meru*) and Southern Polar (*Bhadvanala*). The North Pole has been called as *Zenith* and the South Pole as *Nadir*. The South Pole was truly considered as the antipode of the North Pole, i.e. diametrically opposite to it. However, the world was not believed to exist beyond Equator, as the region here was compared to hell of the earth. The Eastern part, on the other hand, was believed to be 'the land of Gods'. This thinking is in consonance with that of the Europeans in the Early Medieval period, when the Dark Ages prevailed and the East in 'T-in-O' Maps (Fig. 4.1) was assumed to be the place of *Adam* and *Eve*. The ancient Indian scholars have also drawn Prime Meridian. These imaginary lines, the position of Sun and various stars have helped them to determine local time at various places. In Rigveda, there is formulated the idea of four main directions, viz. *Purva* (East), *Paschim* (West), *Uttar* (North) and *Dakshin* (South). By adding Zenith (*Meru*) and Nadir (*Bhadvanala*) it was raised to six. But, afterwards, ten directions have been frequently mentioned in the Puranic literature. The designation of these directions in the *Puranas* is significant in the sense that it bears concept of the ruling deities dominating in each of these ten directions (Table 3.5).

TABLE 3.5 Cardinal points and the Ruling Daities as per Puranic Literature

Direction	*Ruling Deity*
Purva (East)	Indra (The God of Rain)
Agneyay (Southeast)	Agni (The God of Fire)
Dakshina (South)	Yama (The God of Death)
Nairitya (Southwest)	Niriti (The God of Disaster)
Paschim (West)	Varuna (The God of Water)
Vayavya (Northwest)	Marut/Vayu (The God of Air)
Uttar (North)	Kubera (The God of Wealth)
Isana (Northeast)	Isa (The God of Power)
Urdhava (Zenith*)	Brahma (The Creator of Universe)
Adhoh (Nadir^)	Sesanaga (The Universal Ocean)

*The point directly above the observer.
^ The point directly below the observer.

The knowledge regarding the earthquakes, atmosphere, weather, climate and seasons in this period is excellent. For 'earthquakes' the term *bhukamp* has been used in *Puranas*. It was assumed that the deities of Air, Fire and Water caused earthquakes. The ancient Indian scholars have identified the vacuum between the earth and the heaven as *antariksha*. But it is not clear whether it referred to 'space' or 'atmosphere'.

However, they were undoubtedly aware of its vast extent or thickness and the occurrence of various weather phenomena here. *Ramayana* furnishes a lot of this kind of information. Bhaskaracharya has conceived the thickness of this *antariksha* around the earth to be 12 *yojans* (or 96 kms). As far as the knowledge about the seasons is concerned, it is based largely on the studies in India. Rigveda mentions five seasons. In *Valmiki Ramayan*, however, six seasons have been identified. They are listed below (Table 3.6):

TABLE 3.6 The Seasons as mentioned in Ancient Indian Literature

Season	*Months*
Basant (Spring)	*Chaitra-Baisakh* (March-April)
Grishma (Summer)	*Jyestha-Asadh* (May-June)
Prourit/Varsha (Rainy)	*Shravan-Bhadrapad* (July-August)
Sharad (Autumn)	*Ashvin-Kartik* (September-October)
Hemant (Winter)	*Mangsir-Paus* (November-December)
Shishir (Severe Winter)	*Magh-Phagun* (January-February)

The earth studies of Chinese scholars pertain to hydrological cycle, forests, seismology and weather and climate. There is record of progress in the understanding of natural processes, particularly hydrological cycle, work of running water in wearing down mountains and forming down alluvial plains, erosion of mountain slopes, and study and measurement of the amount of silt carried by the rivers. The studies on forests and effects of forest clearing were going on in China around 4th century B.C. The observations of Chinese scholars on weather and climate are commendable. They are known to have prepared a number of weather reports on the basis of empirically based facts. Some of these reports date back to 13 centuries before Christ. Shen Kuo devised a geological hypothesis for land formation, based upon findings of inland marine fossils, knowledge of soil erosion, and the deposition of silt. He also proposed a hypothesis of gradual climate change, after observing ancient petrified bamboos that were preserved underground in a dry northern habitat that would not support bamboo growth in his time. Early speculation and hypothesis pertaining to what is now known as meteorology had a long tradition in China before Shen Kuo. For example, the philosopher Wang Chong (27-97) accurately described the process of the water cycle.[33] Shen wrote vivid descriptions of *tornadoes*—the first known description of them in East Asia—and gave reasoning that rainbows were formed by the shadow of the Sun in rain, occurring when the Sun would shine upon it. Paul Dong writes that Shen's explanation of the rainbow as a phenomenon of atmospheric refraction "is basically in accord with modern scientific principles."[34] Shen hypothesized that rays of sunlight refract before reaching the surface of the earth, hence people on earth observing the Sun are not viewing it in its exact position. Shen also hypothesized that the altitude of the apparent Sun is higher than the actual altitude of the Sun.

Known for excellent record keeping, the Chinese kept accurate records of not only celestial events, but of earthquakes as well. From the earliest times, the Chinese were concerned with the destructive force of earthquakes. The people of the ancient Zhou Dynasty explained them as disturbances with cosmic *yin* and *yang*, along with the heaven's displeasure with acts committed (or the common people's grievances ignored) by the current ruling dynasty.[35] At the time, people believed that seismic events were supernatural in origin: signs from heaven from angry gods designed to punish those below. Heng discounted these superstitions and offered a scientific explanation in its place. He proposed that earthquakes were caused by wind and air, writing: The chief cause of earthquake is air, an element naturally swift and shifting from place to place. As long as it is not stirred, but lurks in a vacant space, it reposes innocently, giving no trouble to objects around it. But any cause coming upon it from without rouses it, or compresses it, and drives it into a narrow space...and when opportunity of escape is cut off, then 'With deep murmur of the Mountain it roars around the barriers', which after long battering it dislodges and tosses on high, growing more fierce the stronger the obstacle with which it has contended.[36]

Physics

The foundations of physics lie in Empedocles' assumption that there are four 'elements' of matter, or 'roots' as he calls them: earth, air, fire and water. These are able to create all things, including all living creatures, by being 'mixed' in different combinations and proportions. Each of the elements, however, retains its own characteristics in the mixture, and each is eternal and unchanging. The four elements correspond closely to their expression at the macroscopic level of nature, with the traditional quadripartite division of the cosmos into earth, sea, air, and the fiery ether of heavenly bodies. These four naturally occurring 'elements' of cosmos clearly represent a fundamental natural division of matter at the largest scale. This division at the macroscopic level of reality is applied at the microscopic level to produce a parallelism between the constituents of matter and the fundamental constituents of the cosmos. Empedocles stresses this parallel between the elements at the different levels of reality by using the terms 'sun', 'sea' and 'earth' interchangeably with 'fire', 'water' and 'earth'. In Stoic physics, Posidonius advocated a theory of cosmic "sympathy", the organic interrelation of all appearances in the world, from the sky to the earth, as part of a rational design uniting humanity and all things in the universe, even those that were temporally and spatially separate.

It would be surprising to know that the concepts of atom (*Anu, Parmanu*) and relativity (*Sapekshavada*) were explicitly stated by an Indian philosopher nearly 600 years before the birth of Christ. From the Vedic times, around 3000 to 1000 B.C., the Indians had classified the material world into five elements, viz. Earth (*Prithvi*), fire (*Agni*), air (*Wayu*), water (*Jala*) and ether or *Akasha*. These five elements or '*Pancha Mahabhootas*' were identified with various human senses of perception; earth with smell, air with feeling, fire with vision, water with taste and ether with sound. Whatever the validity behind this interpretation, it is true that since very ancient times Indians had perceived the material world as comprising these 5 elements. Since ancient times, Indian philosophers believed that these elements were physically palpable and hence comprised miniscule particles of matter. The last miniscule particle of matter which could not be subdivided further was termed *Parmanu*. The word *Parmanu* is a combination of *Param*, meaning beyond, and *anu* meaning atom. This Indian concept of the atom was developed independently and prior to the development of the idea in the Greco-Roman world. The first Indian philosopher who formulated ideas about the atom in a systematic manner was Kanada who lived in the 6th century B.C. Kanada propounded that the *Parmanu* (atom) was an indestructible particle of matter. According to him the material universe is made up of *Kana*. When matter is divided and subdivided, we reach a stage beyond which no division is possible, the undivisible element of matter is *Parmanu*. Another 6th century B.C. Indian philosopher was Pakudha Katyayana,[37] a contemporary of Gautama Buddha. He propounded ideas about the atomic constitution of material world. These philosophers considered the Atom to be indestructible and hence eternal. There are different types of *Parmanu* for the five *Pancha Mahabhootas*, Earth, water, fire, air and ether. Each *Parmanu* has a peculiar property which depends, on the substance to which it belongs. It was because of this conception of peculiarity of *Parmanu* (atoms) that this theory founded by Kanada came to be known *Vaisheshika-Sutra* (Peculiarity Aphorisms). The Vaisheshika School of philosophers believed that an atom was a mere point in space. Indian theories about the atom are greatly abstract and enmeshed in philosophy as they were based on logic and not on personal experience or experimentation. In this context Kanada seems to arrive at conclusions which were surpassed only many centuries after him. Kanada's idea also had shades of relativity in it which was propounded by Einstien in our times. About his ideas on atom, Kanada observed that an inherent urge made one *Parmanu* combine with another. When two *Parmanu* belonging to one class of substance combined, a *dwinuka* (binary molecule) was the result. This dwinuka had properties similar to the two parent *Parmanu*. In the material universe, according to him, *Parmanu* belonging to different classes of substances combine in different combinations giving us a variety of *dwinuka*, which in other words means different types of substances. Thus, it remains a fact that Indian ideas about atom are the oldest. It is only after the 4th century B.C., after the Greeks had come in contact with India do we find references to the idea of an atom in Greek science. Hence, it is quite possible that the Greeks borrowed the ideas about atom from Indian philosophers in the 4th century B.C.

Astrology and Astronomy

Astrology developed in the **Greco-Roman** world during **Hellenistic** period, borrowing many elements from **Babylonian** astronomy, and Ptolemy played a key role in this. Ptolemy's treatise on **astrology**, the **Tetrabiblos**, was the most popular astrological work of antiquity and also enjoyed great influence in the Islamic world and the medieval Latin West. Ptolemy's astrological outlook was quite practical: he thought that astrology was like medicine, that is conjectural, because of the many variable factors to be taken into account: the race, country, and upbringing of a person affects an individual's personality as much if not more than the positions of the Sun, Moon, and planets at the precise moment of their birth. Ptolemy's is the most popular work on astronomy, the scientific study of celestial objects and phenomena that originate outside the Earth's atmosphere. The Ptolemaic explanation of the motions of planets remained the accepted wisdom until the Polish scholar Copernicus proposed a heliocentric view in 1543. In the Almagest, one of the most influential books of classical antiquity, Ptolemy compiled the astronomical knowledge of the ancient Greek and Babylonian world. The Almagest also contains a star catalogue, which is an updated version of a catalogue created by Hipparchus. Its list of forty-eight constellations is ancestral to the modern system of constellations. The Babylonians and the Egyptians merited in the field of observational astronomy. These observations were made with naked eye and with the help of some simple instruments as *gnomon*. Unlike them, the Greeks', more particularly Anaximander's merits lie in that of speculative astronomy. Three of his astronomical speculations include: (1) that the celestial bodies make full circles and pass also beneath the earth, (2) that the earth floats free and unsupported in space, and (3) that the celestial bodies lie behind one another. These propositions, which make up the core of Anaximander's astronomy, meant a tremendous jump forward and constitute the origin of our Western concept of the universe. Anaximander's vision implied *depth* in the universe, that is, the idea that the celestial bodies lie *behind* one another. Anaximander's order of the celestial bodies is clearly that of increasing brightness. Due to the inclination of the axis of the heaven, the celestial bodies do not circle around the earth in the same plane as the earth's–flat–surface, but are tilted. This tilting of the heaven's axis has been one of the biggest riddles of the universe. Hipparchus is considered the greatest astronomical observer of antiquity. He was the first Greek to develop quantitative and accurate models for the motion of the Sun and Moon. For this he made use of the observations and knowledge accumulated over centuries by the Chaldeans from Babylonia. With his solar and lunar theories and his numerical trigonometry, he was the first to develop a reliable method to predict solar eclipses. His other achievements include the discovery of precession, the compilation of the first star catalogue of the western world, and the invention of the astrolabe. With an astrolabe Hipparchus was the first to be able to measure the geographical latitude and time by observing stars. Previously this was done at day time by measuring the shadow cast by a *gnomon*, or with the portable instrument known as *scaphion*. The most intriguing for historians of astronomy is Hipparchus' use of Babylonian astronomical material, including methods as well as observations. Hipparchus' work provides a clear link between Babylonian and Greek astronomy. He was responsible for the direct transmission of both Babylonian observations and procedures and for the successful synthesis of Babylonian and Greek astronomy. The discovery for which Hipparchus is most famous is the discovery of precession which is due to the slow change in direction of the axis of rotation of the earth. This work came from Hipparchus' attempts to calculate the length of the year with a high degree of accuracy. There are two different definitions of a 'year' for one might take the time that the sun takes to return to the same place amongst the fixed stars or one could take the length of time before the seasons repeated which is a length of time defined by considering the equinoxes. The first of these is called the sidereal year while the second is called the tropical year. Hipparchus also calculated the length of the sidereal year, again using older Babylonian data, and arrived at the highly accurate figure of $\frac{1}{144}$ days longer than $365\frac{1}{4}$ days. This gives his rate of precession of 1° per century. Hipparchus also made a careful study of the motion of moon. There are three different periods which one could determine. There is the time taken for the moon to return to the same longitude, the time taken for it to return to the same velocity (the anomaly) and the time taken for it to return to the same latitude. In addition there is the synodic month that is the time between successive oppositions of the sun and moon. Hipparchus not only gave observational data for the moon which enabled him to compute accurately the various periods, but he developed a theoretical model of the motion of the moon based on epicycles. Hipparchus applied his knowledge of spherical angles to the problem of denoting locations on the Earth's surface. Philolaus' astronomical system is the first to consider the earth as an orbiting planet instead that lying at the centre of the universe believed earlier. Copernicus referred to Philolaus as his precursor in his *De Revolutionibus*. Unlike Copernicus, however, Philolaus did not replace the geocentric with a heliocentric universe. The central fire rather than the sun is at the centre of Philolaus' cosmos. The heavenly bodies are arranged in ten concentric circles around this central fire. Beginning from the outside, the fixed stars come first, then the five planets, the sun, the moon, the earth and a mysterious counter-earth. The fixed stars serve as the backdrop against which the other heavenly bodies move. All of the bodies have only one circular motion around the central fire, from west to east. It is the motion of the earth that accounts for the apparent daily motion of the sun across the sky and hence for night and day. Philolaus is the first to incorporate all five planets commonly known in antiquity into an astronomical scheme in the correct order, which indicates that he was aware of the most up-to-date astronomical data.

The knowledge of ancient Indian scholars in astronomy, particularly, was fascinating. Astronomy has been interwoven with astrology since ancient times in Indian literature. The planets *Shani* (Saturn) and *Mangal* (Mars), for instance, were considered inauspicious. In the working out of horoscopes (*Janmakundali*), the position of the *Navagrahas* (nine planets) plus *Rahu* and *Ketu* (mythical demons/evil forces) was considered. The scholars like Aryabhatta-I, Bhaskara-II, Varahamihira, Latadeva (505)[38] and Brahamgupta were associated with astronomy. In Indian language, the science of astronomy is called *Khagola-shastra*. The word '*Khagola*' is derived from the name of renowned astronomical observatory at the University of Nalanda. It was at Khagola that Aryabhatta studied and extended the subject. Aryabhatta discovered astronomical facts thousand years before Copernicus and Galileo, and this makes him an early pioneer of the field. Aryabhatta was the earliest to discover that the orbits of planets around the Sun are ellipses. He made an accurate approximation of the Earth's circumference and diameter. His book, *the Aryabhattiya,* described numerical and geometric rules for eclipse calculations. He gave logical explanation to the theory of solar and lunar eclipses. Solar and lunar eclipses, he believed, occurred not because

Rahu gobbled the sun and the moon, as the Hindu mythology claimed, but because of the shadows cast by the Earth and the moon. Aryabhatta's ideas about eclipses and the sun being the source of moonlight may not have caused much of an impression on European astronomers as by then they had come to know of these facts through the observations of Copernicus and Galileo. However, earlier Arabic translations of the *Āryabhattīya* were available in the Middle East by the 10th century, before the Crusades took place, so it's also likely that Aryabhatta's work had an influence on Copernicus' ideas. Aryabhatta was pioneer to infer that the Earth is spherical and it rotates on its own axis which results in day and night. He even concluded that the moon is dark and shines because of the light of sun. Considered in modern English units of time, Āryabhatta calculated the sidereal rotation (the rotation of the earth referenced the fixed stars) as 23 hours 56 minutes and 4.1 seconds; the modern value is 23:56:4.091.

Similarly, his value for the length of the sidereal year at 365 days 6 hours 12 minutes 30 seconds is remarkably close to the true value. He was aware that the earth spins on its axis, and that it moves round the sun and the moon rotates round the earth. Thus it has been suggested that Āryabhatta's calculations were based on an underlying heliocentric model in which the planets orbit the Sun. He discusses about the location of planets in relation to its movement around the sun. He refers to the light of the planets and the moon as reflections from the sun.[39] Aryabhatta was the first astronomer to make an attempt at measuring the Earth's circumference since Erastosthenes (200 B.C.). His measurement, which remained the most accurate for over a thousand years, was a significant improvement over the computation by Eratosthenes. His working was later adopted by the Greeks and then the Arabs. Another Indian astronomer, Brahmagupta estimated in 7th century that the circumference of the earth was 5000 *yojanas* (A *yojana* being equal to 7.2 kms). Calculating on this basis we see that the estimate of 36,000 kms as the earth's circumference comes quite close to the actual circumference known today. Bhaskara-II and his works also represent a significant contribution to mathematical and astronomical knowledge of antiquity. His main works were *Lilavati* (dealing with arithmetic), *Bijaganita* (*Algebra*) and *Siddhanta Shiromani* consisting of two parts: *Goladhyaya* (sphere) and *Grahaganita* (mathematics of the planets). In '*Siddhant Shiromani*' Bhaskaracharya mentions about force of attraction or gravity, discovered centuries later by Newton[40]. The study of astronomy in Bhaskara's works is based on a model of the solar system which is heliocentric and whose movements are determined by gravitation. Heliocentrism had been propounded in 499 by Aryabhata, who argued that the planets follow elliptical orbits around the Sun. Using this model, Bhaskara accurately defined many astronomical quantities, including, for example, the length of the sidereal year, the time that is required for the Earth to orbit the Sun. There is an old Sanskrit Sloka (couplet):" *Sarva Dishanaam, Suryaha, Suryaha, Suryaha.*" This couplet means that there are suns in all directions. This couplet which describes the night sky as full of suns indicates that in ancient times Indian astronomers had arrived at the important discovery that the stars visible at night are similar to the Sun visible during day time. In other words, it was recognized that the sun is also a star, though the nearest one. The Indian astronomers had also formulated ideas about gravity and gravitation.

Brahmagupta, in the 7th century had said about gravity that "Bodies fall towards the earth as it is in the nature of the earth to attract bodies, just as it is in the nature of water to flow". About a hundred years before Brahmagupta, another astronomer, Varahamihira had claimed for the first time that there is a force which might be keeping bodies stuck to the earth, and also keeping heavenly bodies in their determined places. Thus the concept of the existence of some tractive force that governs the falling of objects to the earth and their remaining stationary after having once fallen; as also determining the positions which heavenly bodies occupy, was recognized. It was also recognized that this force is a retractive force. The Sanskrit term for gravity is *Gurutvakarshan* which is an amalgam of *Guru-tva-akarshan*. *Akarshan* means to be attracted. Thus the fact that the character of this force of attraction was also recognized. The function of attracting heavenly bodies was attributed to the sun. The term *Guru-tva-akarshan* can be interpreted to mean, "to the attracted by the Master". The sun was recognized by all ancient people to be the source of light and warmth. Among the Aryans the sun was deified. The sun (*Surya*) was one of the chief deities in the Vedas. It was recognized as the source of light (*Dinkara*), source of warmth (*Bhaskara*). In the Vedas it is also referred to as the source of all life, the centre of creation and the centre of the spheres; the sun being recognized as the centre of the universe (solar system). The idea that the sun was looked upon as the power that attracts heavenly bodies is supported by the virile terms like Raghupati and *Aditya* used in referring to the sun. While the male gender is applied to refer to the sun, the earth (*Prithivi*, *Bhoomi*, etc.,) is generally referred to as a female. The literal meaning of the term *Gurutvakarshan* also supports the recognition of the heliocentric theory, as the term *Guru* corresponds with the male gender, hence it could not have referred to the earth which was always referred to as a female. Thus, heliocentricism and the concept of gravitation could have existed in India before the Greek astronomers developed it. The heliocentric theory of gravitation was, a thousand years later, articulated by Copernicus and Galileo.

The intellects in the two early worlds, Europe and China, paid great attention to the sky and believed the heavens to be organized in order. Regulating the calendar, drawing the sky map, and investigating omens are the same tasks they undertook. However, the early Chinese and Greek astronomy have fundamental differences: first, Greek astronomy highlighted planetary motions; as the apparent irregularities threatened the very notion of celestial order itself, the Greeks sought to geometrize them and in doing so turn irregularities into regularities. In contrast, the Chinese were more confident in the inherent order of the heavens and more open minded about its possible messages for the earth. Chinese theories seem to have "imposed far less rigid patterns on the order they expected, and they would no doubt have been amazed at the Greek ambition to prove celestial regularities". Secondly, the early Chinese and Greeks developed very different models of the universe: the former primarily with a flat earth, round heaven, free heavenly bodies and infinite cosmos, and the latter with a round earth centred by layers of round heavens, bound heavenly bodies and finite cosmos/heavens. The motions of heavenly bodies were, for the Greeks, the consequence of the rotation of the concentric celestial spheres on a common axis, and, for the Chinese, generated by vapour with each having its own path around the earth.

For the ancient Greek scientists, their aim was to provide a tempo-spatial model of the universe for explaining the apparent motions of the heavenly bodies. In contrast, early Chinese cosmological theories did not give detailed descriptions of the movement of the heavenly bodies, as they are viewed as independent entities from one another, moving freely. It is because, as Needham (1959) commented, the Chinese appeared to concentrate on the polar star (based on their keen observation to the sky) while the Greeks emphasised on the earth (or rather, where man is located) and, much later on, the sun. Ancient Chinese astronomy was primarily a government activity. It was the astronomer's role to keep track of planetary motions as well as divine what astronomical phenomena may mean for the ruling emperor. Solar eclipses were important enough to be recorded in chronicles and on "oracle" bones. The eclipse observations from the Chou dynasty and Warring States period (1050-221 B.C.), and onward, have been reliably dated. The ancient Chinese invented the first planetarium. By 2300 B.C., ancient Chinese astrologers and astronomers already had sophisticated observatory buildings. By about 20 B.C., the Chinese astrologers understood what caused eclipses. By 206 A.D. Chinese astrologers could predict solar eclipses. Observing total solar eclipses was a major element of forecasting the future health and successes of the Emperor. Based on the theories of his predecessors, Zhang Heng systematically developed the celestial sphere theory. An armillary constructed on the basis of his hypotheses bears a remarkable similarity to Ptolemy's earth-centred theory. It is astonishing that the celestial model Zhang Heng constructed was almost a physical model of Ptolemy's earth-centred theory. Su Song created a celestial atlas, which had the hour circles between the *xiu* (lunar mansions) forming the astronomical meridians, with stars marked in an equidistant cylindrical projection on each side of the equator, and thus, was in accordance to their north polar distances. Being the head official for the Bureau of Astronomy, Shen Kuo was an avid scholar of medieval astronomy. Shen is credited with making improved designs of the gnomon, armillary sphere, and clepsydra clock. Much like what Zhang Heng had said, Shen Kuo likened the Moon to a ball of silver, which does not produce light, but simply reflects light if provided from another source (the Sun). He explained that when the Sun's light is slanting, the Moon appears full. He reasoned that celestial bodies were spherical.[41] Shen compared the relationship of the Moon's path to the ecliptic, the path of the Sun, "to the figure of a rope coiled about a tree". Shen is also known for his cosmological hypotheses in explaining the variations of planetary motions, including retrogradation. Shen's hypotheses were similar to the concept of the epicycle in the Greco-Roman tradition.

Mathematics

Among the various fields of Greek interest, mathematics was one and Pythagoras, known as 'the father of numbers', excelled in it. The importance he gave to numbers was upheld by his followers, many of whom thought numbers were divine. Plato was also fond of mathematics because of its idealized abstractions. The heavenly bodies, he believed, exhibited perfect geometric form. This he expresses most clearly in a dialogue called '*Timaeus*' in which he presents his scheme of the universe. One of the most startling discoveries of early Greek science was that the fundamental intervals of music, the octave, the fourth, and the fifth, corresponded to whole number ratios of string length. This discovery that the phenomena of musical sound are governed by whole number ratios played a central role in the Pythagorean conception, first expressed by Philolaus, that all things are known through number. The Egyptian mathematics had also reached its heights in ancient times when he practical skill of land measurement was invented in Egypt because of the necessity frequently to re-measure plots of land after destructive inundations.

As a part of mathematics, Geometry is concerned with questions of size, shape, and relative position of figures and with properties of space. Geometry originated in Egypt where it developed out of necessity; it was adopted by Thales who had visited Egypt, and was introduced into Greece by him. It was from Egypt that Thales acquired the rudiments of geometry. However, the evidence is that the Egyptian skills were in orientation, measurement, and calculation. Thales' unique ability was with the characteristics of lines, angles and circles. Initially a body of practical knowledge concerning lengths, areas, and volumes in 3rd century B.C. geometry was put into an axiomatic form in *'the Elements'* by Euclid. Euclidean geometry set a standard for many centuries to follow. The development of geometry is preserved in a work of Proclus, *A Commentary on the First Book of Euclid's Elements*. Proclus provided a remarkable amount of intriguing information on the subject.

Hipparchus made an early contribution to trigonometry, the study of angles and of the angular relationships of planar and three-dimensional figures. He produced a table of chords, an early example of a trigonometric table. He also introduced the division of a circle into 360 degrees into Greece. Hipparchus was not only the founder of trigonometry but also the man who transformed Greek astronomy from a purely theoretical into a practical predictive science.

In India, mathematics has its roots in nearly 4000 years old Vedic literature during which a number of mathematical traditions were set forth for the first time. It has now been generally accepted that the technique of algebra and the concept of zero originated in India.[42] In addition, trigonometry, having evolved in the Hellenistic world and having been introduced into ancient India through the translation of Greek works, was further advanced in India. Even the rudiments of Geometry, called *Rekha-Ganita,* were formulated here. Also the technique of calculation, called *algorithm,* which is today widely used in designing software programmes (instructions) for computers was derived from Indian mathematics. These mathematical concepts were transmitted to the rest of the world and led to further developments that now form the foundations of mathematics. Vedic literature is replete with concepts of zero, the techniques of algebra and algorithm, square root and cube root. The period 400-1200 A.D. is often known as the golden age (the classical period) of Indian Mathematics. This period saw mathematicians such as Aryabhatta, Varahamihira, Brahmagupta, Bhaskara-I, Mahavira, and Bhaskara-II give broader and clearer shape to many branches of mathematics including, for instance, Arithmetic, Geometry, Algebra, Mathematical logic, General mathematics and Trigonometry. Their contributions would spread to Asia, the Middle East, and eventually to Europe. In fact, mathematics of this period was included in the 'astral science' (*jyotiṣśāstra*) and consisted of three sub-disciplines: mathematical sciences (*gaṇita* or *tantra*), horoscope astrology (*horā* or *jātaka*) and divination (*saṃhitā*). Excavations at Harappa, Mohenjodaro and other sites of the Indus Valley Civilization (IVC) have uncovered evidence of the use of "practical mathematics". The people of the IVC manufactured bricks whose dimensions were in the proportion 4:2:1, considered favourable for the stability of a brick structure. They used a standardized system of weights in regular geometrical shapes, which included hexahedra, barrels, cones, and cylinders, thereby demonstrating knowledge of basic geometry.[43] The inhabitants of Indus civilization also tried to standardize measurement of length to a high degree of accuracy. They designed a ruler—the *Mohenjo-daro ruler*—whose unit of length (approximately 1.32 inches or 3.4 centimetres) was divided into ten equal parts. Bricks manufactured in ancient Mohenjo-daro often had dimensions that were integral multiples of this unit of length. Baudhayana (8th century B.C.) composed the *Baudhayana Sulba Sutra*, best known as *Sulba Sutras*. The *Sulba Sūtras* ("Aphorisms of the Chords" in Vedic Sanskrit) (700-400 B.C.) list rules for the construction of sacrificial fire altars.[44] Jaina mathematicians (6th century B.C.) are important historically as crucial links between the mathematics of the Vedic period and that of the Classical period. A significant historical contribution of Jaina mathematicians lay in their freeing Indian mathematics from its religious and ritualistic constraints. In particular, their fascination with the enumeration of very large numbers and infinities, led them to classify numbers into three classes: enumerable, innumerable and infinite. In addition, Jaina mathematicians devised notations for simple powers (and exponents) of numbers like squares and cubes, which enabled them to define simple algebraic equations (*beezganit samikaran*). Jaina mathematicians were apparently also the first to use the word *shunya* (literally *void* in Sanskrit) to refer to zero. More than a millennium later, their appellation became the English word "zero" after a tortuous journey of translations and transliterations from India to Europe. Important Jaina mathematicians included Bhadrabahu (298 B.C.), Yativrisham Acharya (176 B.C.), and Umasvati (150 B.C.). Among other scholars of this period who contributed to mathematics, the most notable is Pingala (*piṅgalá*) (300-200 B.C.). There is evidence that in his work on the enumeration of syllabic combinations, Pingala stumbled upon both the Pascal triangle and Binomial coefficients.[45] It is well-known that the decimal place-value system in use today was first recorded in India, then transmitted to the Islamic world, and eventually to Europe. The earliest surviving evidence of decimal place value numerals in India and Southeast Asia is from the middle of the first millennium A.D. Decimal numerals recording the years 683 A.D. have also been found in stone inscriptions in Indonesia and Cambodia, where Indian cultural influence was substantial.[46] Besides, there was the Kerala school of astronomy and mathematics, founded by Madhava of Sangamagrama in Kerala, South India. It flourished between the 14th and 16th centuries and included among its members: Parameshvara, Neelakanta Somayaji, Jyeshtadeva, Achyuta Pisharati, Melpathur Narayana Bhattathiri and Achyuta Panikkar. In attempting to solve astronomical problems, the Kerala school astronomers independently created a number of important mathematics concepts. The most important results, series expansion for trigonometric functions, were given in Sanskrit verse in a book by Neelakanta called *Tantrasangraha*. Their discovery of these three important series expansions of calculus—several centuries before calculus was developed in Europe by Isaac Newton and Gottfried Leibniz—was a landmark achievement in mathematics. The invention of the Concept of Zero, Numerals, Arithmetic, Algebra, Geometry, Algorithm, Trignometry and Calculus are major accomplishments of ancient Indian mathematicians.

Numerals are found in the inscriptions of Ashok the Great in 3rd century B.C. This knowledge travelled from here to West and in Arab countries even now numerals are known as *Hindse*, i.e. from India. The concept of zero originated in ancient India. Apart from being a numeral, it is also considered a fundamental concept. The inclusion of zero in mathematics opened up a new dimension of negative numerals and gave a cutoff point and a standard in the measurability of qualities whose extremes are as yet unknown to human beings, such as temperature. Like the technique of algebra, the concept of zero also reached Europe through the Arabs. With the integration of zero into the numerals it became possible to note higher numerals with limited characters. In the earlier Roman and Babylonian systems of numeration, a large number of characters were required to denote higher numerals. Thus enumeration and computation became unwieldy. The first, however, to compute with the zero as a number was Brahmagupta. Brahmagupta's treatise *Brahma-sputa-siddhanta* was translated into Arabic and for several centuries this translation remained a standard text of reference in Arab world. It was from this translation of an Indian text on Mathematics that the Arab mathematicians perfected the decimal system and gave the world its current system of enumeration which we call the Arab numerals.

Arithmetic was discovered by Indians in about 2nd Century B.C. Bhaskaracharya's book *'Lilavati'* is regarded as the first book on modern arithmetic. The Arabs learnt and adopted it from India and spreaded it to Europe.[47] Algebra is the Arabic name for *Bijaganit*. Algebra as we know it today is a corruption of the term *'Al jabr'* which the Arabs gave to *Bijaganitam*. In India around the 5th century A.D. a system of mathematics that made astronomical calculations easy was developed. Astronomical calculations are complex and involve many variables that go into the derivation of unknown quantities. Algebra is a short-hand method of calculation and by this feature it scores over conventional arithmetic. In ancient India conventional mathematics, termed *Ganitam*, was known before the development of algebra. This is borne out by the name—*Bijaganitam*, which was given to the algebraic form of computation. *Bijaganitam* means 'the other mathematics' (*Bija* means 'another' or 'second' and *Ganitam* means mathematics). The fact that this name was chosen for this system of computation implies that it was recognized as a parallel system of computation, different from the conventional one which was used since the past and was till then the only one. It is certain that this technique of computation originated in India and was current around 1500 years back. Aryabhatta has referred to *Bijaganitam* in his treatise on Mathematics, *Aryabhattiya*. Another Indian mathematician-astronomer, Bhaskaracharya, has also authored a treatise, *'Siddhanta-Shiromani'*, on this subject. Thus the technique of algebraic computation was known and was developed in India in earlier times. From the 13th century onwards, India was subject to invasions from the Arabs and other Islamic communities like the Turks and Afghans. Along with these invaders, came critics like Al-Beruni who studied Indian society and polity. The Indian system of mathematics could not have escaped their attention. It was also the age of the Islamic Renaissance. Between the 10th to 13th centuries, the Christian kingdoms of Europe made numerous attempts to conquer the birthplace of Jesus Christ from its Mohammedan-Arab rulers. These attempts (the Crusades) failed in their military objective, but the contacts they created between oriental and occidental nations resulted in a massive exchange of ideas. The technique of algebra passed on to the West at this time. During the Renaissance in Europe, followed by the industrial revolution, the knowledge received from the East was further developed.

In the area of Geometry, as well, the Indian mathematicians had their contribution. There was an area of mathematical applications called *Rekha Ganita* (Line Computation). The *Sulba-sutras*, which literally mean 'Rule of the Chord' give geometrical methods of constructing altars and temples. The brick work of Harappa and Mohenjodaro excavations shows that people of ancient India (2500 B.C.) possessed knowledge of geometry. Aryabhatta formulated the rules for finding the area of a 'triangle', which led to the origin of Trignometry. Baudhayana was the author of one of the earliest *Sulba-sutras* (texts dealing with geometry and mathematical principles). His writings contain several important mathematical concepts that were later attributed to the Greeks. One such example is the Pythagorean Theorem. The now known Pythagorean Theorem is believed to have been invented by Baudhayana. The *Sulba-sutra* of Baudhayana contains geometric solutions.

Algorithm, a process of calculation based on decimal notation numbers, was deduced by Khwarazmi from the Indian techniques of geometric computation which he had studied. Al Khwarazmi was the chief exponent of this Indo-Arab amalgam in mathematics who evolved a technique of calculation from Indian sources. This technique which was named by westerners after Al Khwarazmi as "Algorismi" gave us the modern term *Algorithm*, which is used in computer software. In mathematics, computer science, and related subjects, an *'algorithm'* is an effective method for solving a problem expressed as a finite sequence of instructions. Algorithms are used for calculation, data processing, and many other fields.

Aryabhatta and Bhaskara-II demonstrate the knowledge of trigonometry, a branch of mathematics that studies triangles. Aryabhatta's contribution in trigonometry is unparalleled. He suggested formula to calculate the areas of a triangle and a circle, which were correct. He was the first mathematician to give the 'table of the sines'. The *Siddhanta Shiromani* of Bhaskara-II imparts knowledge on the sine table and relationships between different trigonometric functions. He also discovered spherical trigonometry, along with other trigonometrical results.

Calculus constitutes a major part of modern mathematics. Calculus has widespread applications in science, economics and engineering. The *Siddhanta Shiromani* of Bhaskara-II is an astronomical treatise, but the preliminary concepts of infinitesimal calculus and mathematical analysis, along with a number of results in trigonometry, differential calculus and integral calculus that contained in the work are of particular interest in the field. Madhava (1340-1425) and the Kerala School mathematicians from the 14th century to the 16th century expanded on Bhaskara's work and further advanced the development of calculus in India.

It has been suggested that Indian contributions to mathematics have not been given due acknowledgement and that many discoveries and inventions by Indian mathematicians were known to their Western counterparts, copied by them, and presented as their own original work; and further, that this mass plagiarism has gone unrecognized due to Eurocentrism.

The Chinese mathematics had discovered the use of 'decimal system', which was vastly superior to the sexagesimal system of Mesopotamia and Egypt. It is generally believed that the Indians derived their decimal system from the Chinese. This decimal system was introduced into Baghdad about 800 A.D. from the Indians. As many of the innovations, e.g. that of decimal form of arithmetic, have diffused westward from China, it is believed that the idea of triangulation method of survey and the use of grid-lines to locate places has travelled from the West, particularly from Greeks, to China. Chang Heng, the 2nd century scholar, was the first to introduce the grid system into China. In the broad field of mathematics, Shen Kuo mastered many problems, including various complex formulas for geometry, 'packing' equations for calculus, and chords and arcs problems employing trigonometry. Shen addressed problems of writing out very large numbers, as large as 10. Shen's "technique of small increments" laid the foundation in Chinese mathematics for packing problems involving equal difference series.[48]

History

Herodotus is the first known historian of Classical Antiquity. His account, *the Histories*, is the first example of historical writing in Western World. Before Herodotus the idea of recording past was not widespread and not attempted in a systematic manner. Polybius is another historian among the ancient Greeks to attempt to present history as a sequence of causes and effects, based upon a careful examination of tradition. For Posidonius "history" extended beyond the earth into the sky; humanity was not isolated each in its own political history, but was a part of the cosmos. His *Histories* were not, therefore, concerned with isolated political history of peoples and individuals, but they included discussions of all forces and factors (geographical factors, mineral resources, climate, nutrition), which let humans act and be a part of their environment. For example, Posidonius considered the climate of Arabia and the life-giving strength of the sun, tides (taken from his book on the oceans), and climatic theory to explain people's ethnic or national characters.

Although the style and form of Chinese historical writings varied through the ages, *Shiji* has defined the quality and style from then onwards. Before Sima, histories were written as certain events or certain periods of history of states; his idea of a general history affected later historiographers like Zheng Qiao in writing *Tongshi* and Sima Guang in writing *Zizhi Tongjian*. The historians regard Sima Qian's work as their model, which stands as the "official format" of the history of China. In writing *Shiji*, Sima initiated a new writing style by presenting history in a series of biographies. His work extends over 130 chapters—not in historical sequence, but was divided into particular subjects, including annals, chronicles, treatises—on music, ceremonies, calendars, religion, economics, and extended biographies. The modern historian Hsu Mei-ling states that Ban Gu's written work in geography set the trend for the establishment of geographical sections of history texts, and most likely sparked the trend of the gazetteer in ancient China.[49]

Cartography

Pei Xiu (224-271 A.D.), active during Three Kingdoms Period of China, outlined and analyzed the advancements of cartography and surveying up to his time. There is also evidence that Zhang Heng (78-139 A.D.) was the first to establish the grid reference system in Chinese cartography. However, the Qin State maps and Mawangdui maps of the Han Dynasty discovered by modern archaeologists were far superior in quality than those by Pei Xiu. The makers of Han maps were familiar with the use of scale. Pei Xiu outlined six principles that should be observed when creating a map; the first three principles outlined the use of scale, direction and distance, while the last three principles were related to the calculation of distances and represention on a flat, two dimensional map.[50] Pei Xiu states: In making a map there are six principles observable: (1) the graduated divisions, which are the means of determining the scale to which the map is to be drawn; (2) the rectangular grid (of parallel lines in two dimensions), which is the way of depicting the correct relations between the various parts of the map; (3) pacing out the sides of right-angled triangles, which is the way of fixing the lengths of derived distances (i.e. the third side of the triangle which cannot be walked over); (4) measuring the high and the low; (5) measuring right angles and acute angles; (6) measuring curves and straight lines. Pei Xiu was the first in China to describe in full the geometric grid reference for maps that allowed for precise measurements using a graduated scale, as well as topographical elevation. Historian Howard Nelson states that there is ample written evidence that Pei Xiu derived the use of the rectangular grid reference from the maps of Zhang Heng.

Meteorology

The studies in meteorology stretch back millennia, though significant progress in meteorology did not occur until the 18th century. Anaxagoras claimed that the earth is flat, rests on air, and remains where it is because of its size. A number of meteorological views attributed to Anaxagoras were also held by Anaximenes. Despite the manner of its formation, the earth is stationary, not spinning. The (relative) flatness of the earth allows water to spread over the earth, with mountains and plains rising above the level of the water. The levels change as water is evaporated or added by rain, and as water that has been trapped in the earth by the rotation makes its way out through rivers to the sea. The extant sources report views on thunder and lightning, the source of the Nile, the first correct account of the nature of hail, and inquiry into why the sea is salty.

Biology and Botany

Aristotle's most successful scientific writings were those on biology, the science of living organisms. He was a careful and meticulous observer who fascinated by the task of classifying animal species and arranging them into hierarchies. His formation of a hierarchy of living things led him towards the idea that animals represented a chain of progressive change, a sort of evolution. Other Greek philosophers groped similarly in this direction. However, barring any knowledge as to the physical mechanism whereby evolutionary changes could be brought about, such theories invariably became mystical. A rational theory of evolution had to await Darwin, 2200 years after the time of Aristotle. Botany, as a branch of biology and the scientific study of plant life and its development, began with tribal efforts to identify edible, medicinal and poisonous plants, making it one of the oldest sciences. From this ancient interest in plants, the scope of botany has increased to include the study of over 550,000 species of living organisms. Greek philosopher, Theophrastus, a pupil of Aristotle, is regarded as the founder of botany. Theophrastus classified plants into trees, shrubs, under shrubs and herbs.

Chemistry and Metallurgy

Parallel to the development of the concepts of atom and atomic combinations in physics there also was a similar development of ideas in the area of Chemistry. In ancient India, chemistry was called *Rasayan Shastra*, *Rasa-Vidya*, *Rasatantra* and *Rasakriya* all of which mean 'science of liquids'. There also existed chemical laboratories and chemical works, which were called *Rasakriya-nagaram* and *Rasakriya-shala* which literally mean 'School where liquids are activated'. Apart from the term *Rasa* which means liquid, another word, *Dravya* which means slurry, was also used to refer to chemicals. Thus, in ancient India, chemistry was evidently developed to a significant level. In the area of metallurgy, Indians had acquired proficiency in the extraction of metals from ore, and also in the casting of metals. In very early times: around 2000 B.C. the idea of smelting metals was known in Mesopotemia and the Near East. Though Indians could have had borrowed the idea of smelting metals from an outside source, they seem to have had used metals in warfare from around 1500 B.C. when the Aryans are said to have invaded the Indus Valley cities. The next definite reference to the use of metals by Indian soldiers is by the Greeks. The Greek historian Herodotus has observed in the 5th century that "Indians in the Persian army used arrows tipped with iron". Indian steel and iron were reportedly being used by the Romans for manufacturing armour as well as cutlery. But these references apart, it is in India itself that we find actual objects that reflect the advancement of the technique of smelting. Ashoka pillar at Mehrauli, New Delhi and another iron pillar in Karnataka stand proof of India's metallurgical heritage. This kind of a rust-proof iron had not been smelted anywhere else in the world, till we invented the stainless steel a few decades ago. Another instance of Indian metallurgy is the copper statue of Gautama Buddha found at Sultan Ganj in Bihar. There are many such examples that bear testimony to the excellence in smelting metals achieved in India in ancient times. This testifies to the fact that Indian metallurgy and chemistry had reached a high stage of perfection more than 1500 years ago. Nagarjuna was one such practitioner of the technique of combining various metals in order to invent a superior metal. The technique of alchemy was borrowed by the Arabs from India. The homeland of the Arabs is not rich in metals, thus alchemy and the smelting of metals could not have been indigenous to the Arabs. The Arabs called the technique Al Kimia which means the 'transformation of metals'. Al means 'The' and Khimia which is derived from the Greek term Khemia means 'to transmute metals'. The word A1chemy is a corruption of the term Al Kimia which the Arabs gave to the technique of converting base metals into gold like substances which they culled out from Indian texts on the subject.

Psychology

Psychology is a discipline involving the systematic, and often scientific, study of human mental functions and behaviour. Psychologists study such phenomena as perception, cognition, emotion, personality, behaviour and interpersonal relationships. Philolaus presents account of psychic faculties: The head is the seat of intellect, the heart of soul (*psyche*) and sensation, the navel of rooting and first growth, the genitals of the sowing of seed and generation. Philolaus thus recognizes four distinct psychic faculties. He makes a clear distinction between intellect (*nous*), which is limited to human beings, and perception, which animals possess as well. This hierarchy of four faculties is strikingly associated with the hierarchy of living beings (human, animal, plant) as well as a hierarchy of parts of the body (head, heart, navel, genitals). Philolaus makes a significant advance over earlier Pre-Socratics by making a firm distinction between thinking and perception.[51]

Linguistics

The learning of Indian curriculum in classical times had at its heart a system of grammatical study and linguistic analysis. The core text for this study was the Aṣṭādhyāyī of Pāṇini, the *sine qua non* of learning. Pāṇini's grammar can be considered to be the world's first formal system. To design his grammar, Pāṇini used the method of "auxiliary symbols," in which new affixes are designated to mark syntactic categories and the control of grammatical derivations. This technique was rediscovered by the logician Emil Post and is now a standard method in the design of computer programming languages.

Science and Technology

The Ancient Chinese are known for not only their trade and travels, but also for their many inventions and technological advancements. These inventions and advancements included, for instance, the development of writing, a calendar, gunpowder, printing, silk, paper, and the compass, besides the use of rain and snow gauges, use of magnetic compass for navigation, surveying and cartographic skills. Zhang Heng is the first person known to have applied hydraulic motive power to rotate an armillary sphere[52], an astronomical instrument representing the celestial sphere. The Greek astronomer Eratosthenes (276-194 B.C.) invented the first armillary sphere in 255 B.C. The Chinese armillary sphere was fully developed by 52 B.C. Zhang Heng's water-powered armillary sphere had profound effects on Chinese astronomy and mechanical engineering in later generations. In 132, Zhang Heng presented to the Han court what many historians consider to be his most impressive invention, the first seismometer. It was named *Houfeng didong yi* (instrument for measuring the seasonal winds and the movements of the Earth), and it was able to determine the exact direction of tremors and earthquakes.[53] The device was considered important enough to be mentioned in the "Annals" chapter of the *Book of Later Han*, detailing the reign of Emperor Shun. Although Heng's original seismograph did not survive time, its description did, and several modern scholars have tried to recreate the device. Su Song was the engineer of a water-driven astronomical clock tower in medieval Kaifeng, which employed the use of an early escapement mechanism.[54] Su Song compiled one of the greatest Chinese horological treatises of the Middle Ages. *Xinyi Xiangfayao* (lit. "Essentials of a New Method for Mechanizing the Rotation of an Armillary Sphere and a Celestial Globe"), written in 1092. In his written work, Shen Kuo made one of the first references in human history to the magnetic compass-needle, the concept of true north, and its use for navigation at sea.[55] Shen Kuo's writing on magnetic compasses has proved invaluable for understanding China's earliest use of the compass for seafaring navigation. An artisan known as Bi Sheng (990-1051) invented ceramic movable type printing.[56] Although the use of assembling individual characters to compose a piece of text had its origins in antiquity, Bi Sheng's methodical innovation was revolutionary for his time. By 15th century, metal movable type printing was developed in Ming Dynasty China and was widely applied in China by 16th century. Zhang Heng is often credited with inventing the first odometer, an achievement also attributed to Archimedes (c. 287-212 B.C.E.) and Heron of Alexandria (fl. C.E. 10-70).[57] Similar devices were used by the Roman and Han-Chinese empires at about the same period. The South Pointing Chariot was another mechanical device credited to Zhang Heng.[58] It was a non-magnetic compass vehicle in the form of a two-wheeled chariot. Differential gears driven by the chariot's wheels allowed a wooden figurine (in the shape of a Chinese state minister) to constantly point to the south, hence its name.

Travels and Explorations

The discovery of the rest of world by Chinese travellers is an aspect of history that is often overlooked in Western writings. Travels beyond the Chinese horizons were undertaken by orders of an emperor or by missionaries or traders. The travel books are known as the *'Travels of Emperor Mu'*, who ruled from 1001 to 945 B.C. Emperor Mu[59], it is recorded, had the ambition to travel all-round the world and to leave the marks of the chariot wheels on every land. These books, which may be compared with that of *Odyssey of Homer*, the ancient Greek scholar, also reveal the story of high adventures. There is no doubt that at a very early date, the Chinese travellers had gone far beyond their original cultural hearth area. The discovery of Mediterranean civilization is credited to Chinese scholar, Chang Chien (Zhang Qian), in 128 B.C. There are a number of other Chinese travellers whose records are enough to ensure their place in history. One of the most distinguished was the Buddhist monk, Huan Tsang. In the 7th century after Christ, he was able to cross the high windswept plateau of Tibet and the world's highest mountain on the way to India. After studying in the centres of the Buddhist faith for several years, he returned to China, carrying back a large collection of Buddhist relics and manuscripts. He was also the Chinese discoverer of India. In the same century, another Buddhist monk, I Ching, reached India by sea. When he returned to China he carried with him more than 10,000 rolls of Sanskrit Buddhist texts, which he undertook to translate into Chinese. In 1296 A.D. the Chinese traveller, Chou Ta Kuan, visited Combodia and wrote a detailed account of the strange customs of the Combodians. By 13th century A.D. the Chinese merchants were sailing their junks to Java and Malaya and even as far as India. Cheng Ho, a Chinese admiral, carried out major work of exploration in this direction. His voyages opened regular trade routes to Java, Sumatra, Malaya, Ceylon and the west coast of India. He also reached the Persian Gulf, the Red Sea, and the east coast of Africa, south of the equator. In the east he went as far as Taiwan. Zhang Qian, a 2nd century B.C. explorer, visited directly the kingdom of Dayuan in Ferghana, the territories of the Yuezhi in Transoxiana, the Bactrian country of Daxia with it remnants of Greco-Bactrian rule, and Kangju. He also made reports on neighbouring countries that he did not visit, such as Anxi (Arsacid territories), Tiaozhi (Mesopotamia), Shendu (India) and the Wusun. The inhabitants of Dayuan, in Ferghana, west of the Tarim Basin, are considered by Zhang Qian as sophisticated urban dwellers, on the same footing as the Parthian and the Bactrians. Zhang Qian also describes the origins of the Yuezhi, explaining they came from the eastern part of the Tarim Basin, a momentous explanation which has encouraged historians to connect them to the Caucasoid mummies, as well as to the Indo-European-speaking Tocharians that have been identified from precisely the same area. Zhang Qian witnessed the last period of the Greco-Bactrian kingdom, as it was being subjugated by the nomadic Yuezhi. Zhang Qian also reports about the existence of India southeast of Bactria. The name *Shendu* comes from the Sanskrit word "Sindhu", used for the province of Sindh (now a province of Pakistan) by its local people. Sindh was one of the most advanced regions of India at the time. Although it was part of India, it practiced separate authority over itself. Because of its coastal borders with Persia and the Arabian Sea, it invited great wealth from these regions. Parts of Northwestern India (modern Pakistan) were ruled by the Indo-Greek Kingdom at the time, which explains the reported cultural similarity between Bactria and India. Zhang Qian identifies "Anxi" as an advanced urban civilization, like Dayuan (Ferghana) and Daxia (Bactria). Zhang Qian also visited the area of Sogdiana (Kangju), home to the Sogdian nomads. From his missions he brought back many important products, the most important being alfalfa seeds (for growing horse fodder), strong horses with hard hooves, and knowledge of the extensive existence of new products, peoples, and technologies of the outside world. Zhang Qian's journeys had promoted a great variety of economic and cultural exchanges between the Han Dynasty and the Western Regions, because silk was the dominant product traded this trade link became known as the Silk Route.

GEOGRAPHICAL INHERITANCE

Geography in ancient world has originated out of the disciplines as discussed above. The initial contribution of ancient scholars and travellers goes to the fields of regional, mathematical and practical, historical, physical, economic and theoretical geography.

Regional Geography

The field of regional geography in Europe was largely aided by the works of Homer, Hecateus and Strabo, and the explorations of Alexander the Great and Pytheas. As derived from the work and experience of these people, the ancient Greeks divided the world into three continents, Europe, Asia, and Libya (Africa). Hellespont formed the border between Europe and Asia, and the Nile River between Asia and Libya. 'Ges Periodos', the literary work of Hecateus, is actually regional description of these two earliest known continents. Besides, parts of northern Africa have also been discovered. Homer describes a circular world ringed by a single massive ocean. The works show that the Greeks, by the 8th century B.C., had considerable knowledge of geography of the eastern Mediterranean. 'Illiad' and 'Odyssey', the epic poems of Homer, contain a large number of place names and descriptions. The relatively narrow habitable band was considered to run from the Atlantic Ocean in the West to India in the east. The southern portion of Africa was unknown, as was the northern portion of Europe and Asia. It was believed that they were circled by a sea. These areas were generally considered uninhabitable. The military and armed explorations of Alexander the Great added substantially to the field of regional descriptions. His conquests pushed Greek knowledge of the earth as far as Indus River in the east and to Egypt southward, where he established his rule. A large part of the Mediterranean region became the part of his kingdom. He founded the city of 'Alexandria' in 332 B.C., which later emerged as one of the great commercial and intellectual centres of ancient world. Alexander's staff, while travelling, included writers to describe the lands they crossed, and astronomers to make observations. Although, the primary objective of Alexander's conquests was to expand Greek territory, he sent back to the Greek world a wealth of new observations concerning what it was like beyond and they proved highly beneficial to the field of ancient geography. With his accounts the Greeks became familiar with the natural features of Persian Empire, Central Asia, Afghanistan, India and coastal parts of Iran. While Alexander was extending Greek geographic horizons to the east, Pytheas was voyaging to the northwest, covering parts of Western and Northern Europe. He is known to have made remarkable voyages between 330-300 B.C. Pytheas described about the customs of people of Britain in detail. He observed the change in character of agriculture from south to north in Great Britain. His writings contain a reference of a sea full of ice that could neither be traversed on foot nor in a boat. He has mentioned a place north of Norway, where the length of the longest day was between 17 and 19 hours, and where the Sun remained above horizon during whole of day. He was also the first Greek to tell about ocean tides and their relation to various phases of moon. Thus, Pytheas reported things that were so contrary to Greek experience that the scholars of his day discredited him and treated his important observations and information as pure fantasies. However, today we acknowledge him as a great scholar of the ancient period. In the view of Strabo, geography acquainted man with the habitable land, with oceans, with vegetation, fruits, and with the various socializations of earth's surface, etc. that helped in identifying the various related problems and happiness of the man. He called all these elements as 'natural attributes of a place', and established their relationship with man and other places of the surface of earth. This concept is still the basic in all spatial studies and has a firm place in present day geography. These natural attributes of the place distinguish it from other places and regions. These ideas underlie the "areal differentiation" viewpoint, as presented later by Richard Hartshorne, a 20th century American scholar. Strabo considered 'place' in two different perspectives, viz. in terms of its astronomical (abstract) location and in terms of its terrestrial (on land) position.

In ancient period, the knowledge about various parts of the world was limited. It was due to the poor means of communication and transportation. Even then, the attempts were made to divide the world into several regions. Such descriptions exist

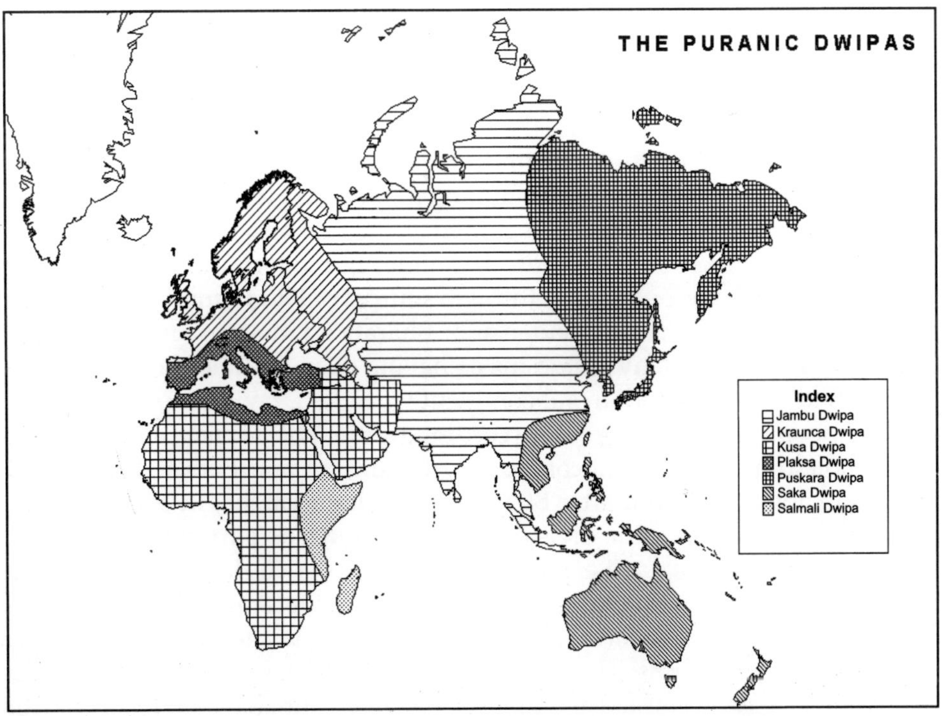

FIGURE 3.7 The Puranic Dwipas (*Source:* Hussain, M. (1995), p. 105)

in Puranas. Although incorrect, the term *Dwipa* has been used to designate various realms (continents) of the earth. Different scholars have interpreted this term in several ways. Some conceived its meaning as 'an island', others as 'peninsula', 'doab' or even a 'natural region'. Accordingly, the known world during the Puranic period was divided into seven *Dwipas* or 'regions' (Fig. 3.6). These divisions exclude the American Continents, Greenland, England and Antarctica, since they were discovered only during the Age of Discovery in the late medieval period. The regions, as depicted in Fig. 3.7, seem to have derived their names from the existing popular vegetation (trees or grasses) here.

These seven regions were known as Jambu, Krauncha, Kusha, Plaksha, Pushkara, Shaka and Shalmali. Jambu dwipa lies in the centre of all these continents. In relation to the present day context, Jambu covers present Central Asia from North to South, including India, or the region north of Salt Sea. Jambu, in fact, is a bush found in Himalayan region. Kusha extends over present Middle East and most of Africa. The name is taken from a sacred grass, *Kusa*, used in brahmanical ceremonies. The present Eastern Asia and adjoining lands form Pushkara. The present Mediterranean region forms Plaksha. Shalmali represents the region of Eastern Africa and Madagascar Island. This region is rich in *Salmala*, the silk-cotton tree, found on the margins of Equatorial regions of monsoon lands with moderate rainfall. Most of present Europe was Kraunca. Lastly, Shaka forms South-East Asia and adjoining Island groups. Hot and moist climate and thick evergreen forests characterize the region.

The geographical knowledge about Indian Sub-continent is related to its identification, people and culture and relief and drainage. In Vedic and Puranic literature, the entire country from Himalayas to Kanyakumari has been referred to as *Bharatvarsha* by the Aryans. This name has both geographical and historical significance. According to one school of thought, the name is derived after the name of *Bharat*, the ruling king. Bharata, according to Hindu legend was the first King to conquer all worlds as known to the Vedic Aryans, uniting it into a single entity which was named after him as *Bharatvarsha*. Bharata is regarded as the first and only emperor to rule all of India. Bharata's paramount position as founder of modern India is imbibed in the consciousness of Indians, but mainly through sources of Hindu mythology and Hindu religious works. Archaeological evidence of Bharata's kingdom and reign is scarce. Bharata is construed by many historians as having been an Indo-Aryan king, and as king he unified the entire Indian subcontinent with the Dravidian peoples and other indigenous peoples as his subjects. According to the *Mahabharata*, Bharata's empire covered whole of the Indian subcontinent, Afghanistan and Persia. The Republic of India is also known as Bharat after Bharata. This Bharatvarsha, in ancient times was divided into *Nav-Khandas* (nine divisions)—eight shown as part of Greater India and the 9th surrounded by the sea (i.e. present day Sri-Lanka). Certain parts of the country are very distinctively mentioned in the ancient Indian literature. They are, e.g. *Sapta-Sindhu* (Punjab Plains), *Aryavarta* (the Aryan domain) and the region of Indus valley or the Upper Gangetic Plains. The Vedas, Epics and Puranas make mention of a series of mountains in *Bharatvarsha*. They are, for instance, Himalayas *(Himavat)*, existing like a bow in its northern part and divided into *Antagiri* (Inner Himalayas) and *Bahyagiri*

(Outer Himalayas); *Kailash Parbat*, the abode of *Apsaras* (nymphs) and *Devas* (Deities) and rich in diamonds, minerals and other precious stones; *Vindhayans*, the extensive mountains with hundreds of peaks, variegated trees and creepers; *Mahendra Mali*, the Eastern Ghats; Sahyadri, the Western Ghats; *Rika*, the mountain range from Ken to Ton rivers north of Vindhayans; and *Suktiman*, the mountains of Khandera, Ajanta and Golkunda. The descriptions are also available for a number of Himalayan and other Inland rivers. Rigveda has mentioned various rivers originating from Himalayas, viz. Ganga, Yamuna, Brahamputra, Saraswati, Satudri (Sutlej), Asikni (Chenab), Vitasta (Jhelum), Arjikeya (upper part of Indus), Susoma (Savan), Sindhu (Indus), Kubha (Kabul), Gomati (Gomala), Krumu (Kurrum), etc. Among the inland rivers the important ones that get a mention are Narmada, Tapti (Tapi), Godavari, Krishna, Cauvery and Tungbhadra. However, the most elaborate descriptions exist about Ganga and Brahamaputra. The religious flavour is very strong in these descriptions, as the rivers have been considered sacred, to be worshipped as Goddesses in the Hindu mythology.

While the invaders ruthlessly destroyed the architectural achievements of this country and imposed their despotic rule, they also transmitted Indian sciences to the outside world. Along with Mahmud of Ghazni came scholars like Al Beruni who studied Indian texts and translated them into Arabic. Many Indian ideas of medicine were incorporated into the Unani system of medicine of the Arabs. Nagarjuna's works could not have escaped their attention.

In China, the earliest known geographical writing dates back to 5th century B.C., and this belongs to the Warring States period (481 B.C.–221 B.C.). It is in the form of the book *Shu Jing* (*Classic of History*), which describes the traditional nine provinces, with their human, physical and economic characteristics. These regional descriptions pertain to areas of the Yellow River, the lower valleys of the Yangtze, the Shandong peninsula, and the parts of the Wei and Han Rivers in the west. In this ancient geographical treatise that later greatly influenced geographers and cartographers, the Chinese used the mythological figure of Yu the Great to describe the known world. Although the Chinese geographical writing in the time of *Herodotus* and *Strabo* were of lesser quality and contained less systematic approach, this changed from the 3rd century onwards when Chinese methods of documenting geography became more complex than found in Europe. From the time of the 5th century B.C. *Shu Jing* forward, Chinese geographical writing provided more concrete information and less legendary element. In later periods of the Song Dynasty (960-1279 A.D.) and Ming Dynasty (1368-1644 A.D.) there were much more systematic and professional approaches to geographic literature. The Song Dynasty scholar and government official Fan Chengda (1126-1193 A.D.) wrote the geographical treatise known as the *Gui Hai Yu Heng Chi*. It focused primarily on the topography of the land, along with the agricultural, economic and commercial products of each region in China's southern provinces. The polymath Chinese scientist Shen Kuo (1031-1095 A.D.) devoted a significant amount of his written work to geography. The Ming Dynasty geographer Xu Xiake (1587-1641) travelled throughout the provinces of China to write his enormous geographical and topographical treatise (compiled posthumously into The Travel Diaries of Xu Xiake), documenting various details of his travels, such as the locations of small gorges, or mineral beds such as mica schists. From the snowy passes of Szechuan, to the semi-tropical jungles of Guangxi, to the mountains of Tibet, Xu Xiake wrote of all his experiences and provided enormous amounts of written detail from his observations. In Guizhou, he made the discovery of the true source of the West River (Xi Jiang) in Guangdong. He was also responsible for the discovery that the Mekong and Salween rivers were in fact separate. Local gazetteers featured a wealth of geographic information in ancient China. The ancient Chinese historian Ban Gu (32-92) started the trend of the gazetteer.

The Chinese were also concerned with documenting geographical information of foreign regions outside China. They have been writing of civilizations of Middle East, India, and Central Asia since the time of traveller Zhang Qian (2nd century B.C.). The Tang Dynasty (618-907 A.D.) Chinese diplomat Wang Xuan-ce travelled to Magadha (modern northeastern India) during 7th century A.D. and wrote the book *Zhang Tian-zhu Guo Tu* (Illustrated Accounts of Central India), which included a wealth of geographical information. Chinese geographers such as Jia Dan (730-805) wrote descriptions of places far abroad, including a sea route through the Persian Gulf. The geography of medieval Cambodia (the Khmer Empire) was documented in the book *Zhen-La Feng Tu Ji* of 1297 A.D., written by Zhou Daguan. Zhou Daguan is most well known for his accounts of the customs of Cambodia and the Angkor temple complexes during his visit there. Zhou's portrayal is today one of the most important sources of understanding of historical Angkor and the Khmer Empire. Huang-tsang's book, *Great Tang Records on the Western Regions*, was one of the primary sources for the study of the regions of Central and South Asia. He was also the Chinese discoverer of India. In the same century, another Buddhist monk, I-Ching, reached India by sea. The records of I-Ching's travels contributed to the world knowledge of the ancient kingdom of Srivijaya, as well as providing information about the other kingdoms lying on the route between China and the Nalanda Buddhist University in India. Zheng He led seven expeditions (1405-33) to what the Chinese called "the Western Ocean" (Indian Ocean) and brought back information on more than thirty kingdoms, covering, for instance, Champa, Java, Palembang, Malacca, Aru, Sumatra, Lambri, Ceylon, Kollam, Cochin, Calicut, Siam, Kayal, Pahang, Kelantan, Hormuz, Maldives, Mogadishu, Barawa, Malindi, Aden, Muscat, Dhufar, Quilon, Kaya, Coimbatore, Puttanpur, Sharwayn, East Africa and countries of the Arabian Peninsula (Table 3.3). His voyages opened regular trade routes to Java, Sumatra, Malaya, Ceylon and the west coast of India. He also reached the Persian Gulf, the Red Sea, and the east coast of Africa, south of the equator. In the east he went as far as Taiwan. Besides, Zhang Qian visited directly the kingdom of Dayuan in Ferghana, the territories of the Yuezhi in Transoxiana, the Bactrian country of Daxia with it remnants of Greco-Bactrian rule, and Kangju; he also made reports on neighbouring countries that he did not visit, such as Anxi (Arsacid territories), Tiaozhi (Mesopotamia), Shendu (India) and the Wusun.

Mathematical and Practical Geography

The development of mathematical and practical geography is linked primarily to Crates, Thales, Anaximander, Pythagoras, Eratosthenes, Hipparchus, Posidonius, Aristotle, Ptolemy and Eudoxus of Cnidus. The studies were conducted in relation to the shape of earth, the size of earth, the location over earth's surface and the observations on earth's surface. Thales and Anaximander, particularly, are known as the originators of mathematical tradition in geography. Greeks gave the system of latitudes and longitudes and proved that the earth is a sphere. Pythagoras was the first to propose a spherical world, arguing that the sphere was the most perfect form. This idea was embraced by Plato and Aristotle presented empirical evidence to verify this. The arguments given in support of his statement about the sphericity of the earth were twofold: (1) he deduced from the law of gravitation the tendency of all things towards the centre and proved that by the action of this, when the earth was in the course of formation, and the component elements were coming together, the mass thus formed by acceleration was so constituted that its entire circumference must be equidistant from its centre; and (2) He inferred the shape from what is seen to take place in lunar eclipse, i.e. the shadow falls on moon is spherical. He noted that the Earth's shadow during an eclipse is curved, and also that stars increase in height as one moves north. Aristotle, whose ideas were later adopted by Eratosthenes and others, was of the view that the Earth lies in the centre of universe around which the celestial bodies revolve. This misconception prevailed for centuries. Eudoxus of Cnidus used the idea of a sphere to explain how the sun created differing climatic zones based on latitude. This led the Greeks to believe in a division of the world into five regions. Crates of Mallus is famous for constructing the earliest known globe of earth. Thus, during the time of Aristotle, the sphericity of earth had been proved, latitudes and longitudes of many places had been calculated and the practice of plotting world map on a grid had begun. Eratosthenes and Posidonius, both, further extended the mathematical tradition in geography, but the works of the former were more elaborate and substantial. The size of earth was an important question to Eratosthenes, who attempted to calculate its circumference. That some of the figures Eratosthenes had used in his calculation were considerably in error became known, and Posidonius set out to get a more accurate measurement. The first of Greek scholars concerned about the measurement and the location of things on the face of the earth was Thales. He was very sharp in understanding mathematical and geographical regularities. Thales was first to apply various geometrical concepts (circles, diameter, angles, etc.) in geographical studies. Anaximander is known for the introduction into Greek world of a Babylonian instrument called 'Gnomon' (an instrument having functions like a sundial). With the help of Gnomon, it was possible to make a variety of observations regarding noontime, duration of sunshine, time of solstices and equinoxes, the time of sunrise and sunshine and so on. Anaximander is also said to have made the first map of the world. Hipparchus, another mathematician and astronomer of this period, attempted to determine exact positions of every point on the earth's surface. He was first to divide circle into 360 degrees, based on Assyrian mathematics. Hipparchus defined a grid of latitudes and longitudes. The equator, he pointed out, was a great circle (one that divides the earth into two equal parts) and meridians, drawn converging at the poles, are also great circles. The parallels become shorter and shorter as they approach the poles. Since the earth makes one complete revolution in 24 hours and there are 360 meridians drawn from equator to poles, each hour the earth turns through 15 degrees of longitudes.

Among all the ancient scholars of Greek and Roman period Ptolemy was most influential. His contribution in the field of mathematical geography is highly commendable and has been acknowledged throughout ages. Ptolemy developed the sound principles of mathematics and applied to the study of practical geography. He recalculated the circumference of earth. He redesigned a graticule and plotted the world using the concept of conical projection with remarkable accuracy. He worked out the dimension of inhabitable world and plotted the places of physical and human interest of contemporary times. Of the two great books Ptolemy authored, one is *'the Syntaxis'*, popularly known as *'Almagest'*, which is a great work in classical astronomy. It long remained the standard reference work on the movement of the celestial bodies. The other book is *'Geographia'* and it is devoted to the exact determination of the position of places by means of latitudes and longitudes. In the opinion of Ptolemy, geography is a science that deals with the art of map-making. Maps based on scientific principles had been made since the time of Eratosthenes (3rd century B.C.), but Ptolemy improved on projections. Ptolemy used geometric models to predict the positions of the sun, moon, and planets, using combinations of circular motion known as epicycles. Based on his observations of solstices and equinoxes, Ptolemy found the lengths of seasons and proposed a simple model for the sun which was a circular motion of uniform angular velocity, but the earth was not at the centre of the circle but at a distance called the eccentricity from this centre. This theory of sun forms the subject of *Almagest*. The planetary theory which Ptolemy developed here is a masterpiece. He created a sophisticated mathematical model to fit observational data which before Ptolemy's time was scarce, and the model he produced, although complicated, represents the motions of the planets fairly well. The most significant contribution of Ptolemy and his maps is the first uses of longitudinal and latitudinal lines and the specifying of terrestrial locations by celestial observations. When his *Geographia* was translated from Greek into Latin and introduced into Western Europe at the beginning of 15th century, the idea of a global coordinate system revolutionized European geographical thinking and put it upon a scientific and numerical basis. The map distinguishes two large enclosed seas, the first one being Mediterranean, and the second one being Indian Ocean (*Indicum Pelagus*), which extends into the China Sea (*Magnus Sinus*) in the East. The major geographical locations are Europe, the Middle-East, India, Sri Lanka (*Taprobane*), the Southeast Asian peninsula (*Aurea Chersonesus* or "Golden Peninsula") and beyond it China (*Sinae*). The *Geographia* and the maps derived from it played an important role in the expansion of Roman Empire to the East. Trade throughout the Indian Ocean was extensive from 2nd century, and many Roman trading ports have been identified in India. From these ports, Roman embassies to China are recorded in Chinese historical sources from around 166. The basic objective of Ptolemy's attempts was to reform the map of world, on the basis of astronomical principles. Therefore, he was able to prepare the first complete map of the known world. Ptolemy's concept of Universe agreed with that of Aristotle: "the earth was a sphere that remained stationary in the centre while the celestial bodies moved around it in circular courses". This remained an accepted doctrine until the Age of Discovery. Another important contribution of Ptolemy was "Guide to Geography", the largest geographical gazetteer of the times, which contained 8 volumes and the basis on which he revised the world map. In the art of map making, the great contribution made by Ptolemy was his improvement over the previously drawn maps and this he did by adopting projections for the world map, showing the graticule of latitudes and longitudes. In fact, in regard to the mathematical construction of the projection of his maps, Ptolemy was far ahead of his predecessors. The writings and maps of Ptolemy inspired the geographers and explorers of Great Age of Discovery. With the death of Ptolemy, the geographic horizons that had been widened both physically and intellectually by the Greeks closed in again.

In the field of mathematics, Shen Kuo developed techniques that laid the foundations for spherical trigonometry and high-order arithmetic progressions. In his *Dream Pool Essays* (*Mengxi Bitan*) Shen was the first to describe the magnetic needle compass used for navigation. Shen discovered the concept of true north in terms of magnetic declination towards the North Pole, with experimentation of suspended magnetic needles and this was the decisive step in human history. The Chinese mathematics had discovered the use of 'decimal system', which was vastly superior to the sexagesimal system of Mesopotamia and Egypt. It is generally believed that the Indians derived their decimal system from the Chinese. As many of the innovations, e.g. that of decimal form of arithmetic, have diffused westward from China, it is believed that the idea of triangulation method of survey and the use of grid-lines to locate places has travelled from the west, particularly from Greeks to China. Chang Heng, the 2nd century scholar, was the first to introduce the grid system into China. He also devised an armillary or celestial globe to track the movement of planets and stars, proposed the concept of the lunar eclipse, developed longitude and latitude grids for maps and invented the odometer. The odometer was a "mileage cart". A visionary mathematician, Heng computed the value of pi as the square root of 10, or approximately 3.162, not far off from 3.14. He constructed a sundial to measure the position of sun.

The earliest extant maps found in China date to the 4th century B.C. and were made in the ancient State of Qin. The earliest known reference to the application of a geometric grid and mathematically graduated scale to a map was contained in the writings of the cartographer Pei Xiu (224-271). Pei is best known for his work in cartography. He lived during Western Jin Dynasty (265-316), and was claimed to be "the father of China's Scientific Cartography." He was equally famous as Claudius Ptolemy, the ancient Greek cartographer. He was the first to mention a plotted geometrical grid reference and graduated scale displayed on the surface of maps to gain greater accuracy in the estimated distance between different locations. Pei Xiu, in turn, derived the idea of the grid reference from the map of Zhang Heng (78-139 AD), a polymath of the Eastern Han period. Zhang Heng invented quantitative cartography, applying a grid system to maps, from which positions, distance, and itineraries could be calculated. His book, *Discourse on New Calculations*, established the basis for the mathematical use of the grid with maps. He perfected the science of latitude and longitude, and his grids were said to form a "net over the Earth." Pei's major contribution to cartography lies in his establishment of a theoretical foundation for the making of traditional Chinese maps for the first time by summing up the experiences of his predecessors. He put forward the six principles in map-making, representing a milestone in the development of cartography theories. The first three principles outlined the use of scale (*fan lu*), direction (*zhunwang*), and road distance (*daoli*), while the last three principles were used to properly calculate distances on uneven terrain as represented on a flat, two dimensional map. Beside Pei Xiu and Zhang Heng, Su Song also had in-depth knowledge of cartography. He made a celestial atlas which was a work of great precision. The unique characteristic of his star maps was that the maps were laid out in printed form. He created many terrestrial maps as well.

Historical Geography

Mainly Herodotus and Strabo contributed to this field. Although Herodotus is usually described as the first great historian, a large part of his work is easily identified as geography. In fact, he is credited with the very old idea that all history must be treated geographically and all geography must be treated historically. The notion of geography as "the hand-maiden of history" came from Herodotus. According to him Geography provides the physical background, the stage setting, in relation to which the historical events take on meaning. He provided excellent examples of what we would call today 'historical geography', i.e. the re-creation of past geographies and the tracing of geographical change through time. The contributions of Herodotus are based on his own personal observations during many years of travel. Strabo, on the other hand, compiled the writings of his predecessors, particularly Homer, Herodotus, Aristotle, Eratosthenes, Posidonius and Hipparchus, and wrote a 17-volumed book entitled '*Geographie*'. The major part of this work is devoted to detailed description of various parts of the known world. First two volumes are general; eight books cover Europe; six Asia; and one covers present Africa. This work is mainly historical, but Strabo also described in it—the character, physical properties, surface configuration and natural production of each country known in his times. Strabo considered geography essentially as an encyclopaedic description of the known inhabited world.

Physical Geography

The Greeks made a remarkable development in the field of physical geography.[60] Their works contain numerous references to studies, for instance, on mountains, delta-building, tides, change of weather, rain, earthquakes, volcanoes and the plant studies. The scholars like Homer, Posidonius, Herodotus, Theophratus, Plato and Aristotle made valuable contributions in this direction. From the Aegean Sea the Greeks expanded the horizon of knowledge in the study of seas and oceans distinguishing the varying properties of their coastlines, salinity, waves, tides and winds. Posidonius was the most outstanding physical geographer. He was considered an authority on oceanography, on which he contributed a book, entitled *"The Ocean"*. The studies on tides and the measurement of the depth of the sea and related aspects go to his credit. Interestingly, Herodotus was a historian by profession, but he always placed and studied historic events in geographic setting. He introduced Greek alphabet 'delta' to geographical dictionary, on the basis of his observations of the black soil along the Nile River. He linked the shape of silt-deposits on the flood plain to Greek alphabet 'delta' (Δ). He also corrected the misconceptions regarding Caspian Sea, for which the contemporary identification was that of a 'sea'. He supported evidences for it's being a 'lake'—with no contacts with the outside world. Physical geography was the field of interest for Aristotle and Plato as well. Aristotle, for instance, explained the phenomenon of expansion of land in the shallow seas and formation of delta. He correctly pointed out the process of alluvial deposition by which so many places of land area being eroded and added into the sea. He affirmed this with the study of Sea of Azov, which was continuously becoming shallower and he feared that one day it would be filled up completely and would convert into a dry land. The delta formation of Nile was also attributed to the enormous silt carried by the river from its upper reaches (Ethiopia). Plato discussed some of the barren lands of Attica (Greece) and explained that such waste tracts in the past were full of vegetative covers and fertile soils. Under the impact of the external forces, the forests have been depleted and soils leached resulting into barren topography. Such waste lands he said are like the skeleton of sick man, all the fat and soft earth having been wasted away, and only the bare framework of the land being left. Plato considered man also an active agent who changed the face of the earth. Theophratus was a student of Aristotle. He examined the relationships of plants and climate, and this marked the beginning of the study of plant geography. Besides, the Greeks, right from the Homerian period, have recognized four major winds, having different properties, directions and the associated weather type. They are: *Boreas*: the north wind—strong, cool, with clear skies; *Eurus*: the east wind—warm and gentle; *Notus*: the south wind—wet and sometimes violent; and *Zephyrus*: the west wind—balmy (fragrant) but with gale force. Moreover, the Greeks also divided the world into: Torrid, Temperate and Frigid zones. They established a relationship between temperature and ecumene regions of the world. The physical studies of Chinese scholars, on the other hand, pertain to climatology and paleoclimatology, earthquakes and seismology, hydrological cycle and geomorphology.

Economic Geography

The travel reports of Chinese stimulated trade relations with the West. Particularly, the Chinese discoverer, Zhang Qian (172-114 B.C.), travelled as far west as Samarkand and Usbekistan almost 1400 years before Marco Polo. On a second trip he visited Bactria and Sogidiana in Parthia. His travels led to the introduction into China of a superior breed of horses and new plants, such as grapes and alfafa. He was the first official diplomat to bring back reliable information about Central Asia to the Chinese imperial court, then under Emperor Wu of Han, and played a pioneering role in the Chinese colonization and conquest of the region now known as Xinjiang. Today Zhang Qian's travels are associated with the major route of transcontinental trade, the Silk Road. In essence, his missions opened up to China the many kingdoms and products of a part of the world then unknown to the Chinese. Following Zhang Qian's embassy and report, commercial relations between China and Central as well as Western Asia flourished, as many Chinese missions were sent throughout the end of the 2nd century B.C., initiating the development of Silk Road. Although Zhang Qian's journeys had promoted a great variety of economic and cultural exchanges between the Han Dynasty and the Western Regions, because silk was the dominant product traded this trade link became known as the Silk Route. In the 40-volume work that Jia completed in 801, Jia wrote of two common sea trade routes in his day: one from the coast of the Bohai Sea towards Silla in Korea and another from Guangzhou through Malacca towards the Nicobar Islands, Sri Lanka and India, the eastern and northern shores of the Arabian Sea to the Euphrates River. A small branch of this extensive second trade route led all the way to Dar-es-Salaam in Tanzania, East Africa. In his work written between 785 and 805, he described the sea route going into the mouth of the Persian Gulf, and that the medieval Iranians (whom he called the people of Luo-He-Yi) had erected 'ornamental pillars' in the sea that acted as lighthouse beacons for ships that might go astray. Confirming Jia's reports about lighthouses in the Persian Gulf, Arabic writers such as al-Masudi and al-Muqaddasi a century after Jia wrote of the same structures.

Theoretical Geography

Two great philosophers, namely Plato and Aristotle, made contributions to the development of geographical ideas. They respectively were the masters of deductive and inductive reasoning. Plato insisted that the observable things on earth were only poor copies of ideas, or perfect predicates from which observable things had degenerated or were in process of degenerating. According to him, the existing patterns related to man's activities are the result of his thinking, ideas or imagination. He argued that every system is gradually moving towards chaos, i.e. there is always degeneration of things from their original perfect state. He argued that in seeking explanations one should proceed from general theory to particular situation, i.e. to explain things in the context of some general rule or theory. Because he believed that there is a hidden cause behind every occurrence (existence). This kind of thinking is called deductive reasoning. But, according to Aristotle, the best way to build theory was to observe facts and the best way to test theory was to confront it with observations. Whereas, Plato believed in building theory by reasoning from the general to particular, Aristotle did it in the reverse order, i.e. by reasoning from the particular to general. The latter approach is inductive. Aristotle recognized that observations made through senses can never provide explanations. Our senses, he said, can tell us that, e.g., 'fire is hot' but cannot tell us that 'why it is hot'. Unlike Plato, Aristotle assumed that things were in process of physical change leading to a final perfect shape. This model for scientific explanations constituted the world's first paradigm for the guidance of scholars. The multifarious works of this great philosopher created a keen desire in his pupils to make direct observations. One of his outstanding and adventurous pupils was Alexander, who made his own heroic efforts for travelling.

Thus, the major contribution of ancient scholars was a body of philosophical and cosmological ideas, besides a number of terminologies that still constitute the basic vocabulary of geography. The ancient Greeks were the originators of literary and mathematical traditions in geography. Academically, these ancient scholars formed strong ties of geography with history and this inter-disciplinary bond could not be broken for several centuries to come, and has been intensely represented in the writings of geographers till the modern times. The roots of the concept of 'areal differentiation' as adopted by Hartshorne in the modern period lie in the writings of Strabo, the 1st century B.C. Roman scholar. Ptolemy's geocentric view of the universe, i.e. "the earth was a sphere that remained stationary in the centre, while the celestial bodies moved round it in circular courses", remained an accepted doctrine until the Age of Discovery. The Greeks are also credited for the introduction of *Gnomon*, one of the first scientific instruments ever made. The *Gnomon* is an instrument that functions as a sundial and can be used for finding the declination of the Sun through the year, among other things.

One of the major contributions of the Greek scholars was towards suggesting the methods of theory building and hypothesis testing. Plato, a well known Greek scholar, favoured theory building by intuition and reasoned from general to particular (the deductive approach). On contrary to it Aristotle suggested the method of reasoning from particular to general (the inductive approach). These methods of deductive and inductive reasoning were introduced to geography as early as 4th and 5th centuries B.C.

To conclude, there were certain positive and negative gains from the Greek and Roman Schools of thought. Belief in 'Geo-centric Concept of the Universe' was one of the negative views prevailing during this period. This was contrary to the present 'Helio-centric view'. Moreover, no attempts were made to translate the works of Greek or Roman writers into other languages in use. Therefore, the interest in philosophical and scientific writings declined and Geography had to face a period of recession in the centuries to come. However, the positive gains of the period were more numerous. Geography as a discipline got enriched by a number of viewpoints and techniques and various terminologies were added to its vocabulary. Besides coining the term 'geography' for our discipline, a number of other terminologies are also credited to ancient Greeks. The important ones include, for instance, 'cartography', 'chorography', 'longitude', 'latitude', 'torrid', 'frigid', 'temperate' and many more. All these terms are still in use today. The contribution of ancient Greeks and Romans to various fields of study covered almost entirely the gamut of geographical knowledge.

NOTES

1. Crawford, O.G.S. (1927). *Antiquity*. [Gloucester, Eng.].
2. Rana, L. (2011), pp. 46-74.
3. Rana, L. (2011), pp. 164-226.
4. Moriz Winternitz (1863-1937), was a German Indologist. He was an outstanding scholar who wrote on religion, folk-lore, the epics and Sanskrit literature. He is best remembered for his *"History of Indian Literature,"* a work in three volumes, which is considered indispensable for teachers and students alike.
5. Rana, L. (2011), pp. 227-292.
6. Breasted, J. H. (1916), *op. cit.*
7. Needham (1986), Volume 3, 500.
8. Needham (1986), Volume 3, 512-17.
9. Hsu (1993), 98.
10. Eves, Howard (1969), p. 31.
11. The Levant is a geographical term that denotes a large area in Western Asia, roughly bounded on the north by the Taurus Mountains, on the south by the Arabian Desert, and on the west by the Mediterranean Sea. Levant refers to an area of cultural habitation rather than to a specific geographic region. The Levant forms the middle part of the Fertile Crescent, between the Nile Valley (Egypt) to the south-west, and Mesopotamia (Iraq) to the east. It consists of Syria, Palestine, Lebanon and Jordan.
12. *Ancient History*, www.infoplease.com.
13. Devlin, Keith, 1996.
14. The period from 800 BCE to 200 BCE, during which, similar revolutionary thinking appeared in China, India and the Occident. The period is also sometimes referred to as the axis age.
15. Durant, Will (1939), p. 538.
16. John Dalton FRS (1766-1844), was an English chemist, meteorologist and physicist. He is best known for his pioneering work in the development of modern atomic theory, and his research into colour blindness (sometimes referred to as Daltonism, in his honour).
17. His book on arithmetic is the source of interesting legends that assert that it was written for his daughter, Lilavati. In one of these stories, which is found in a Persian translation of *Lilavati*, Bhaskara II studied Lilavati's horoscope and predicted that her husband would die soon after the marriage if the marriage did not take place at a particular time. To alert his daughter at the correct time, he placed a cup with a small hole at the bottom of a vessel filled with water, arranged so that the cup would sink at the beginning of the propitious hour. He put the device in a room with a warning to Lilavati to not go near it. In her curiosity though, she went to look at the device and a pearl from her nose ring accidentally dropped into it, thus upsetting it. The marriage took place at the wrong time and she was soon widowed.
18. Rana, L. (2011), pp. 239-262.
19. Ru "Confucian"—originally term denoting "ritual dancer"; school/sect of early Chinese thought that emphasizes correct social and ethical relations through literary and ritual self cultivation based on models derived or idealized from the early Zhou era.

20. Nelson (1974), 359.
21. Needham, Volume 4, 446.
22. Sivin (1995), III, 22.
23. Bowman (2000), 105.
24. Yao (2003), 545.
25. Needham, Volume 3, 524.
26. Vlastos, Gregory (1975).
27. Radhakrishnan and Moore, (1957), pp. 453-487.
28. Dasgupta, Surendranath (1992). p. 258.
29. The Great Proletarian Cultural Revolution, commonly known as the Cultural Revolution was a social-political movement that took place in the People's Republic of China from 1966 through 1976. Set into motion by Mao Zedong, then Chairman of the Communist Party of China, its stated goal was to enforce socialism in the country by removing capitalist, traditional and cultural elements from Chinese society, and to impose Maoist orthodoxy within the Party. The movement politically paralyzed the country and significantly affected the country economically and socially.
30. Husain, 1988.
31. Huang (1997), 64.
32. Needham (1986), Volume 4, Part 2, 479.
33. Sivin (1995), III, 17.
34. Needham (1986), Volume 3, 413-414.
35. Needham (1986), Volume 3, 625.
36. Neehdam (1986), Volume 4, Part 2, 484.
37. Pakuda Katyayana was one of the seven well-known heterodox thinkers of 6th century B.C. He put forward a theory known as *Anuvada* (Atomism). Very little is known of this man's personal life or his overall philosophy. Apparently, he was obsessed with getting to "the root of things." He theorized that man is made up of seven *eternal elements,* which are nothing but atoms ("that which cannot be split"): four basic elements, namely fire, air, earth, water, (which Charvakas also believed made up the body), pleasure and pain (*Dwandwam,* which Upanishadist considered as the main function of mind—likes and dislikes, pleasure and pain, gain and loss), and the soul (which Brahmins and Upanishadists called *Atman*).
38. Latadeva the author of "*Soorya Siddhanta*", talked about the earth's axis and called it SUMERU. "That the earth is a sphere and it rotates on its own axis", was known to Varahamihira and other Indian astronomers much before Copernicus published this theory. Latadeva in his Surya Siddhanta text has divided the year into 12 months. Seven planets of the solar system effect the earth's atmosphere and their names were added to the seven days of the week, which was accepted all over the world.
39. Boyer, Carl B. (1991). p. 207.
40. Jewish Encyclopaedia.
41. Ebrey *et al.* (2006), 162.
42. Ifrah 2000, p. 346.
43. Katz 1995, pp. 173-174.
44. Seidenberg 1978.
45. Fowler 1996, p. 11.
46. Filliozat 2004, pp. 143-144.

47. *Encyclopaedia Britannica.*
48. Hsu (1993), 96-97.
49. Hsu (1993), 98.
50. Sivin (1995), III, 23-24.
51. Huffman, Carl A. (1993).
52. An armillary sphere is basically a skeletal celestial sphere with a model of the Earth or, later, of the Sun placed in the centre. Armillary spheres were developed by the Greeks in antiquity for use as teaching tools. Armillary spheres became popular again in the late middle ages.

The Armillary sphere consists of two major components, the sphere and the stand, as seen in the figure above. The central body in the sphere represents the Earth, which was, of course, considered the centre of the Universe. The colures and the Equator (the rings defining the sphere) represent the firmament, that is, the sphere upon which the fixed stars reside. The band going around the sphere, at an angle to the equator, represents the zodiac. The line running through the middle of this band defines the ecliptic, or the path followed by the Sun through the sky.
53. Crespigny (2007), 1050; and Loewe (1968), 107.
54. Bodde (1991), 140.
55. Needham (1986), Volume 1, 135; and Fraser & Haber (1986), 227.
56. Wu (1943), 211-212.
57. An odometer (milometer) indicates distance travelled by a car or other vehicle. The device may be electronic, mechanical, or a combination of the two. The word derives from the Greek words "hodós", meaning "path" or "way", and "métron", "measure".

58. Needham (1986), Volume 4, Part 2, 40.
59. King Mu was the fifth sovereign of the Chinese Zhou Dynasty. King Mu was the most pivotal king of the Chinese Zhou Dynasty, reigning nearly sixty-six years, from ca. 976

B.C. to ca. 922 B.C.. He was reputed to have lived until the age of 105. He liked to travel, and in particular visited the Kunlun Mountains several times during his reign, and is said that he traveled 90,000 kilometers.

60. Husain, 1988.

REFERENCES

Ancient History, www.infoplease.com.

Bodde, Derk (1991). *Chinese Thought, Society and Science.* Honolulu: University of Hawaii Press.

Bowman, John S. (2000). *Columbia Chronologies of Asian History and Culture.* New York: Columbia University Press.

Boyer, Carl B. (1991). "The Mathematics of the Hindus". *A History of Mathematics* (Second Ed.). John Wiley & Sons, Inc. ISBN 0471543977.

Breasted, J. H. (1916). *Ancient times, a history of the early world: an introduction to the study of ancient history and the career of early man,* Boston: Ginn and Company.

Crawford, O. G. S. (1927). *Antiquity.* [Gloucester, Eng.]

Crespigny, Rafe de. (2007). *A Biographical Dictionary of Later Han to the Three Kingdoms (23-220 CE).* Leiden: Koninklijke Brill.

Dasgupta, Surendranath (1992). *A History of Indian Philosophy,* Volume 1. Motilal Banarsidass Publ.

Devlin, Keith, (1996), *Mathematics: The Science of Patterns: The Search for Order in Life, Mind and the Universe* (Scientific American Paperback Library) 1996, ISBN 978-0-7167-5047-5.

Durant, Will (1939) *The Life of Greece.* New York: Simon and Schuster, p. 538.

Ebrey, Patricia Buckley, Anne Walthall, and James B. Palais (2006). *East Asia: A Cultural, Social, and Political History.* Boston: Houghton Mifflin Company. ISBN 0-618-13384-4.

Encyclopaedia Britannica.

Eves, Howard, (1969) *An Introduction to the History of Mathematics,* Holt, Rinehart and Winston.

Filliozat, Pierre-Sylvain (2004), "Ancient Sanskrit Mathematics: An Oral Tradition and a Written Literature", in Chemla, Karine; Cohen, Robert S.; Renn, Jürgen *et al., History of Science, History of Text (Boston Series in the Philosophy of Science),* Dordrecht: Springer Netherlands, pp. 137-157, 360-375.

Fowler, David (1996), "Binomial Coefficient Function", *The American Mathematical Monthly* 103 (1): 1-17.

Fowler, Jeaneane (2005). *An Introduction to the Philosophy and Religion of Taoism: Pathways to Immortality.* Brighton: Sussex Academic Press. ISBN 1-84519-085-8.

'History of China', Wikipedia, the free encyclopaedia.

Holt-Jenson, A. (2009), *Geography: Its History and Concepts,* Fourth Edition, Sage, London.

Hsu, Mei-ling (1988) "Chinese Marine Cartography: Sea Charts of Pre-Modern China," *Imago Mundi,* Volume 40, pp. 96-112.

Hsu, Mei-ling, (1993), "The Qin Maps: A Clue to Later Chinese Cartographic Development," *Imago Mundi,* Volume 45, pp. 90-100.

Huang, Ray (1997). *China: A Macro History.* New York: An East Gate Book, M.E. SHARPE Inc.

Huffman, Carl A. (1993) *Philolaus of Croton: Pythagorean and Presocratic,* Cambridge University Press.

Husain, M. (1988), *Evolution of Geographical Thought*, Second Revised Edition, Rawat Publications, Jaipur.

Ifrah, G. (2000): *The Universal History of Numbers*. John Wiley & Sons, New York.

Jewish Encyclopedia.

Katz, Victor J. (1995), "Ideas of Calculus in Islam and India", *Mathematics Magazine (Math. Assoc. Amer.)* 68 (3): 163-174.

Needham, Joseph (1986). *Science and Civilization in China: Volume 1, Introductory Orientations*. Taipei: Caves Books, Ltd.

Needham, Joseph (1986). *Science and Civilization in China: Volume 3, Mathematics and the Sciences of the Heavens and the Earth*. Taipei: Caves Books, Ltd.

_______ (1986). *Science and Civilization in China: Volume 4, Physics and Physical Technology, Part 2: Mechanical Engineering*. Taipei: Caves Books, Ltd.

_______ (1986). *Science and Civilization in China: Volume 4, Physics and Physical Technology, Part 3, Civil Engineering and Nautics*. Taipei: Caves Books, Ltd.

Nelson, Howard. "Chinese Maps: An Exhibition at the British Library," *The China Quarterly* (Number 58, 1974): 357-362.

Radhakrishnan, S.; Moore, C. A. (1957). *A Source Book in Indian Philosophy*. Princeton, New Jersey: Princeton University Press. ISBN 0-691-01958-4.

Rana, L. (2008), *Geographical Thought: A Systematic Record of Evolution*, Concept Publishing Company, New Delhi.

Rana, L. (2011), *Ancient Tradition in Geography: The Western and Oriental Perspective*, Project Report (unpublished), ICSSR, New Delhi.

Seidenberg, 1978. *The origin of mathematics. Archive for the history of Exact Sciences*, vol. 18.

Sivin, Nathan (1995). *Science in Ancient China: Researches and Reflections*. Brookfield, Vermont: VARIORUM, Ashgate Publishing.

Vlastos, Gregory (1975). *Plato's Universe*. Seattle: University of Washington Press.

Wu, Kuang Ch'ing. "Ming Printing and Printers," *Harvard Journal of Asiatic Studies* (February 1943): 203-260.

Yao, Xinzhong. (2003). *Routledge Curzon Encyclopaedia of Confucianism:* Volume 2, O–Z. New York: Routledge.

CHAPTER-4

Medieval Period

The Medieval phase is a very long period of development in the history of geographical thought, i.e. beginning from the end of Roman Empire (5th century A.D.) to the beginning of Industrial Revolution (1750). The developments during this period may be studied under three phases, viz.: Early Medieval period: Dark Age in Europe (500-1100 A.D.); High Medieval period: Rise of Arab School of Thought (800-1400 A.D.); and Late Medieval period: Age of Discovery (1400-1750A.D.).

EARLY MEDIEVAL PERIOD

The Early Medieval period is associated with the Dark Age in Europe. By 500 A.D., the Roman Empire in the west had declined and this marked the onset of Dark Age in Europe which continued up to 1100 A.D. The political unity of the Roman Empire dwindled after 3rd century and its territories were settled by succeeding waves of barbarian tribal confederations, which rejected the classical culture of Rome. Prominent among these peoples in the movement were the *Huns*, *Avars* and *Magyars* with the large number of Germanic peoples. The barbarian invasions brought the end of the Roman Empire. Christianity influenced the social, economic, literary and philosophical developments during the early medieval period. The Christian era was marked by the loss of ancient scientific concepts about the world and their replacements by unscientific, uncritical cosmogonies based largely on scriptures. During this period, travelling and explorations, owing to political instability was hazardous. However, the missionary travels were the only source of knowledge for the regional account of different nations of that period. The approach to studies of any kind became theocratic. During Dark Age, Geography made a little progress as the organized body of knowledge. The quest for scientific learning stopped; instead, there was regression of the already acquired scientific knowledge. At best, the scholars made accurate but sterile copies of the works of ancients, rejecting anything that did not conform to the dogmas of Church. Such an intellectual environment stifled any development of critical scientific analysis. Concepts, which had been developed in Greek and Roman times, were reshaped to conform to the teachings of Church. The image of world was moulded to fit Biblical references. The earth was considered as a flat disc with Jerusalem at its centre. This kind of thinking had an adverse impact on the art of map-making and this resulted in its great distortions. The tradition of making fairly accurate maps, as started by the Greeks and Romans got lost. Instead, there emerged maps, which were purely fancy and unscientific. The ideal map of this period was 'Orbis-Terrarum' or: 'T-in-O' Map (Fig. 4.1). 'Orbis-Terrarum' means 'wheel of the world'; 'world', of course, meaning 'the World as it was known at the Time'. This originally included the Mediterranean Sea and its surrounding territories. In the Early Middle Ages, the religious map-makers wished to illustrate geography of faith rather than the terrestrial territory. The 'Orbis-Terrarum' maps literally used 'O' with a 'T' in it—a circular ocean surrounding the earth, which itself surrounding a T-shaped body of water (the stem of T representing the Mediterranean Sea and its top showing the Aegean and Black Seas on the one hand and the Nile River and Red Sea on the other). The three divisions—Europe, Asia and Africa—were accepted as standard. The centre of the inhabited world, just above the centre of the T, was Jerusalem. At the Far East, beyond the limit of the inhabited world, was conceived the location of paradise (the house of *Adam* and *Eve*). Moreover, in all these maps were inserted the mythical places, beasts and dragons. This type of cartography was in practice for a long time. These maps were made very beautiful and extremely decorative, but have less relation to actual facts, as they were far from reality with no scientific base. As compared to modern standards, the orientation of the maps was totally erroneous; as there was vagueness in the east-west and north-south extension of world.

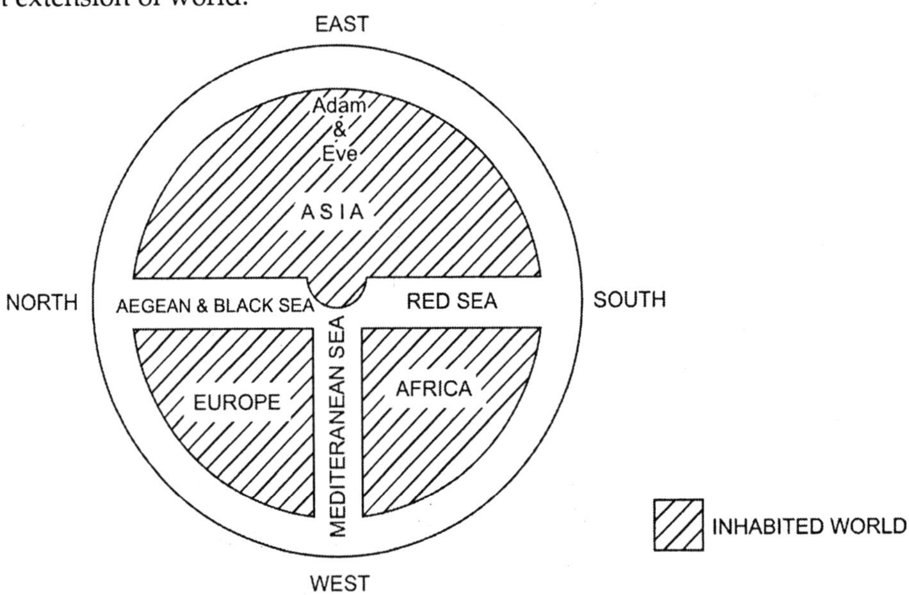

FIGURE 4.1 'T-in-O' Map of Early Medieval Period

So far the sprawl of the habitable world is concerned many wrong ideas were put forward. The sphericity and the nearly correct distances of places, latitudes and longitudes were ignored. As for the travels and journeys, they continued but geography was hardly benefited. Because, the purpose of these travels was not scientific or speculative; these were mainly religious or missionary movements made by pilgrims to holy shrines. Moreover, the descriptions made were also purely aesthetic (beautiful without logic). Most of the classical correct concepts were forgotten, and the old errors appeared again about the world maps and habitable portions of the earth. Moreover, the involvement of Christians and Muslims (Arabs) in crusade (holy) wars further aggravated the problem. Against this background, there emerged an intellectual curiosity of the Arab scholars that led to the great achievements of Islamic civilization. Thus, when Europe was plunged into darkness, the Arab world was getting enlightened. Now the new zeal of learning spread through the world of Islam. The Arab civilization, in fact, became the main link between the ancient civilization and the modern world. The rise of Arab school marks the end of the recession in development of geographic literature and return to more scholarly research.

HIGH MEDIEVAL PERIOD

Arab civilization became the most advanced during medieval period. In the early years of 7th century, Prophet Mohammad founded Islam. This new religion, within a very short period of time, not only united the warring tribes but also led to the establishment of a big empire and the building of a civilization, which in many respects was the finest civilization of time. Nearly all Arabia had accepted the new religion and became a new unified state before the death of Prophet Mohammad in 632. After his death, his successors, known as 'Caliphs' or 'Khalifas', held the religious and political authorities. From Arabia, Islam spread very fast to many other parts of the world. Within about a hundred years, the Khalifas and their generals had conquered Iran, Syria, Egypt, Central Asia, North Africa and Spain. The Arabs had also come to India. The greatest contribution of Islam was that it unified the scattered Arab empire under one faith and spreaded a lesson of equality and brotherhood among all the Muslims. Wherever, the Arabs went, their religion went with them. This astonishingly rapid conquest was due to several reasons. It was partly missionary zeal, but economic and social causes favoured the expansion. The Arabs, who were the inhabitants of a poor country, wanted richer lands. They were excellent fighters. The people of the neighbouring countries tired of oppressive rule and fascinated by simplicity of the teachings of Islam, welcomed the invaders. The Arab empire was the largest that the world had so far seen. In the pre-Islamic period of their history, the Arabs were illiterate and unfamiliar with the art of writing. United by Islam, the Arabs started on their quest for knowledge. They sought knowledge wherever they could find it. In fact, the Arab Civilization was far superior to that of Europe in these centuries. In 762, Muslims founded the new city of Baghdad and for more than a century it was the centre of intellectual world. With the patronage of Caliph Harun-al-Rashid an academy, entitled

'Baitul-Hikma', was established. At this academy, scholars from all over the world were invited to teach and assist the Arab academicians and to help them in the translation of the Greek, Latin, Persian and Sanskrit works into Arabic. Arabs invited the Indian scholars at Baghdad to learn Indian mathematics and numerals. They studied Surya-Sidhant, the ancient Indian astronomical treatise, and the works of Aryabhatta. The Arabic translation of the Surya-Siddhanta was a landmark in the history of Arab astronomy[1]. In this manner, the Arabs collected geographical ideas from the Greeks, Romans and Indians. On the basis of their observations, explorations and studies they developed their own concepts and theories with great vigour and produced innumerable books on various aspects of geography and allied sciences; and carried their knowledge to far-off lands from the shores of Atlantic to the Pacific and Indian Ocean. The establishment of vast empire facilitated the coming together of intellectual and scientific traditions of disparate civilizations, from the Western to the Oriental worlds. The Arabs made all knowledge their own and developed it further. They borrowed, for instance, the knowledge of medicine from Greece and the system of numerals (Hindsa) from India. Their system of medicine is still called '*Yunani*' that means 'Greek'. Besides, the Arabs also developed a decorative style of writing called 'calligraphy'. The Arab traders travelled widely and brought back new information that the scholars compared with Greek ideas and incorporated into Ptolemy's maps. Besides Baghdad, the main centre of the Muslim intellectual world, the other major centres of learning that emerged during this period included: Jerusalem in Israel; Cairo in Egypt; Toledo in Spain; Istanbul in Turkey; Damascus in Syria; and Shiraj in Iran. Various factors that helped to stimulate this interest in learning may be summarised as: (i) The large extent of Arabic Empire which stretched over so many different areas from the Atlantic to the borders of China; (ii) The excellent system of roads, a legacy of Rome; (iii) The desert routes which encouraged movement and facilitated the pilgrimages to Mecca demanded by the faith; (iv) Trade, which naturally arose out of the diversity within the Empire and was further stimulated by the high rank of traders in Muslim society; (v) The conquest of Syria, Persia and Bactria—where the Greek culture still flourished and gave to the Muslims a rich fruit of Greek learning; and (vi) The cultural and economic ties with the Indian Sub-continent.

In the West, on the other hand, from the beginnings roughly about 1000 A.D., greater stability was observed in Europe. Here too, the 'High Middle Ages' describe the expansionalist culture and intellectual revival from the late 11th to the 14th century.

CONTRIBUTION

Many Arab scholars, most of whom were also travellers, contributed to the development during High Medieval Phase of geographic development. They included, for instance, Al Battani, Al Masudi, Al Beruni, Ibn Sina, Al Idrisi, Ibn Batuta, Ibn Khaldun, etc. Al Battani (858-929) was an astronomer, astrologer and mathematician. He made important observations and measurements regarding celestial bodies. In astronomy, his well-known discovery is the determination of solar year as being

365 days, 5 hours, 46 minutes and 24 seconds, which is very close to the latest estimates. Al-Battani also determined with remarkable accuracy the obliquity of the ecliptic, the length of the seasons and the true and mean orbit of the sun. He proved, in sharp contrast to Ptolemy, the variation of the apparent angular diameter of the sun and the possibility of annular eclipses. He rectified notions about the orbits of moon and planets and propounded a new theory to determine the conditions of visibility of the new moon. He also provided solutions by means of orthographic projection for various problems of spherical trigonometry. In mathematics, he was the first to replace the use of Greek chords by sines, with a clear understanding of their superiority. He also developed the concept of cotangent and furnished the table in degrees. He wrote a number of books on astronomy and trigonometry. His most famous book was his astronomical treatise with *"De scienta stellerum–De numeris stellerum et motibus."* His treatise was extremely influential in Europe till the Renaissance, with translations available in several languages. Abul Hasan Ali Ibn Husain Ibn Ali Al-Masu'di (888-957), the Arab discoverer, was born in Baghdad and is known as the 'Herodotus of the Arabs' because he was the first Arab to combine history and scientific geography. He travelled extensively in India, Middle East and Africa. Al-Masudi wrote a 30-volume history of the world, *Al-Tanbih,* recounting the experiences of his travels from Europe to India. By presenting a critical account of historical events, he initiated a change in the art of historical writing, introducing the elements of analysis, reflection and criticism, which were later on further improved by Ibn Khaldun. In particular, in his *Al-Tanbih* he makes a systematic study of history against a perspective of geography, sociology, anthropology and ecology. Masu'di had a deep insight into the causes of rise and fall of nations. Al Biruni (973-1032), whose full name was Abu Arrayhan Muhammad Ibn Ahmad al-Biruni, was a Persian mathematician, physicist, scholar, philosopher, astronomer, astrologer, traveller, historian, pharmacist, and teacher. He contributed greatly to the fields of mathematics, philosophy, medicine, and science. Al-Biruni crater, on the Moon, is named after him. He was a colleague of the Iranian Muslim philosopher and physician, Ibn Sina. He also travelled to South Asia with Mahmud of Ghazni, who also became his patron. Based on his travels and observations, he wrote *Ta'rikh al-Hind* ("Chronicles of India"). He knew Greek, Sanskrit and other languages; but, he wrote his books in Persian and Arabic, although his native language was Khwarezmian. At the age of 17, he had calculated the latitude of Kath, Khwarazm, using the maximum altitude of the sun. He recalculated the Earth's radius from approximately 6,314 km (measured by Eratosthenes in 240 A.D.) to 6,339.6 km. This feat was not surpassed in the western world until the 16th century. Abu Ali al-Husayn Ibn Abdullah Ibn Sina (980-1037) was one of the foremost philosophers of the golden age of Islamic tradition. He is also known as 'Al-Sheikh al-Rais' (Leader among the wise men), a title that was given to him by his students. In the west he is also known as the "Prince of Physicians" for his famous medical texts. Ibn Sina he became the most influential of all the Islamic philosophers and earned royal favour for treating the Kings of Bukhara. Though trained as a physician, Ibn Sina made important contributions to philosophy, mathematics, chemistry, and astronomy. His philosophical encyclopaedia, *Kitab al-Shifa* ("Book of Healing"), brought Aristotelian and Platonian

philosophy together with Islamic theology. It divided the field of knowledge into two major streams, viz. theoretical knowledge and practical knowledge. Al Idrisi (1099-1166) was one of the leading scholars of the 12th century. He was a historian, politician and geographer. He was the master of several languages as Persian, Arabian, Latin, Greek and Spanish. Idrisi's major contribution lies in the form of his book, *Kitab al-Jami-li-Sifat Ashtat al-Nabatat,* which is a study and review of all the literature on medicinal plants. In addition to this, he made original contributions to geography, especially related to economic, physical factors and cultural aspects. He made a planishere in silver for King Roger II, and described the world in *Al-Kitab al-Rujari* (Roger's Book), also entitled *Nuzhat al-Mushtaq fi Ikhtiraq al-Afaq* (The delight of him who desires to journey through the climates). This is practically a geographical encyclopaedia of his time, containing information not only on Asia and Africa, but also on Western countries. Al-Idrisi also compiled another geographical encyclopaedia, larger than the former; it was entitled as *Rawd-Unnas wa-Nuzhat al-Nafs* (Pleasure of men and delight of souls) also known as *Kitab al-Mamalik wa al-Masalik.* Apart from botany and geography, Idrisi also wrote on fauna, zoology and therapeutical aspects. His work was translated into Latin and, especially, his books on geography remained popular both in the East and the West for several centuries. Abdullah Muhammad Ibn Batuta (1304-1378) was of Negro origin, but was educated in Muslim religion. He was the most remarkable traveller, whose voyages extended east as far as Northern China and south along the coast of Africa well beyond the Equator, covering Egypt, Syria, Persia, Arabia, Zanzibar Isles, India, Maldives, Ceylon, Sumatra and China. Ibn Batuta visited China sixty years after Marco Polo and in fact travelled 75,000 miles, much more than Marco Polo. Ibn Batuta's contribution in geography is unquestionably great. His writings, including the book '*Travelogue*' (1357), are an important source of knowledge on large parts of the known world of medieval times. Ibn Batuta was the only medieval traveller who is known to have visited the lands of every Muslim ruler of his time. He is still considered a most reliable source for the geography of his period and an authority on the cultural and social history of Islam.

Ibn Khaldun (1332-1406) was the last great Moslem scholar, and a great historical geographer. He is seen as one of the forerunners of modern historiography, sociology and economics. He is best known for his book "*The Muqaddimah*" (The introduction to History). Ibn Khaldun is said to have been the first social scientist. He was a Muslim philosopher and statesman who provided a useful way of looking at the political and social conditions of the pre-modern world. He lived in the era when the Western world was experiencing the renaissance. This was also the beginning of the decay of Islamic Civilisation. There were crusades and invasions from East. Besides, disputes of Caliphs were common. Among all this background, Ibn Khaldun represents the climax of history of Islamic economic thought. The Muqaddimah of Ibn Khaldun records an early Muslim view of 'universal history'.

Through their travels and extensive works on history, botony, mathematics, astronomy and astrology the Arabs made outstanding contribution to various fields of geographical learning as Mathematical and Practical Geography, Regional and Historical Geography, Physical Geography and Climatology and Human Geography.

Mathematical and Practical Geography

In Arab School, the Greek science was the starting point. The works of Aristotle and Ptolemy were studied elaborately. It is not surprising, therefore, that mathematical geography received the maximum attention. The factors like semi-desertic conditions, open atmosphere, clear skies and vast distances to scan facilitated the astronomical observations and also helped in promotion of studies in this field. The Arabs, who were largely influenced by the Greek tradition, adopted the same ideas about shape and size of earth. The early Arabs considered the earth as the centre of Universe, around which revolved seven planets. These planets, in order of their distance from the earth were: Moon, Mercury, Venus, Sun, Mars, Jupiter and Saturn[2]. It was imagined that each of these planets occupied a particular 'sky' or rather using the expression of Ptolemy a separate 'sphere' and that each sky was under the authority of the 'planet' to which it had been assigned. In his hypothesis all the stars formed the seven skies or the 'highest heaven'. They believed that the first of the seven skies, closest to the earth is of green-emerald, the second of white silver, the third of red ruby, the fourth of white pearl, the fifth of red gold, the sixth of yellow topaz and the seventh of a mass of fire (*nur*). Similarly, they imagined of seven earths, one inside the other, the lowest of them being hell. There were several attempts to ascertain the size of earth. Al-Battani, for instance, calculated the length of equator at 27000 miles, whereas Ptolemy in 157 A.D. had established it at 24000 miles. Sea-Charts and maps based on cylindrical projection were made. The Arab geographers borrowed from the Greeks, the division of earth into five zones; each zone represented particular temperature conditions. The five zones are: (i) the Torrid Zone—situated between the two tropics, (ii) the two Frigid Zones—placed near the poles, and (iii) the two Temperate Zones that occupied the intermediate positions. For the determination of latitudes, the Arabs like Eratosthenes and other Greeks, made use of the sun's shadow when it happened to be on the meridian; the shadow was marked by a column (gnomon). The Arabs improved over the works of Greeks. Besides, another greatest contribution of the Arabs to the advancement of mathematical knowledge was the introduction of Hindu invention of the zero in arithmetic (*Hindsa*). Before this, the Roman numerals were in use and their range was limited. The zero liberated arithmetic and people could now count in millions, billions and trillions. In mathematics, the Arabs learned the Indian numerals and spread their use far and wide, so that in the West they are to this day called "Arabic numerals". They improved upon the existing knowledge of Algebra, trigonometry and chemistry. However, surprisingly enough, the attempts of Arab scholars did not bring about any marked improvements in cartography and map-making. Because, it was distorted by their love of ornamentation in speech and writing. Moreover, they still believed in geocentric view of the planetary system.

REGIONAL AND HISTORICAL GEOGRAPHY

A wealth of regional descriptions exists in the works of Arab scholars, particularly Al Idrisi and Ibn Batuta. Both physical and human aspects are covered in detail. According to Arabs the western limit of the habitable world is at the end of the Mediterranean Sea, the eastern at the Sila (Japan), the northern in the land of Yajuj Majuj (Siberia) and the southern to the south of the equator.[3] The writings of Ibn Batuta described the ecumene areas of world. The travels of Ibn Batuta took him to many parts of Arabia never visited earlier. His travels extended as much as 10° south of the Equator and more. This venture brought empirical proof that Aristotle, the ancient Greek scholar, had been wrong in asserting that the Torrid Zone was too hot and unfavourable for human habitation. Batuta, on the contrary, confirmed that numerous tribes occupied it. The major focus of the writings of Al Idrisi, on the other hand, was on Arabia and India. He has described the country of India as an agriculturally rich country, and mentioned the presence of about 150 towns and cities at that time. Al Idrisi also provided the detailed descriptions of these regions.

Physical Geography and Climatology

The physical geography finds a special place in the writings of the scholars like Al Biruni, Ibn Sina, Al Idrisi, Al Masudi, Al Balakhi and Al Aaqdisi. Their important observations are related to the study of: (i) the processes shaping landforms; (ii) alluvial deposits; and (iii) phenomenon of tides. Al Biruni in his book entitled *'Kitab-al-Hind'* describes the fluvial processes and alluvial deposits of Himalayan region. In this book he recognized the significance of the rounded stones he found in the alluvial deposits south of the Himalayas. The stones became rounded, he pointed out, as they rolled along the torrential mountain streams. Furthermore, he recognized that the alluvial material dropped close to the mountain was relatively coarse in texture and that it became finer in texture farther away from the mountain. Ibn Sina wrote about the processes shaping landforms, viz. weathering and denudation. According to his findings, the streams play an important role in this and the high mountains are more exposed to the process of weathering than any other landform. He keenly observed the work of agents of denudation and weathering in the mountains and ascertained that the mountain streams erode the slopes. He formulated the idea that mountains were being constantly worn down by rivers and that the highest peaks occurred where the rocks were especially resistant to erosion. Mountains are raised up, he pointed out, and are immediately exposed to this process by wearing down, a process that goes on slowly and steadily. Ibn Sina also noted the presence of fossils in the rocks of high mountains. Various Arab scholars and navigators observed the phenomena of tides, and proved that they are caused because of the gravitational pull of the sun and the moon. Al Masudi who navigated in the Caspian Sea recorded the spring and neap tides. Besides, Al Idrisi corrected the erroneous ideas about the Indian Ocean and the Caspian Sea. The Indian Ocean was supposed to be enclosed, whereas the Caspian Sea was shown as an arm of the World Ocean in 'T-in-O' maps of Early Medieval period. He also tried to redefine more accurately the courses of many of the rivers including that of Denube and Niger rivers.

However, the most notable were the contributions made to climatology. The important contributions in the field have come in the form of coining of the term *mausim* (Monsoon); preparation of the first Climatic Atlas; and division of the world into Climatic Regions. The term 'monsoon' has an Arabic origin, from *mausim*. Arabs were the first to give the idea of periodic nature of monsoons. The studies of Al Masudi, particularly, deal with the Indian Monsoons. The first Climatic Atlas of the world, entitled '*Kitabul Ashkal*' was prepared by Al Balakhi in 921 A.D. He gathered the climatic data and information from the Arab travellers. Al Aaqdisi attempted to divide the world into 14 Climatic Regions in 985 A.D. He also observed that climate varies not only by latitude, but also by the positions east and west. He presented the view that the Southern hemisphere has most of the water and that most of the land area is in the Northern hemisphere and this has reflection in the climate as well. Undoubtedly, this idea had been conceived from the Greeks, basically.

Human Geography

The studies carried on under Human Geography by Arab scholars are dominated by the writings of Ibn Khaldun. They deal mainly with the influence of physical factors on life forms, including human beings. Ibn Khaldun, in his monumental work, 'Muqaddimah', for instance, has described and discussed various aspects of human society. The work has been divided into 6 sections, viz.: (1) On civilization, geography and anthropology; (2) The nomadic culture and its contrast with the sedentary culture, sociological and historical causes and consequences of the conflicts arising continually from that of fundamental opposition; (3) Dynasties and Kingdoms; (4) Life in villages and cities; (4) Professions or the Means of Livelihood; and (5) Classification of Sciences. The prime focus of the writings of Ibn-Khaldun was on human society, culture and its development. Modern geographers are particularly impressed with his cultural interpretation of the physical environment; and his analysis of the role of the city in regional economy. The work of Ibn-Khaldun begins with a discussion of man's physical environment and its influence on the living styles of people. He investigated various stages of social organization, identifying the desert nomads as the most primitive and the purest. He suggested that the sedentary city dweller is dependent on luxuries and becomes normally soft. He discussed the forms of government, describing a sequence of stage that marks the rise of dynasty to power, followed by its decline through corruption to its fall. Thus, he explained, in a way, the rise of Arabian power, and also foresaw its fall. Khaldun maintained that the northern hemisphere is more densely populated than the southern and also that the population is meagre along the equator; but away from equator, there are greater concentrations of population up to 64th parallel. Further away, there is once again very little or no population at all. The intense heat of the equatorial belt, he considered as the discouraging factor for the concentration of population. A harmonious blending of heat and cold in the temperate region, according to him, is conducive to human growth and settlements. Away from the temperate regions, the excessive cold of the Polar areas is again deterrent to human growth. About the origin of settlements, he stated that 'those who settle are attracted by fertility of land and by the seas with which they can defend themselves against invaders'. In course of time the population increases, and so also the pressure on land,

and the habitation becomes a permanent settlement. The settlement grows in the course of time and becomes city—which shows a definite socio-economic structure, in ways of division, dearth and abundance, supply and demand. The roots of all cities have always been these small settlements at sometimes or the other. Moreover, he tried to explain various human activities with the help of environment. He explains why the Arabs are nomads and why the Negroes are of easy disposition. Ibn-Khaldun tried to correlate man and environmental relationships in a scientific way. According to him, the people of warmer climates are noted for their passionate nature. Those of the colder climate incline to stolidity and lack of vivacity. Those in the temperate climates excel in wisdom and are neither excessively passionate nor markedly stolid. He also explained that the Negros are black because they live in the warm and humid climatic zones while the people of temperate and cold regions are whitish in colour. Similarly, people try to build their houses and settlements on the southern slopes and close to the springs and water sources. In fact, the roots of the concept of climatic determinism lie in the writings of Ibn-Khaldun, who believed that the climate has a tremendous influence on the nature of human beings, their colour and settlements. According to him, in warm climates the people are more passionate; in colder climates people are stolid and are of harsh nature; whereas in temperate climates they are full of wisdom as they occupy the middle position. Similarly, the settlement patterns are also influenced by the climates, according to him.

As a result of expanding colonialism and imperialism, the Arab Empire started disintegrating after 1400 A.D. and the Arab School also came to a decline. However, the important contributions of the Arabs may be summarized as follows: (1) Laying roots for the philosophy of environmental determinism (climatic); (2) Generating a voluminous literature on the ecumene and non-ecumene areas of the known world; and (3) Presenting a newer outlook of the world that later led to the onset of Age of Discovery. The scholarly works of Arabs provided a new and more realistic picture of the world. But, on the whole, the literary developments among the Arabs were hampered by the over-refined state of language and love of story-telling in which the geographical nature of facts was over-shadowed. Nevertheless, the Arabs did keep the Greek learning alive, and it was from their hand the Europe received it, as the Dark Age drew to a close. The Renaissance Movement had begun in the times of Arabs and got momentum in the Age of Discovery. Starting in the 8th century the renewed study of Greek revived much of the forgotten learning during the Arab period. But the classical spirit was not revived, i.e. no new concepts were established or formulated by Arab scholars in a recognizable manner. The renaissance remained largely a religious movement, more concerned with preserving the knowledge of the past than expanding it through scientific investigation.

LATE MEDIEVAL PERIOD

The Age of Discovery (1400-1750 A.D.) in late Medieval period was the age of colonization and exploration, a period from the early 15th century and continuing into the 18th century. During this period, the European ships travelled around the world in search of new trading routes and partners to feed burgeoning capitalism in Europe. In the process, Europeans encountered peoples and lands previously unknown to them. Among the most famous explorers of the period were Christopher Columbus, Vasco da Gama, Pedro Álvares Cabral, John Cabot, Yermak, Juan Ponce de León, Fernand Magellan, Willem Barentsz, Abel Tasman, Captain Cook, Amerigo Vespucci and Willem Jansz (Table 4.1). The Age of Exploration was rooted in new technologies and ideas growing out of the Renaissance and these included advances in cartography, navigation, and shipbuilding, besides the discovery of new lands. Obviously, geography was the most benefitted field of learning during this period. This post-Arab phase was not only the Age of Discovery, but also a period of transition from medieval to modern, from Old to New. The important developments of this period include: Exploration of New Lands; Scientific and Technological innovations; and Improvements in Cartography and Map making.

Exploration of New Lands

Although the travels had become very popular in Arab period itself, a number of other developments gave further impetus to the explorations; and they included: (i) colonization, (ii) use of magnetic compass, and (iii) improvement in the art of navigation. Great explorers emerged in the scene. Numerous journeys of geographical exploration were commissioned by a variety of nation states in Europe. Most of these voyages were financed because of the potential commercial returns from resource exploitation. The voyages also provided an opportunity for scientific investigation and discovery. These voyages also added many significant contributions to geographic knowledge. Most of the explorers of this period were of Italian, Portuguese and Dutch origin, besides a few also being Spanish, Russian and English.

TABLE 4.1 Explorers in Late Medieval Period

Name	*Time*	*Nationality*	*Area Explored*
John Cabbot	(1450-1499)	Italian	Newfoundland
Christopher Columbus	(1451-1506)	Italian	America
Amerigo Vespucci	(1454-1512)	Italian	America
Juan Ponce de Leon	(1460-1521)	Spanish	Florida, USA
Pedro A. Cabral	(1467-1520)	Portuguese	Brazil (South America)
Vasco-da-Gama	(1469-1524)	Portuguese	Indian Ocean
Fernand Magellan	(1480-1521)	Portuguese	Circumnavigation of Earth
Yermak Timofeyevich	(1532-1585)	Russian	Siberia
Willem Barents	(1550-1597)	Dutch	Northern Shores of Europe
Willem Janszoon	(1570-1630)	Dutch	Coast of Australiaand New Zealand
Abel Tasman	(1603-1659)	Dutch	Tasmania, New Zealand and Fiji Islands
James Cook	(1728-1779)	English	Pacific Ocean

The Italian discoverers, namely John Cabbot, Christopher Columbus and Amerigo Vespucci explored lands across the Atlantic Ocean. John Cabbot was an Italian-born-English. Born in Gaeta and later a citizen of Venice, Italy, his Italian name of Giovanni Caboto became shortened to the English style John Cabot after he had lived in Bristol. In Italian, 'Caboto' means 'a coastal seaman' and was quite a common name given to sailors and navigators. He was the son of a spice merchant (Guilo Caboto). The Caboto family moved to Venice in 1461 where this young Giovanni worked with his father in a spice shop for many years. John Cabot was attracted to the treasures of the Orient from a relatively early age. Whilst still a young man he moved to Venice and it was there that he began learning of the value of Eastern spices and materials as it became his trade. He was the first European to search for the Northwest Passage. John Cabot discovered the island of Newfoundland in 1497. Christopher Columbus was also an Italian explorer and trader. He sailed across the Atlantic Ocean, hoping to find a route to India in order to trade for spices. Instead, he crossed Atlantic Ocean and reached the America on October 12, 1492 under the flag of *Castile*. History places great significance on his landing in America in 1492, with the entire period of the history of the Americas before this date usually known as Pre-Columbian. The anniversary of this event, *Columbus Day*, is celebrated in many parts of America. Columbus is commonly credited as the first European to see the Americas because of the profound impact his contact wrought on history. His voyage marked the beginning of European exploration and colonization of the Americas. Although Columbus was the first European to land America, the continent has actually been named after another Italian explorer, Amerigo Vespucci. He explored the east coast of South America between 1499 and 1502. His exploratory journeys along the eastern coastline of South America convinced him that a new continent had been discovered, a bold contention in his day when European explorers crossing the Atlantic Ocean, including Christopher Columbus, thought they were travelling to East Asia. Vespucci's voyages became widely known in Europe. In 1507, this new continent was named as "America"after Vespucci's first name, Amerigo.

Juan Ponce de León was a Spanish explorer and soldier who was the first European to set foot in Florida. He also established the oldest European settlement in Puerto Rico and discovered a current in the Atlantic Ocean, the Gulf Stream. Born in Santervas, Spain, Ponce de Leon was a soldier fighting Muslims in southern Spain in the early 1490s. He sailed on Christopher Columbus' second expedition to the Americas in 1493, but did not return to Spain with Columbus. Instead, he stayed in Santo Domingo (now called the Dominican Republic). He was appointed governor of the Dominican province of Higuey. He later heard of gold in the neighbouring island of Borinquen (now called Puerto Rico) and brutally conquered the island, claiming it for Spain. He was then appointed governor of this island. Ponce de Leon was then given the right to find and take the island of Bimini (in the Bahamas); he was searching for riches and the fountain of youth (a legendary spring that gave people eternal life and health). He sailed from Puerto Rico on March 3, 1513, and reached the east coast of Florida (St. Augustine) in April 1513. Ponce de Leon named the land "Pascua de Florida" (feast of flowers). He also claimed this land for Spain.

Pedro Álvares Cabral was a Portuguese navigator and explorer, generally regarded as first European discoverer of Brazil (1500). He was born in Belmonte, in the Beira Baixa province of Portugal. He had excellent training in navigation and large experience as a seaman. Vasco da Gama was also a Portuguese explorer who discovered an ocean route from Portugal to the East. He was one of the most successful in the European Age of Discovery, and the first person to sail directly from Europe to India, through Cape of Good Hope in Africa. Another Portuguese maritime explorer, Ferdinand Magellan, led the first successful attempt to circumnavigate the Earth. But he could not complete his final voyage as he was killed. However, his voyage made him one of the first individuals to cross all meridians of the globe. He became the first person to lead an expedition sailing westward from Europe to Asia and to cross the Pacific Ocean.

Yermak Timofeyevich, a Russian by birth, was the Cossack leader and explorer of Siberia. His exploration of Siberia marked the beginning of the expansion of Russia towards this region and its colonization. Yermak's exploration of Siberia is reflected in the so called Siberian Chronicles[4]. His life and death have been subjects for numerous Russian songs, books, and paintings since the 16th century.

Willem Barents was a Dutch navigator and explorer, a leader of early expeditions to the far north. Near the end of 16th century, several expeditions tried to find a northeast passage to Asia. No one pursued this quest with such persistence and skill as this Dutch navigator, for whom the Barents Sea is named. Seeking the illusive Northeast Passage, Barents made three voyages in 1594, 1595, and 1596-97 along the frozen and dangerous northern shores of Eurasia. Another Dutch navigator and colonial governor, Willem Janszoon, was the first European known to have seen the coast of Australia. Janszoon was born in Amsterdam, the Netherlands. He is first recorded as entering into the service of the *Oude compagnie,* one of the predecessors of the Dutch East India Companyr (VOC), as a mate aboard the *Hollandia,* part of the second fleet dispatched by the Dutch to the Dutch East Indies in 1598. He again sailed from the Netherlands for the East Indies in December 1603 as captain of the *Duyfken,* part of a fleet of twelve ships. He proceeded to chart some 320 km of the coastline, which he thought to be a southerly extension of New Guinea. Finding the land swampy and the people inhospitable, Willem Janszoon headed home in 1606. He called the land he had discovered "Nieu Zelandt" after the Dutch province of Zealand but this name was not adopted, and was later used by Abel Tasman to name 'New Zealand'. Abel Janszoon Tasman was also Dutch by birth. He is best known for his voyages of 1642 and 1644, in the service of the VOC (Dutch East India Company). His was the first European expedition to reach the islands of Van Diemen's Land (now Tasmania) and New Zealand and the first to sight the Fiji islands in 1643. He also mapped substantial portions of Australia.

Captain James Cook was an English explorer and cartographer. He made three voyages to the Pacific Ocean, accurately charting many areas and recording several islands and coastlines on European maps for the first time. His most notable accomplishments were the British discovery and claiming of the east coast of Australia; the European discovery of the Hawaiian Islands; and the first circumnavigation and mapping of Newfoundland and New Zealand. James Cook's eleven years sailing around the Pacific Ocean contributed much to European knowledge of the area. Several islands such as Easter Island and the Sandwich Islands (Hawaii) were encountered for the first time by Europeans, and his more accurate navigational charting of large areas of the Pacific was a major achievement.

With the discovery of new lands across the Atlantic, Pacific and Indian Ocean, the Age of Exploration is generally said to have ended in the early 18th century. However, the Arctic and Antarctic seas were not explored until the 19th century. It also took much longer for Europeans to reach the interior of continents such as Africa and North America than it did the coasts. Columbus, Vasco da Gama, Magellan, Cook and almost all the expeditionary leaders relied on Ptolemy's calculations about latitudes and circumference of the earth. As a result of these voyages huge knowledge was accumulated. There was a huge mass of literature that had to be tested, to be made accessible to others.

Scientific and Technological Innovations

Mainly the invention of printing machine and the works of scholars like Copernicus, Kapler, Galileo and Newton brought the scientific and technological developments in geography during the Age of Discovery. The printing press was invented at a time when it was really desired. The older form of printing machine had already come into existence before the Age of Discovery; it was invented in China by Bi Sheng between 1041 and 1048. But, the use of movable type to mass produce printed works was popularized by a German printer, Johannes Gutenberg in 1450. While there are several local claims for the invention of the printing press in other parts of Europe, including Laurens Janszoon Coster in the Netherlands and Panfilo Castaldi in Italy, Gutenberg is credited by most scholars with its initial invention. The invention of 'Printing Machine' in Europe aided the scientific developments by publishing the accounts of explorations, voyages, discoveries and the geography of newly discovered lands. Up to 15th century most of the scholars took earth as the centre of universe and tried to study the universe with a geo-centric approach. But during the Age of Discovery, the heliocentric idea was strengthened. The Polish Scholar, Nicholus Copernicus (1473-1543), carried out numerous observations about the stars, planets and moon for 32 years (1497-1529). His experiments provided the first modern formulation of a heliocentric (sun-centred) theory of the solar system in his epochal book, *De revolutionibus orbium coelestium* (On the Revolutions of Celestial Spheres). In 1618, Kepler proved that planetary motions were elliptical rather than circular. Johannes Kepler (1571-1630), a German mathematician and astronomer, was one of the most important scientists in the field of astronomy. He was the founder of "celestial mechanics", having been the first to explain planetary motion. He gave fundamental laws of planetary motion and postulated that the Earth and planets travel about the sun in elliptical orbits. In 1623, Galileo presented the proof that Copernicus was right about the heliocentric Universe. Galileo Galilei (1564-1642) was an Italian physicist, astronomer and philosopher closely associated with the scientific revolution. Galileo's career coincided with that of Johannes Kepler. Galileo's achievements include improvements to the telescope, a variety of astronomical observations, the first and second laws of motion, and effective support for Copernicanism. He has been referred to as the "father of science". In 1686, Newton presented the 'Laws of Gravitation'. Isaac Newton (1642-1727) was an English physicist, mathematician, astronomer, alchemist, inventor, and natural philosopher generally regarded as one of the most influential scientists in history. Newton wrote the *Philosophiae Naturalis Principia Mathematica* wherein he described universal gravitation and the laws of motion, laying the ground for classical mechanics. By deriving Kepler's laws of planetary motion from this system, he was the first to show that the motions of bodies on Earth and of celestial bodies are governed by the same set of natural laws. The unifying and deterministic power of his laws was integral to the scientific revolution and the advancement of heliocentrism.

Improvements in Cartography and Map Making

The field of geography that was most negatively affected during Dark Age was the most benefited one during the Age of Discovery. It was Cartography. Ptolemy's maps were unearthed and restudied. The missing details were filled in and the discrepancies were removed. Ptolemy's calculations about the latitudes, longitudes and circumference of the earth were corrected. The shape of earth (i.e. elliptical and not sphere or round) was finally ascertained. Missing details being filled up, the world map became almost complete. On this basis the first accurate 'globe' of world was made. Martin Behaim (1459-1507), a Portuguese navigator and geographer, created the spherical globe depicting the Earth in its three-dimensional form in 1492. Behaim's invention was a significant advance over two-dimensional maps because it created a more realistic presentation of Earth's shape and surface configuration. Mercator's Projection was devised (1569 A.D.) and the new maps were prepared on it. Gerardus Mercator (1512-1594) was a Flemish[5] mapmaker and geographer best known for the map projection that bears his name. Mercator's projection is based on the concept of 'loxodrome' or 'rhumb line', the line of constant bearing. It makes this projection particularly useful to navigators.

With all the above developments, there came a scientific revolution not only in materialistic terms but also in the ideas and thinking of man. All the disciplines, along with geography, were benefited. Many of the earlier erroneous ideas were eliminated gradually giving way to more reliable accounts. The quality of geographic literature, as such, improved.

THE IMPACT

The impact of the developments during Age of Discovery on Geography was tremendous. A number of Geographical Societies were constituted in different parts of the world to promote further research in the discipline. The 18th and 19th centuries were the times when geography became recognized as a distinct academic field and became part of University curriculum in Europe. This development took place with the foundation of, for instance, Societie de Geographie, Paris (1821); Royal Geographical

Society, London (1830); Russian Geographical Society, St. Petersburg (1845); American Geographical Society, New York (1851); and National Geographical Society, Washington D.C. (1888). Due to vast amount of accumulated knowledge there began branching of geography into different specialized fields of study. Earlier geography was 'all-in-one'. Many old theories or concepts were corrected and re-established; and many new theories about the origin of the earth, continents and man-nature relationships were advocated. The German scholar, Bernard Varenius (1622-1650) made the most fruitful use of all available information and knowledge about the earth during Age of Discovery; and gave the first scientific literature in geography in the form of *Geographia Generalis*. The positive gains of this period may be summarized as follows:

- Discovery of new lands and completion of world-map;
- The revival of interest in geography from a scientific perspective;
- 'Academic Freedom' of geography, i.e. identification as an independent field of study;
- Accumulation of huge amount of data and knowledge of geographical nature and its publication the form of *Geographia Generalis;*
- Advancements in Cartographic technique of data representation and map-making;
- Beginning of scientific or speculative investigations and generalizations in geography;
- Emergence of various modern schools of thought in geography (as German, French, American British and Soviet);
- Geography acquiring a new definition and methodology; and
- Split of geography in various specialized fields of study, e.g. physical, human, etc.

In spite of large-scale and worldwide explorations and voyages during the Age of Discovery, certain parts of the world still remained unexplored and untouched. They were the Polar areas, more specifically the parts of Arctic and Antarctic regions. The scientific explorations of these regions began only in the 20th century. Further, due to vast knowledge and information as accumulated during the Age of Discovery, the field of geography could not remain unified. The scholars did not have a common consensus on the definition, content, scope and approach of geography as a scientific discipline. Consequently, the discipline entered into a long period of dualism, as physical *versus* human, idiographic versus nomothetic, or systematic versus regional. Then, with the passage of time there was more and more use of theoretically deductive approaches. The empirically inductive methods became less popular and there emerged a whole generation of armchair geographers.

However, looking on the positive side of development, the foundations of Modern period of geographical thought were already laid before the Medieval period came to a close after the Age of Discovery. This was in the form of the writings of Bernard Varenius, the scholar of Pre-Classical phase of German School of Thought.

NOTES

1. The Surya Siddhanta was an astronomical treatise written in India, which had rules laid down to determine the true motions of the luminaries, which conform to their actual positions in the sky.
2. Husain, 1988, p. 83.
3. Husain, 1988, p. 84.
4. A chronicle is the historical account of events. The Siberian Chronicles are the Russian chronicles of the late 16th–18th centuries on the *History of Siberia*.
5. Flemish is the territorial name for the Dutch language spoken in historic Flanders, a region mostly comprising the northern part of Belgium, but also including a southern part of the Netherlands and a small area of northern France.

REFERENCES

Husain, M. (1988), *Evolution of Geographical Thought*, Second Revised Edition, Rawat Publications, Jaipur.

Rana, L. (2008), *Geographical Thought: A Systematic Record of Evolution*, Concept Publishing Company Pvt. Ltd., New Delhi.

CHAPTER-5

Modern Period

The "modern" era is generally associated with the European Enlightenment, which began in the middle of 18th century. The thinkers of Enlightenment saw themselves as continuing the work of the great 17th century pioneers—Galileo, Isaac Newton, Kapler, Copernicus, etc. who extended scientific methods into every field of inquiry, thus laying the groundwork for the development of modern social sciences. Precisely, the modern period begins from 1750s onwards. In fact, the scientific geography as we know today is outcome of the product of last 250 years, i.e. after the emergence of German School of Thought. It was during this period that a number of scholars tried to define the field, content and scope of geography in the way we accept today. Notable among them were, for instance, Bernard Varenius, Emmanuel Kant, Alexander von Humboldt, Carl Ritter, Friedrich Ratzel, Vidal de la Blache, Jean Bruhnes, W.M. Davis, Carl Sauer and many more. They discussed and elaborated on the methods of collecting, organizing and presenting the information of geographical nature. Geography before mid-18th century was not more than encyclopaedic compilations of places, without orderly presentation (general principles) of areal differentiation. The new form of geographic descriptions was primarily due to German writers, followed by French, American, British and Soviet. It is generally said that as far as geography is concerned, the Greeks laid the foundation and the Germans built the structure.

GERMAN SCHOOL

In spite of all the adversities and other developments, it was clear, at least, that geography is a science of areal distribution. But the principles and methods to study areal distributions were not at hand. Because, the knowledge of geography was not organized up to this point. The principles behind spatial arrangement of phenomena, their causal connections, inter-relationships, which could give geography a true scientific status, were yet to be established. The credit for making attempt in this direction for the first time goes to the German School of Thought. For the next three centuries after the Age of Discovery, geography was largely a German science. The German scholars organized the large amount of empirical knowledge accumulated during Age of Discovery in a systematic manner. Geography became very popular under the German School. This was a period when the relationships between the physical and biological phenomena were explored in a new perspective. The scientific academies were established and attempts were made to formulate generalizations and laws. However, this long period was not a period of steady and smooth development in geographical thought, as soon the discipline entered into a number of controversies. The roots of geography, as a field of learning, reach back to antiquity, but its establishment as a modern science was essentially the work of decades from 1650 to 1950. The period since early 19th century to the end of it may be termed as the 'Classical Phase' (1800-1880) of German School. This was mainly the period of Alexander von Humboldt (1769-1859) and Carl Ritter (1779-1859). Humboldt and Ritter, together, are popularly known as the 'Fathers of Modern Geography'. Although, the real beginnings of German School are believed to have been made with the writings of these two founders of Modern Geography, contribution of the School had actually began with the works of few other scholars before them and also continued even after their death. Therefore, the period of German School before Humboldt and Ritter may conveniently be called as "Pre-Classical" (1650-1800) and that after them as "Post-Classical" (1880-1950).

Pre-Classical Phase

The pre-classical phase of the German School (1650-1800) was a phase at the close of Medieval period and before the beginning of Modern period. It was the time when a large mass of accumulated empirical knowledge was organized on systematic lines. Bernard Varenius and Emmanuel Kant are considered as the most important personalities of the 'Pre-Classical Phase'. Bernard Varenius (1622-1650) was highly influenced by the works of Copernicus, Kepler and Galileo. Varenius made two significant contributions to the development of geography. Firstly, he brought together the contemporary knowledge of astronomy and cartography and made their constructive use in geography. Secondly, he divided the nature of enquiries in geography into two: 'general' and 'specific'. Later, this led to the development of 'systematic' and 'regional' geography. Varenius believed in two-fold division of geography, i.e. 'systematic' (general or topical) and 'regional' (specific or special). The former approach emphasizes on one aspect of study throughout all the regions. The regional studies, on the other hand, deal with a particular area in detail covering all its relevant and inter-related elements. The systematic studies lead to the formulation of generalizations, laws and theories. General geography provides the fundamentals that need to be applied to regional studies. These generalizations help to understand the regional geographies. All these ideas are summarized in *Geographia Generalis* of Bernard Varenius. This work continued to be a standard geographic reference for about

100 years. The book is divided in two parts, viz. General Geography and Regional Geography. General Geography consists of descriptions of three types of features of earth. They are: (i) *Absolute Part*: covering terrestrial characteristics of the earth, independent of others, as size, shape, form, etc.; (ii) *Relative Part*: covering the planetary characteristics of the earth in relation to Sun or other planets, as revolution, rotation, etc.; and (iii) *Comparative Part*: dealing with absolute and relative locations of various places, features and areas over the surface of the earth. Similarly, Regional Geography also consists of descriptions of three types of properties of the constituent areal units. These include: (i) *Celestial Properties*: related to the study of climate, atmosphere, etc.; (ii) *Terrestrial Properties*: related to location, relief, soil and land characteristics; and (iii) *Human Properties*: concerned with population, political divisions, communication, socio-cultural traits, religion, diet, etc. Varenius' *Geographia Generalis* became so popular that Newton, his contemporary mathematician and astronomer, got it translated into English as well. However, the scientific trend started with the writings of Varenius was interrupted by Kantian philosophy. Kant considerably modified the scientific structure of geography as proposed by Varenius.

Emmanuel Kant (1724-1794) was a Professor in Konigsberg University, where he taught geography for about 40 years. He tried to secure the foundations of geography within the framework of traditional philosophy. As per the tradition, geography was being considered as a discipline concerned with the description of phenomena distributed over space. In this process, he overlooked the scientific beginnings marked by the work of his predecessor, Bernard Varenius. Kant's main interest was in physical geography, which he considered as the basis (preliminary) of all geographical studies. Kant's philosophical contribution goes primarily to the method of acquiring knowledge; classification of knowledge; and classification of geography. Kant agreed that there are two ways to acquire knowledge, viz. through sense perceptions and pure reasoning. The former approach is 'inductive' and the latter is 'deductive'. The sense perceptions, in turn, are of two types, viz. 'inner' and 'outer'. However, Kant's approach to geographical studies was purely deductive. He pointed out that there are two ways of organizing knowledge for the various phenomena on the surface of earth, viz. logical and physical. The Logical Classification is grouping according to the nature of inherent characteristics of the phenomena. It deals with organizing knowledge according to the type of objects studied, and accordingly covers a series of systematic sciences that are devoted to the study of particular kinds of phenomena, e.g. Botany, Zoology, Physics, Chemistry, etc. The Physical Classification, on the other hand, is grouping according to the position or occurrence of phenomena in time or space. It deals with organization of knowledge in its temporal or spatial dimensions and, therefore, includes historical and spatial sciences. The historical sciences are concerned with the arrangement of phenomena in chronological order; and the spatial sciences are concerned with the arrangement of phenomena in chorological order. The disciplines falling within this category, according to Kant, are different from those in the previous classification. In simple words, history is the description in terms of time; and geography in terms of space. In this manner, Kant considered both history and geography descriptive in their nature. He also regarded both of these disciplines complementary to each other. Besides laying emphasis on the study of Physical Geography, Kant also identified some other branches, namely: (i) *Mathematical Geography*—studying the form, size and movement of earth, and its place in the Solar system; (ii) *Political Geography*—the study of relationship between the political units and their physical background; (iii) *Moral Geography*—the study of customs and characteristics of man in relation to environment; (iv) *Mercantile Geography*—the study of trade and commerce; and (v) *Theological Geography*—the study of distribution of religious traits in different types of environment, e.g. Christians in Europe, Muslims in Central Asia, Confucians in Hwang Ho Valley in China, and Vaishnavs in Sutlej-Ganga Plains, etc.

Kant's views have been criticized on various grounds. Particularly since the later part of 19th century, the questions have been raised about the validity of his concepts and classification. Kant became highly unpopular, later, and came to be known as the 'Father of Exceptionalism' in geography. It is because he accorded geography a purely descriptive status, bisecting it from the other scientific disciplines. He made this exceptionalist claim not only for geography but also for history. The most notable opposition was put forward by F.K. Schaefer, a 20th century British scholar. According to Schaefer, there is only one form of explanation for all disciplines, i.e. to recognize phenomena as instances of laws. Even if the explanations are considered of two types, viz. descriptive and scientific, this holds true for all disciplines. It is illogical to state that some disciplines are purely descriptive and others purely scientific. The very basic distinction of knowledge between 'scientific' (systematic) and 'descriptive' is wrong in principle. Because, the so-called systematic sciences also have spatio-temporal dimension. Moreover, every study begins with some kind of description. The explanations, causal connections, laws and theories follow later. Schaefer considered Kant a poor and immature geographer in comparison to his predecessor, Bernard Varenius, who died 150 years ago. He called Kantian philosophy as pre-Newtonian. Kant was not influenced by the ideas of Newton, who established a functional similarity among the formally diverse things. However, the subsequent writers, namely, Humboldt and Ritter, revived the tradition of Varenius in geography, again giving the discipline a scientific twist.

The Classical Phase

The writings of Humboldt and Ritter characterize the Classical Phase of German School. In spite of being short spanned, this phase of German School turned out to be a very influential period of development. Humboldt and Ritter lived and worked at Berlin for more than 3 decades and tried to mould the subject matter of geography into scientific form. Both were outstanding scholars of the first half of 19th century and were also contemporary. Both died in the same year, i.e. 1859. Coincidently, 1859 is also the year of publication of Charles Darwin's "*Origin of Species*". Their ideas, although formulated before the great impact of Darwin's evolutionary thought, and before the enormous exploration and mapping in the latter half of the 19th century, have continued to serve to the day as 'guide posts' in the field of geography.

Alexander von Humboldt (1769-1859) was a great traveller, explorer, naturalist and cosmographer. He led the way in the expansion of geography in and outside Germany. He was a scholar of versatile genius, who contributed richly to a variety of fields, as geology, botany, zoology, physics, chemistry, anatomy, physiology, history, climatology, geomorphology, plant geography, orography, oceanography, cartography, vulcanism and magnetism. He travelled about 45,000 thousand miles and in all his travels, however short, he made multitudinous observations. He remained a living model for subsequent generation of researchers. The adventurous nature of Humboldt did not permit him to stay at one place. His travels covered the regions around Mediterranean in Europe, Mexico and Caribbean Islands in the West and Siberia in the East. He travelled across the Llanos, sailed up to the Orinico, visited Cuba and Mexico, journeyed extensively in the Northern Andes, and climbed many mountain peaks, including Chimbarajo (the highest peak of Andes, in Equador, South America). The last 3 decades of his life he devoted in the writing of his best-known work, *The Cosmos.* The prevalent philosophies at the time of Humboldt were those of 'natural' and 'moral'. The natural philosophies were related to the physical world, envisaging the earth as an organic whole, inseparable and mutually interdependent; whereas, the moral philosophies were related to man's inner world and human thinking. Although, Humboldt's basic inclination was towards natural, he did not ignore the other view. Instead, he regarded both (natural and moral) philosophies in synthesis—and considered through the scientific analysis of the ways in which things and phenomena on the surface of the earth depended upon each other—an idea which runs right through Cosmos. Thus, he envisaged the earth as a whole, an organic whole, of all parts of which being mutually interdependent. Although Humboldt contributed richly to a variety of fields, he was always concerned with the areal associations of natural phenomena. Humboldt included man and his works in the concept of nature and natural areas, but he did not consider man as a primary determinant. The division of knowledge related to natural world, he contemplated falling into three groups, viz.: (i) *Systematic Sciences*: studying the phenomena in terms of their similar characteristics, e.g. Botany, Geology, Physics, Chemistry, etc.; (ii) *Historical Sciences*: studying the history of development of existing groups of phenomena, e.g. History, Anthropology, etc.; and (iii) *Earth Sciences*: concerned with the distribution and arrangement of phenomena on the earth's surface. The objective of such sciences, e.g. Geography, is to study the phenomena in their spatial distribution, spatial relationships and interdependence. The basis of Humboldt's classification was close to that of Kant's; but, the former treated all kinds of subject-matter (including history and geography) as scientific and accorded them the status of sciences. While dealing with the subject matter of earth sciences, Humboldt coined the term 'cosmography' (study of the Universe) and divided it into *Uranography* and *Geography*. In his opinion, *Uranography* is the descriptive astronomy that deals with the celestial bodies. In other words, *Uranography* is the celestial cartography, the branch of astronomy concerned with mapping the stars, galaxies, or other celestial bodies. Geography, on the other hand, deals with the terrestrial part. Geography, accordingly, is the description of earth, dealing with interrelationships of phenomena that exist together in an area. Humboldt's approach was holistic. Unlike Kant, he followed empirical and inductive method of study. It was based on self-observation and experimentation, because he himself collected enormous data related to plants, animals, rocks and other aspects. Humboldt's methods of recording and describing in the field are important in themselves. He noted, measured and mapped changes in relief, plant life, crops, tree and snowline in a way that had never been done before. To generalize these features (particularly important since no maps were available) he drew cross-sections of the areas he travelled. While travelling he used 40 different types of instruments for observation and recording. They included, e.g. telescope, sextants, quadrants, barometer, cyanometer, etc. As far as the analysis of recorded information was concerned, he used systematic and comparative methods. Because the purpose of his travels was not only to measure and record various kinds of phenomena, but also to find out the ways in which the great variety of observable phenomena of landscape are associated and inter-connected with each other at different places. In order to bring about precision and accuracy in the descriptions he also used various cartographic techniques. Although Humboldt's approach was systematic, he could not ignore the importance of regional studies due to his belief in inter-relationships and causal connections. According to him, there is an order under chaos. The systematic distribution of every phenomenon at individual level produces some areal associations or assemblages. Although each of these associations appears different in outlook from the other, it is not difficult to establish an order to find logic behind this distribution. Therefore, the systematic and regional studies are not different; rather they are complementary to each other. Humboldt conceived man and his activities as the part of nature. Man and nature relationships produce different types of areal associations. No two landscapes exhibit exactly same characteristics; but, there are some similar features; some common cause of their existence may be explored. For example, Llanos (South America), Pampas (Argentina), the Russian Steppes, the Tropical Savannas, the American Prairies, the Tundra Grasslands, etc. are unique in themselves and exist in different regions; but still some common genetic causes of existence or common features (nature of vegetation) may be found among them. Thus, relationships among unique landscapes may be established. This kind of approach became core of all geographic studies. In other words, the world is ordered, and the geographer's task is to discover that order. '*Cosmos*' (the Ordered Universe) is Humboldt's best-known work. It is a comprehensive survey of the world. Running in five volumes, published during 1845-62, the work is translated into eight languages. It is a monumental compilation, based on conclusions drawn from his lifetime travel and research. In '*Cosmos*', Humboldt regards man as an integral and constituent part of the Universe, an item in the balanced unity of nature. Some important concepts of geography may be derived from the work of Humboldt. They are: (i) Geography considers earth's surface as a home of man; (ii) Geography is a science of the spatial distribution of world-phenomena (the basic concept of geography); (iii) General geography is Physical geography, and includes man wherever he is part of nature; (iv) Geography is the study of inter-relationships, i.e. relation between living (organic) and non-living (in-organic) worlds; (v) Geography is the comprehension (understanding) of the world phenomena; (vi) There exists heterogeneity of phenomena in geography, and the geographers' task is to search for similarities or general (common) rule of their existence; and (vii) The 'unity of nature' is a characteristic feature of geographical studies. Thus, Humboldt tried to make geography "as original (distinctive) science" rather than a collection of facts from the physical and biological sciences. Charles Darwin described him as "the greatest traveller who ever lived". Humboldt's travels, experiments and knowledge transformed the Western science in 19th century.

Carl Ritter (1779-1859), a scholar of diversified interest, was one of the founders of modern school of geographical thought. He started his career as Professor in Geography at the University of Berlin and remained there till his death, spending a total of 40 years of teaching here. During these years he was able to exercise profound influence on geographical thought. He was not only contemporary to Humboldt, but his work was also complementary to that of his. Where Humboldt's inclination was towards nature, Ritter emphasized more on human. Humboldt followed systematic approach and Ritter was in favour of regional. Although Ritter did make some travels in Europe, he was basically an armchair geographer. Just like '*Cosmos*' of Humboldt, Ritter also authored a great work entitled *Die-Erdkunde*. Die Erdkunde is Ritter's best-known work and is based on the regional descriptions of the earth covering largely Asia, Africa and Europe. Ritter was deeply a religious man and thought that divine has created the earth as a school of man and looked upon it as a home of man. He, therefore, is famous as a regional geographer and anthropogeographer. He also had a vision of an ordered and harmonious universe. But, his approach was teleological and he made it clear to his students how God's plan revealed in the harmony of man and nature. He founded the Berlin Geographical Society. Ritter identified geography as an empirical and descriptive science. According to him, geography deals with the local conditions and embraces three attributes of a place, viz. (i) *Topographical*: dealing with the natural divisions of the earth; (ii) *Formal*: concerned with systematic or general features as, e.g. atmosphere, movement of water, etc.; and (iii) *Material*: which describes the geographical aspects of natural history and cover distribution of human beings, minerals, plants and animals. These three attributes constitute the basic content of *Die-Erdkunde*. This work is a survey of the known world, i.e. based on the study of three continents, viz. Asia, Africa and Europe. It was written with a distinct theme, i.e. comparative and inter-relationships, showing connections between man's history and nature. On the basis of his work he proved that geography also has a right to be distinctively considered as a science, with the same dignity as other sciences. Ritter claimed that the central principle of geography is "the relation of all phenomena and forms of nature to the human race, are examined and organized within the framework of unique geographical associations of land and man on the earth's surface". Through 'Erdkunde', Ritter has made a rhetorical (super-natural) claim for geography as 'the science of the earth', 'the description of the earth as a home of the man'. Ritter believed that the earth to be an organism made, even in its smallest details, with divine intent, to fit the needs of man to perfection. 'As the body is made for soul, so is the physical globe made for mankind'. This kind of explanation has been termed as teleological interpretation of man-nature relationships. Like *the Cosmos* of Humboldt, Ritter's *Erdkunde* was also a professional work—a product of his lifetime reading, selection, interpretation, and organization of a mass of available data. Thus, like Humboldt, Ritter also believed in the unity of nature, and considered the earth as an organic whole. But, his approach to the study of man-nature relationship was anthropocentric. Ritter sought to develop the concept of terrestrial unity. He divided the earth into major continental units (*Erdteile*). According to him the major continental units were to be arrived at deductively; and the smaller units within each continent to be derived and studied with an inductive approach, i.e. to divide the individual continental units into smaller areal units on the basis of empirical observations. But, he could not attempt this himself due to the lack of maps and related data. Therefore, he was forced to involve himself in teleological speculations. However, he has always argued that the study must advance by collection and generalization of data about particular localities (synthetic approach). Ritter is often accused of making geography "a handmaiden to history"; and mixing religious philosophy of divine purpose with objective investigation. Writers of later period, attempted to discredit the importance of Ritter's work by attacking his philosophical concepts, in particular his teleological view of the universe. But, if one overlooks this, there still remains a treasure of geographic insights. His main contribution is that he conceived each individual units of the earth as inter- related complex of elements. His plan of study became the model for regional presentations and had far-reaching influences. Ritter became popular among geographers for his Regional Concept. His division of the Earth consists of four levels in hierarchical order, viz.: Earth (*Erdkunde*); Continental Units (*Erdteile*); Orographic regions at continental level; and smaller areal or terrestrial units arrived at inductively at the lowest level. After the Earth, the second order units, according to Ritter, are arrived at deductively. The conceptual basis of this division is traditional, i.e. each continent to be sub-divided by its orography (relief characteristics). He held that the continent is not a generic complex, but a complex of individual units. The fourth order areal units are arrived at inductively from the detailed configuration and terrestrial mosaic. Ritter appears to have envisaged each continent as a complex of individual units whose whole structure and content could be arrived at by inductive procedures from the smallest areas of associated phenomena (synthetic approach). In this way one can classify the regions of various orders (micro to macro and *vice versa*). Ritter's main geographical concepts may be discussed as: (i) Geography is an empirical science and not the one based on deductions from rational principles or from *a priori* theories. In other words Ritter saw geography proceeding from observation to observation; (ii) There exist regional-associations of terrestrial phenomena over the surface of earth; (iii) Geography finds coherence (inter-relationships) in the spatial arrangement of terrestrial phenomena and this view has been termed later as "areal-differentiation'; (iv) The boundary lines, whether wet or dry (rivers or mountains) are only a means towards the real purpose of geography, which is understanding of the coherent areas; (v) Physical (or any other) classification of the earth's surface is not an end-product; it is just the beginning for geographical studies and not the aim; and (vi) Geographical studies must proceed both chorologically and chronologically. The fundamental principal evolved by Ritter was 'Unity in Diversity'. According to him there is a fundamental unity in the biotic and abiotic components of habitat in which man sculptures his cultural environment. In such an approach all the physical and cultural components of environment are taken into consideration and their interrelationship is established in understanding the geography of an areal unit. This is regional approach. Unity in diversity means that every naturally bounded area is in unity in respect of climate, production, culture, population and history. Thus, Ritter's main contribution to geography was his concept of regional associations of terrestrial phenomena at various levels on the earth's surface. But, he never recognized any basic difference between the geographical study of particular areas and the general geography. This mixing of two approaches led to dualism later. Further, Ritter has been also criticized for his teleological or anthropocentric interpretations. Finally, he could not break away from the thought that united geography with history, and here he appeared to have been influenced by Kantian Philosophy, although extreme generalizations were not there. In sum, Ritter's preparation for research covered an extremely wide field, in which on the one hand natural sciences, and particularly the observational methods of nature study, predominated, but in which, on the other hand, his interest was increasingly in human problems. His first principle, frequently repeated, was that geography must be an empirical science, rather than one deduced from rational principles—from philosophy—or from *a priori* theories of 'general geography'. "The fundamental rule should assure truth to the whole work", he wrote, "is to proceed from observation to observation and not from hypothesis to observation".

If the works of both Humboldt and Ritter are compared, the writings of Varenius acted as springboard in the thinking of both the scholars. Inspired by Varenius, they worked towards providing geography the status of a scientific discipline. Although, they did not lay any exceptionalist claim for geography, Kantian thought influenced both. This influence can be seen in the classification of sciences by Humboldt and in the chorological perspective of Ritter. Both had a holistic approach to geography, as they believed in the unity of nature. This concept of unity of nature presumed a causal interrelation of all the individual features in nature. Accordingly, the phenomena of nature were required to be studied in order to establish this coherence. For both the scholars, it was axiomatic that the unity of nature included organic as well as inorganic, human as well as non-human, immaterial as well as material. The exclusion of any part would destroy the coherence and unity of the whole. The writings of Humboldt and Ritter have a thrust on inter-relationships, spatial association or integration of phenomena. The Physical Geography was considered as the base of all geographical studies by Humboldt and Ritter. Both emphasized on the importance of empirical method of research in geography. As far as the differences between the two are concerned, Humboldt and Ritter were different from each other in their personality, temperament and family background. Humboldt was born in a wealthy family and was educated privately till the age of 18. But this was not true for Ritter, who belonged to a normal middle class family. Humboldt was a student of science and we know him as a natural scientist; whereas, Ritter was basically a religious man, a human and regional geographer, an anthropogeographer. The sphere of research of both differed. Humboldt worked on general or systematic studies with a major emphasis on physical geography. Ritter, on the other hand, mainly explored the human world with regional approach. Although both believed in the unity of nature, they differed in their philosophical approach to geographical studies. Humboldt's work has deterministic flavour since he considered man as a part of nature, whereas Ritter considered nature as the habitat of man, exhibiting a possibilistic tint. For Humboldt, this unity of nature was a balanced unity of the whole of nature, of which man was a part; whereas for Ritter, it was teleological, as he spoke of discovering the 'divine secrets'. Thus, Humboldt and Ritter were not only contemporary; their works were also complementary to each other. None of the personalities were complete in themselves as far as their contribution to geography is concerned. The great difference in temperament and in general outlook on life gave different colour to the work of the two founders of modern geography, and this difference became fundamental in the works of their followers as well. Because, they could not leave a clear framework for our subject. Therefore, after their death in 1859 geography entered a long phase of dualism. Coincidently, both of them died in the same year. The year 1859 marked not only the death of Humboldt and Ritter but also the publication of Darwin's *"On the Origin of Species"*. Charles Darwin (1809-1882) was an English naturalist who established the theory of organic evolution. In spite of all the diffferences, the works of Humboldt and Ritter together, particularly *Cosmos* and *Erdkunde* did provide an almost complete explanation for geography, in terms of its definition, content, scope, approach and method of study.

The Post-Classical Phase

The Classical Phase of German School terminated with the death of Humboldt and Ritter. The post-Classical period, however, in many respects may be regarded as the critical period. The period from late 19th to mid-20th century was full of debates, controversies, differences and dualisms. As no clearly unified field of geography was left either by Humboldt or by Ritter, the emphasis of geographical study kept on shifting like a pendulum. This situation became more complex with the publication of the work of Charles Darwin. The decades after 1859 saw not only the impact of Humboldt and Ritter, but also of the revolutionary theory of Darwin. The followers exaggerated certain aspects of the views of each of the founders, Humboldt and Ritter. Consequently, geography for a time kept on splitting in several directions and its position as a branch of knowledge thereby brought into serious question. The major problem of this phase of development was to overcome the apparent disunity in the methodology of the field and thus definitely to establish its position as a single field of science. The shift in geographical thinking during this period came primarily due to the works of Oscar Peschel, George Gerland, Ferdinand Richthofen, Friedrich Ratzel, Albrecht Penck, Alfred Hettner, Otto Schluter, Alfred Wagener and Walther Penck. Most of these scholars, except Ratzel, Hettner and Schluter, promoted physically biased studies. Ratzel was an anthropogeographer, whereas, Hettner and Schluter advocated for chorological perspective in geography.

Oscar Peschel (1826-1875) is described as the last great geographer before the discipline was finally overtaken by the impact of Darwinian ideas[1]. He raised fundamental questions concerning the nature of geography, and was critical of the approaches of both Humboldt and Ritter. Peschel was very critical of Ritter's method and noted that since continents and large areal units contained great deal of internal diversity (one quite different from the other) they could not be compared in the proper scientific sense. Comparison was possible only in respect of clearly identifiable features or characteristics, such as particular type of landforms, climatic types, or distribution of plant cover. Peschel insisted that comparative geography should have a definite method and purpose: the geographer pursuing comparative study should seek (with the help of large-scale maps) similar physical features in different parts of the earth's surface, compare their characteristics and origins, and try to relate them all genetically (following the method of comparative anatomy). Peschel thus laid a sound basis for comparative research in geography. His method required the student to begin by studying topographic maps with a view to identifying "homologies '(i.e., landforms of similar types), and then try to trace their origins in each of the areas where they occur. Peschel believed that physical and human geography constituted two entirely separate domains of knowledge, and as such the two could not form part of a single science. But, he was strongly of the view that geography should be pursued primarily as a study of the physical phenomena of the earth[2]. He was critical of Ritter for having neglected physical geography, and he criticized Humboldt for not having attempted a scientific classification of landforms. He was also critical of Humboldt for creating the impression that general geography could be equated with natural science. He further decried Ritter for holding a teleological view about the nature and origins of earth phenomena, and for having subordinated geography to history. Peschel's exhortation that in order to project its status as a science, geography should identify itself more and more as physical geography was further advanced by two other contemporary scholars, namely George Gerland (1833-1919) and Ferdinand Richthofen (1833-1905). Gerland was professor of geography at the University of Strasbourg, and had been the supervisor of the doctoral thesis of Alfred Hettner. In a long essay published in 1887 Gerland had gone so far as to suggest that since man could not be put to scientific analysis, study of man should be taken outside the purview of geography. Richthofen was primarily interested in geomorphology. He had received an advanced training in geology. In geography, Richthofen is best known for his epoch-making inaugural address delivered at the time of accepting the chair of geography at Leipzig in 1883. In this address, he took over the idea of Humboldt and Ritter of geography as science that is distinguished by its spatial-distributional perspective. However, given the fact that he was himself trained as a geologist, Richthofen was in greater sympathy with Humboldt than Ritter, and like Humboldt, he endeavoured to restore the close connection of geography with natural sciences. According to Richthofen, it was the distinctive purpose of geography to focus attention on the diverse phenomena that occur in interrelation on the face of earth. But, at the same time he also emphasized that in order to reach useful and reliable conclusions, geographical study of any part of earth's surface must start with a careful description of its physical features, and from there the student should move on to examine the interrelationships of other features of the earth's surface, and before anything else the physical geographic framework should be described at the outset. But, Ratzel, a younger contemporary of Richthofen, was not in much agreement to this view. He again brought human component in the fold of geographical studies. Ratzel and his work continued to be influential in the studies of modern geography.

Friedrich Ratzel (1844-1904) was a remarkable human geographer. The thinkers like Charles Darwin and the zoologist like E.H. Haekel, in turn, influenced Ratzel's writings. *'Anthropogeographie'* and *'Politische Geographie'* are his two most important literary contributions. He is also notable for advocating the doctrines of 'determinism' and 'geopolitics' and coining the term 'lebensraum' (living space). After he finished his education, Ratzel began the travels that converted him into a geographer. Upon his return from the travels in 1875, Ratzel became a lecturer in geography at the Technical High School in Munich. In 1876, he was promoted to assistant professor, and then rose to full professor in 1880. While at Munich, Ratzel produced several books and established his career as an academician. In 1876, he accepted an appointment at Leipzig and continued to work here till his death. The main interest of Ratzel was in geology, zoology and comparative anatomy. He became popular among geographers because of his work on Man-Nature relationships and Political Geography. Ratzel explored human relationships with physical surroundings in a systematic manner (a mix of the ideologies of both Humboldt and Ritter). The findings of his work are based on empirical observations as he approached all human-nature relationships through travel and direct contact with the realities.

In *Anthropogeographie*, Ratzel explores the influence of physical environment on mankind systematically and empirically. The work runs in two volumes. The basic theme of the work in first volume, published in 1882, is that the man and his actions are not free from the influences of the surroundings. They are controlled by nature. He believed that man is the creature of his environment in much the same way as Darwin had demonstrated the adaptation and survival of the fittest in animal world. The second volume of the book, published in 1891, is written with more liberal views on man-nature relationships. Because, he added that the survival of man and his adaptation to the environment also depends on fact that how fit and skilful he is in moulding the natural circumstances. Man's reactions to the nature differ sharply accordingly to his culture and historical background. If belonging to primitive culture then nature has a stronghold; and if advanced technologically the nature's hold becomes weaker. But, even then the nature is dominant. However, according to Ratzel, the man, whether he adapts or modifies the surroundings, in any case is the creature of his environment; because even the act of moulding is a kind of adaptation prompted by nature in an indirect manner.

In 1897, Ratzel published *Politische Geographie*, the first work written with systematic approach to political geography[3]. In this book, Ratzel develops the concept that views the state as "a particular spatial grouping on the earth's surface." The state, as defined by Ratzel, consists of "a human group with definite organization and distribution"[4]. From these ideas, Ratzel developed the concept of In: 'Lebensraum' or living space, Ratzel hypothesized that the state naturally seeks to increase its size. If the state's neighbours are weak, the state will grow larger and spread into other states. As evidenced, Ratzel believed that space was a great political force. Unfortunately, the people as Karl Haushofer and Adolf Hitler, who used these ideas to formulate their own theories about world domination, misinterpreted these ideas. Ratzel's work, in a way, was reworking of the theme developed by Ritter in his *'Erdkunde'*, but with a newly established scientific method. The difference lies in two respects, viz. (i) Ratzel treated Human Geography systematically and not regionally as done by Ritter; and (ii) Ratzel's views were highly influenced by Charles Darwin. Ratzel saw man as the end-product of evolution, in which the mainspring was natural selection of types according to the human capacity to adjust to the physical environment. Thus, where Ritter considered man and nature as part of harmonious whole serving the creative purpose of God, Ratzel tended to see man as a product of his environment, molded by physical forces that surround man, and succeeding only in so far as he made the correct adjustment to their demands; the ultimate purpose of these adjustments, if any, lay outside the core of his enquiry. A deterministic tint, thus, colours most of his writing. Ratzel is known as the propounder of environmental determinism in geography. He systematically postulated the viewpoint in the later half of 19th century. According to this viewpoint, man is the product of nature. It is the nature that governs man and his activities, and determines his way of life, living styles and also physical appearance. Because of this postulation, Ratzel became very popular particularly in Germany and U.S.A. The influence of revolutionary biology of Darwin led Ratzel to adopt the organic theory of state and society as well. The political entities (nations or states) expand or shrink like organisms. Based on this viewpoint, he postulated the 'Organic Theory of State', the concept of state as an organism, a piece of humanity on the surface of earth. He believed that state (nation) is not only a spatial unit, but also a human entity, and therefore, it is a living unit. The state is not only a man-made, but also a spatial unit. However, during the Second World War, the 'Organic Theory of State' attained its extreme form of explanation, and out of Political Geography of Ratzel emerged a new concept, i.e. the concept of *Geopolitiks*[5]. Rudolf Kjellen, a Swedish student of Ratzel, further elaborated on the 'Organic theory of State' and coined the term *'geopolitics'*. It became the governing principle in Germany during the Second World War (the period of Hitler), when Germans considered themselves as the supreme race. Thus, Ratzel's contribution to geography was both philosophical and methodological. His doctrine of 'determinism', besides being a strong viewpoint, also provided a method for understanding the man-nature relationships. Ratzel viewed man-nature relationship in a balanced way. He gave a new definition to human studies in a systematic way. Out of his works, particularly *'Anthropogeographie'* and *'Political Geography'*, there emerged two new doctrines, viz. 'environmental determinism' and 'geopolitiks'. Whatever the final verdict on Ratzel's work may be, it was of major importance in that it corrected the prevailing tendency to overstress the physical aspects in geography and establish a more balanced viewpoint. Indeed his work served to fix interest still more firmly on systematic studies, although regional geography continued to receive scant attention. In his book, *The Makers of Modern Geography*, Robert E. Dickinson (1961, 1964) states, "There is no doubt that Friedrich Ratzel has been the greatest single contributor to the development of the geography of man." But, this humanistic trend in geographical studies was again broken by the writings of two subsequent scholars, Penck and Wagener.

In the name of Penck, there were two scholars, Albrecht Penck (1858-1945) and Walther Penck (1888-1923), former being the father of latter. Albrecht Penck was basically a geologist. He was professor first at the University of Vienna (1885-1906) and then at Berlin (1906-26) and was also the director (1906-22) of the Institutes of Oceanography and of Geography, Berlin. He is noted for his study on glaciation (in the Alps), for his pioneer classification of landforms, and for his work in the development of modern regional geography. Outstanding among his many works is '*Morphologie der Erdoberfläche*' (Morphology of the Earth's Surface) (1894, Rev. Ed. 1928). Penck particularly dedicated itself and raised the Viennese school of physical geography to international meaning. Following the footsteps of his father, Walther Penck was also an outstanding geomorphologist. Walther Penck was born in Vienna, Austria. From an early age he knew would pursue a career in earth sciences. He became a geologist in 1912 and was offered a professorship in geology at Leipzig University. Research was the lifeblood of Penck and he studied all over Europe. He was offered many other lucrative positions but refused them on the account that it would prevent him from continuing his research. Walther Penck's life was cut extremely short when he died of cancer in 1923. After his death, his father, Albrecht Penck, gathered together his research and published the book '*Morphological Analysis of Landforms*'. Walther Penck contributed considerably to the field of geomorphology and more specifically to the idea of slope development. The main opponent to Penck's view of landform development was William Morris Davis. The theories and ideas of Walther Penck have provided many debates concerning the development of slopes. During 1920s the ideas of Davis had already captivated many in the field. Penck disagreed strongly with Davis' idea that land is rapidly uplifted, followed by a prolonged period of no tectonic movement during which erosive features wear down the uplifted land to a flat surface called a peneplain. Penck believed that the uplift of the ground was continuous and long-lived. He then assumed that through erosive powers, such as rivers, the slope would form a concave, convex or straight profile depending on the intensity of the tectonic activity of the area. After the formation of a certain slope type, Penck ascertained that in some cases a rather low-lying, featureless plain would form called a primarrumpf. Penck's ideas were very controversial and, thus, were never extremely popular. But, the physical geographer who emerged after the death of Penck became highly popular among geographers. He was Alfred Wagener, who gained popularity among geographers for postulating the theory of Continental Drift. Born in Berlin, Alfred Wagener (1880-1930) was a climatologist and geophysicist. He studied natural sciences at the University of Berlin, receiving a doctorate in astronomy in 1904. He did not pursue a career in astronomy, however, but turned instead to meteorology, where the telegraph, Atlantic cable and wireless were fostering rapid advances in storm tracking and forecasting. In 1905, Wegener went to work at the Royal Prussian Aeronautical Observatory near Berlin, where he used kites and balloons to study the upper atmosphere. He also flew in hot air balloons; in 1906 he and his brother Kurt broke the world endurance record by staying aloft for more than 52 hours. Then, Wegener was invited to join a Danish expedition to Greenland's unmapped northeast coast in1906 itself. During this expedition Wegener became the first to use kites and tethered balloons to study the polar atmosphere. When he returned to Germany, Wegener's Arctic research earned him a position at the University of Marberg where, beginning in 1909, he lectured on meteorology, astronomy, and "astronomic-geographic position-fitting for explorers." Both students and professors were impressed by the clarity of the young meteorologist's thinking, by his ability to explain difficult concepts in simple terms, and by the intuitive leaps of his nimble mind. In 1911, still only 30, Wegener collected his meteorology lectures into a book, *'The Thermodynamics of the Atmosphere'*, which soon became a standard text throughout Germany. In 1912, Wagener carried on another expedition, and this time to Greenland. During this expedition, he collected volumes of unique scientific data. The resulting publications established him as one of the world's leading experts on polar meteorology and glaciology. When he returned to Marberg, he worked on continental drift; marshaling all the scientific evidence he could find to support his theory. Using this pioneering interdisciplinary approach, Wegener wrote one of the most influential books in the history of science: *'The Origin of Continents and Oceans'* published in 1915. This work was one of the first to suggest continental drift and plate tectonics. He suggested that a supercontinent called 'Pangaea' had existed in the past; it broke up starting 200 million years ago and the pieces "drifted" to their present positions. He cited the fit of South America and Africa, ancient climate similarities, fossil evidence (such as the fern *Glossopteris* and *Mesosaurus*), and similarity of rock structures. Because of the First World War, Wegener's book went unnoticed outside Germany. In 1922, however, a third (revised) edition was translated into English, French, Russian, Spanish, and Swedish, pushing Wegener's theory of continental drift to the forefront of debate in the earth sciences. Wegener also noted that the continents move up and down to maintain equilibrium in a process called 'isostasy'. He also noted that when the continents of Africa and South America are fitted together, mountain ranges (and coal deposits) run uninterrupted across both continents, writing: It is just as if we were to refit the torn pieces of a newspaper by matching their edges and then check whether the lines of print ran smoothly across. If they do, there is nothing left but to conclude that the pieces were in fact joined in this way.

In spite of the fact that German School is famous for its physically-biased studies in geography, by the mid of 20th century, the School showed a gradual inclination towards the use of chorological viewpoint in the discipline. Interestingly, this development was parallel to the one going in other modern schools of geography as well, and more particularly in France and America. Two German scholars are known to have brought this change and they are Alfred Hettner and Otto Schluter. Alfred Hettner (1859-1941) was the geographer and a teacher. His methodology and his materialistic philosophy, grounded in the work of Immanuel Kant, had a great influence. He founded the journal *'Geographische Zeitschrift'* (1895). Hettner subscribed to a view of classification of sciences, placing geography as a science that was similar to Kant's view on the subject. Hettner noted that on the one hand we have a 'logical classification' of the fields of learning, according to which we have a series of subject-sciences, each concerned with the study and analysis of a logically defined circle of facts. On the other, we have a 'physical classification' of fields of knowledge according to which fields of study are defined not on the basis of logical unity of the subject matter concerned, but on the basis of the physical association of phenomena of diverse origin existing together. The physical association of phenomena may be viewed in two different dimensions: Time and Space. Accordingly, physical classification of knowledge gives us two groups of fields of study, namely, 'historical sciences', which study integrations of diverse phenomena occurring together in the context of time; and 'chorological sciences' which study diverse phenomena existing together in segments of the earth's surface. Hettner has put this idea forward in the first of his methodological essays published in the inaugural issue of his journal *Geographische Zeitschrift*. Hettner vigorously encouraged the development of regional studies through publication in the *Geographische Zeitschrift*. Under the leadership of Hettner, chorology became the guiding principle of Geography in Germany, and soon all over the English-speaking world[6]. An alternative view was built around 'the concept of landscape' by Otto Schluter (1872-1952). Schluter suggested that geographers should concentrate on the study of phenomena on the surface of earth that could be perceived through the senses; and the focus should be on the totality of perception in each area. This totality of "visual perception of area" was termed as landscape. According to him, acceptance of landscape as the subject matter of geography had raised the discipline to the level of the other logically defined fields of science. Schluter incorporated the concept of process, i.e. development through time, in his concept of geography as the science of landscape by means of the derivative concepts of cultural and natural landscapes. He called them as 'Kulturlandschaft' and 'Naturlandschaft'. For him the purpose of the study of landscapes was "not only of classifying categories of phenomena and determining their distribution and associations, but of examining their characteristics through the process of change through time." The concept of landscape and the related methodological principles of Schluter have also been widely used by geographers other than German before the Second World War.

Thus, after the death of Humboldt and Ritter in 1859 till the First World War, the emphasis of geographical studies kept on shifting in Germany. Although this shift was evident in the form of Physical and Systematic (Peschel, Gerland), to Physical and Regional (Richthofen, Penck), to Systematic and Human (Ratzel), to primarily Regional (Hettner, Schluter) and predominantly Physical (Wegener, Penck) during the post-Classical phase of German School of Thought, the physical bias was very strong in most of these works. Altogether the contribution of German school goes to three major fields of geography, viz. Physical Geography, Regional Geography and Human Geography. As far as the literary works are concerned a special mention may be made of *Geographia Generalis* (Varenius); the *Cosmos* (Humboldt); Die Erdkunde (Ritter); and Anthropogeographie and Politische Geographie (Ratzel). Besides, Hettner is credited to begin the publication of a geographical journal named '*Geographische Zeitschrift*' since 1895 onwards. The concepts of Determinism, Geopolitics and Lebensaraum were also put forward by Germans thereby adding new terms to the vocabulary of Geography. On the negative side, Geography also became dualistic in nature during this phase of development. Two forms of dualism were found rooted clearly in the works of Post-Clasical period German scholars. They were 'Systematic *versus* Regional' and 'Physical *versus* Human'. To overcome this separation of the field into distinct directions, Hettner regarded it necessary to make chorological concept dominate. Consequently, the geographical development in Germany largely showed a tilt towards chorology (the regional/landscape studies) in the late 19th and early 20th century.

FRENCH SCHOOL

France entered the modern era shortly after Germany. While the German tradition quickly filled with new scholars, the French owe much of their tradition to one man, Paul Vidal de la Blache. Blache is credited for establishing an entire generation of geographers, who made valuable contributions to various fields of geography. The awakening of interest in geography in France dates from the country's losses in the Franco-Prussian (German) War in 1870-01[7]. Smarting under this defeat in Europe, France sought compensation in new lands in Africa and South-East Asia. Besides the Colonial expansion of France in African and South-East Asian countries after the Franco-German War, there were two other reasons for the evolution of French School of Geographical Thought, viz. the impact of Darwinian Evolutionary thought and big advances in the study of human societies. The newly founded geographical societies in the provinces of France were the leaders of the movement in favour of colonial expansion. It caused the people to be interested in the discipline of geography. It was out of this background that in the last quarter of 19th century there emerged a new school of modern geographers in France under the leadership of Blache. But, geography could not emerge as an independent discipline in France till the end of 19th century. It is because Geography in France in the latter half of 19th century was in the hands of few historians, who taught geography merely as a background to history. Moreover, there was a division between the faculties of Science and Letters (Humanities), so that geographers who were in the Faculty of Letters (Humanities) were not allowed to give lectures on physical geography, which was reserved to the Faculty of Science. Courses on regional, colonial and human geography were prominent in the Faculty of Letters (Humanities) and that on Physical geography in the Faculty of Science. However, it was due to the efforts of Blache that geography in France could get unified and there emerged new a (unified) geography in France in the beginning of 20th century. The beginning of 20th century from this point of view may be considered the watershed between the Old and the New Geography France. Up to that time, the lectures were given either by historians, or were devised as subordinate to the teaching of history. In such circumstances, geography was considered as mere description of the environmental influence upon historical development. But, in the 20th century geography emerged as an independent field in France; and after the First World War (1914) it was well represented in all the universities and nearly all the professors were pupils of Vidal de la Blache.

Contribution

The growth of geography in France, unlike that in Germany with its several distinct schools of thought, has been shaped by the work of one man. For about 4 to 5 decades in the late 19th and early 20th century, geography in France was largely a one-man-show. Vidal de la Blache (1845-1918) was the leader of French School of Geographical Thought. For generations his school was dominant and nearly all the professors and the occupants of chair in geography were either the pupils or the pupils of Blache. No country of comparable age, population, universities and resources has produced such a galaxy of genius, as did France. They altogether have made outstanding contributions to geography primarily in three ways: (i) to the status of discipline, as an integrating link between the humanities and natural sciences; (ii) to the scope of subject, i.e. towards its consequential acceptance in the higher studies as an analytical field and to the use of geographers in the governmental work; and (iii) to the development of its inner framework, particularly through the research in regional, human and historical geography. Blache dominated the scene for nearly 4 decades. The first half of his career covers the last quarter of 19th century when he was a junior contemporary of Richthofen, Ratzel and Wagner, and learned much from them. He was particularly influenced by the second volume of *'Anthropogeographie'* (1891) of Ratzel. The latter half of Blache's career was in the first quarter of 20th century and he was then virtually contemporary with Penck and Hettner. This period also saw the florescence of his school with the publication of massive doctorate studies by his pupils. This tradition also continued even after his death. In the first quarter of 20th century there emerged the first generation of scholars with the inspiration of Blache. Distinguished scholars of this phase included Auguste Himly, Elisee Reclus, Franz Schrader, Lucien Gallois and De Margerie. The outstanding students of Blache who dominated the professional scene in second quarter of 20th century make up the second generation; they include Jean Bruhnes, Camille Vallaux, Albert Demangeon, Em de Martonne, Andre Seigfried, Raoul Blanchard, Jules Sion, Henri Baulig, Rene Musset, etc. The younger scholars as Andre Cholley, George Chabot, Roger Dion and Pierre Deffontaines emerging after Second World War belong to third generation. The fourth generation represents still younger geographers as Jean Gottmann, B. Garnier, Pinchemel, Pierre Vilar, etc. Thus, the influence of Blache penetrated right into the late 20th century. Geography, in France, had a subsidiary position. Blache proceeded to direct geography away from its stultifying place in historical studies. His approach to geography was empirical. He encouraged detailed field studies and their publication as substantial monographs in the *'Annals de Geographie'*, which he founded. Under his vigorous leadership, various university departments undertook field studies. In fact, after Ritter, it was Blache who re-emphasized the regional studies and synthetic approach in geography. Blache put forward two inter-related viewpoints in geography, viz. 'Possibilism' and 'Genre de Vie'. Emm de Martonne collected all the major works and articles of Blache in the form of a book *'Principles of Human Geography'*, published in 1921, after the death of Blache. Blache drew attention to the need for detailed regional studies to elucidate the influence of the rich variety of factors—physical, historical, political and economic—

which affect the existing pattern of an area. From detailed regional studies could be built up a synthesis of general and comparative studies, wherein the causes and consequences of factors would be clarified. This later led him to plan the great 'Geographic Universelle', the best complete study of the world on a regional basis available in any language. Finally, Vidal de la Blache encouraged a number of students to take up regional studies. With the inspiration of Blache, there began the publication of a series of *'Regional Monographs'* in France. This trend continued even after his death and included regional studies not only of France, but also of its colonial territories and other parts of the world. The major thrust of the work of Blache was on the study of man-environment relationship and from the point of view of man. We know him as a great human geographer.

Among other scholars the contributions of Auguste Louis Himly, Élisée Reclus, Franz Schrader, Lucien Gallois, Jean Bruhnes, Camille Vallaux, Albert Demangeon, Emmanuel de Martonne, Andre Seigfried, Raoul Blanchard, Henry Baulig, Jules Sion, René Musset, Pierre Deffontaines, Roger Dion, Pierre Vilar Jean Gottmann and Philip Pinchemel need special mention. Auguste Louis Himly (1823-1906) was a historian and geographer. He wrote, in two volumes, his remarkable *'Histoire de la formation territoriale des etats de l'Europe centrale'*, in which he showed with a firm touch, the reciprocal influence exerted by geography and history. Francois Jean Daniel Schrader (1844-1924), better known as Franz Schrader, was a geographer, mountaineer, cartographer and painter. He contributed to the understanding and mapping of the Pyrenees. While devoting most of his spare time to long hikes in the mountains, during which he collected thousands of data records for its topography, he still used to find time to paint landscapes, including the Pyrenees Alps. In addition to the immense work that left us this great geographer, the scientific committee today that Franz Schrader had established in the French Alpine Club still exists, and the Society of Painters Mountain—Paris. Lucien Louis Joseph Gallois (1857-1941) was a student at the Ecole Normale Superieure in Paris, where he took classes from Paul Vidal de la Blache. Gallois made major contributions to the *Annales de geographie*, a geographical journal that was founded with his mentor, Blache, and after the latters' death in 1918, he assumed directorship of Geographie universelle, a major project involving regional geography of the whole world. Gallois had a keen interest involving the history of geography and cartography, as made evident by an influential 1890 study on German geographers of the Renaissance titled *Les geographes allemands de la Renaissance*. Another noted publication of his was *Regions naturelles et noms de pays: Etude sur la region Parisienne* (Natural Regions and Country Names: A Study of the Paris region). Jean Bruhnes (1869-1930) was a leading exponent of systematic (as opposed to regional) geography in France. Bruhnes is well known because of the English translations of the works of Blache on Human Geography. Bruhnes elaborated a system of "Human Geography" with its raw material in "the visible and tangible facts" of the human activity on earth's surface. He has called these the 'essential facts' of the human geography. Similarly, if one adds the physical facts of the landscape—surface configuration, vegetation, soils in their surface expressions as forms of terrain, then we have 'the essential facts of geography'. The essential facts of geography are divided into three: (i) Facts of unproductive occupation of the soil (houses, roads, settlements, etc.); (ii) Facts of plant and animal conquest (e.g. cultivation of crops, animal husbandry, plantations, etc.); and (iii) Facts of destructive economy or exploitation (i.e. negative use of land, e.g. clearing of forests, hunting and mining, etc.). These were examined in 'regional studies' to show the interdependence of these phenomena in contrasted physical and social environments. According to Bruhnes, study of these three sets of observable phenomena does not set the limit of geography, a fact that is not generally appreciated. Beyond the essential facts, but in association with them, is an area of wider study. This is described as "the geography of history", and it embraces five aspects: (i) The distribution of population (static conditions and migrations); (ii) Geography of work or economic activities (related to production, transport, exchange, etc.); (iii) Geography of political control or power of man (territory, routes and frontiers, groups of states); (iv) Geography of civilization or social geography (nationalities, races, languages, religions, intellectual, artistic and technical phenomena) and social attitudes (mentality, organization, judicial, social) where these are associated with essential facts; and (v) Regional Geography (i.e. a synthesis of all the above). Camille Vallaux (1870-1945) was an eminent geographer and pioneer of oceanography. The originality of Vallaux is to have directed its work towards the geography of sea and oceans. Various publications that go to his credit include: *Low Brittany, Study of Human Geography* (1907), Thesis of Doctorate. Republished in (1980); *Seas and Oceans* (1932); and *General Geography of the Seas* (1933). Albert Demangeon (1872-1940) was an outstanding promoter of regional and human geography. He wrote great number of articles in the *Annales de Géographie* and also collaborated in the sprouting of the magazine *Annales de Economic and Social Historia.* Demangeon was highly influenced by Blache to whom he inherited his inductive and historicist methodology. Jean Gottmann and Pierre Vilar were his disciples. His best-known works include *Le Déclin de L'Europe* (1920), *L'Empire britannique* (1923), and *Les Iles britanniques* (1927). Emmanuel de Martonne (1873-1955) was one of the leading physical geographers of 20th century. He was highly influenced by the American geographer W.M. Davis. Topically, he concentrated on geomorphology, especially the processes of mountain glaciation, peneplain development, and hydrography. As a regional geographer, he contributed numerous studies: on districts in France, Rumania, the Alps, Central Europe, etc. But his most important legacy was his lifelong project, the *Traité de Géographie Physique.* This first appeared in two volumes in 1909, and had expanded to three volumes (including a separate one for biogeography) by the fourth edition of 1925-27, and eventually reached a tenth edition by the time of Martonne's death. Andre Seigfried (1875-1950) was the French political scientist and educator. Siegfried's works were dedicated to the study of geopolitics of its time, to the history of countries like United States, England, New Zealand and France, and to economic and political geography. Raoul Blanchard (1877-1965) was a highly skilled and prolific writer with remarkable personality. He wrote a monograph on Flanders that earned him the chair of geography at the University of Grenoble, a position he held for 50 years. But, later the Alps became his privileged field of study and the object of several works. He established the Institute of Alpine Geography and founded its journal, the *Review of Alpine Geography*, in 1908.

He completed a classic 12-volume work on the French Alps in 1956. Henri Baulig (1877-1962) was a French specialist in geomorphology. While following a course at the University of Harvard in 1904, he came under the influence of the American geographer: William Morris Davis, the founder of geomorphology. Since Baulig was mainly interested in physical geography, he adopted the approach of his master, the genesis of form and its relationship to the cycles of erosion. Pierre Deffontaines (1894-1978) was a disciple of Jean Brunhes and is mainly known for his work in human geography. He founded the Institute of Geography of São Paulo and Rio de Janeiro. Roger Dion (1896-1981) and Pierre Vilar (1906-2003) were great historians of 20th century. Vilar was committed to the idea of history as a fundamental human science. Geography was his first calling, but his approach to the discipline was always historical. Therefore, he contributed regularly to the Annals of Lucien Febvre. Jean Gottmann (1915-1994) was a French geographer who was most widely known for his seminal study on the urban region of Northeast Megalopolis. His main contributions were to human geography and they were in the fields of urban, political, economic, historical and regional geography. His regional specializations ranged from France and the Mediterranean to the United States, Israel and Japan. Philipp Pinchemel (1923-2008) was a disciple of André Cholley. Faced with the evolution of contemporary geography to the social sciences, he had the desire to refocus the geography of what he calls the terrestrial interface. On this interface part two processes: humanize (or transformation of the natural environment) and spatial (or spatial organization in clusters, networks, political or administrative boundaries). *The Face of the Earth* that he wrote is the result of his intellectual journey, a reflection on the focus of geography. It is a manifesto for the unity of the discipline. Geography is defined as the study of writing on human societies the natural interface of the Earth, writing that reflects the geographic reach of men, writing complex made of lines, points, surfaces, shapes, volumes and colour. From there, Pinchemel developed a general conception, ambitious in geography: access to "the intelligence of the terrestrial interface". For him, discipline is knowledge, action and thought. Philippe Pinchemel made known the great works of the "new geography" by the early 1970s. They were *"The Spatial Analysis in Human Geography"* by Peter Haggett and *"The Geography of Markets and Trade Detail"* of Brian Berry. From the 1990s, Philippe Pinchemel published the collection of the Works Committee Historical and Scientific Works of Geographers (often forgotten or ignored by the scientific community) contributing to clarify the basis of geography (epistemological and historical) and testifying to its will continue to define its "essence".

The French school largely contributed to human studies with regional approach, particularly emphasizing on the historical, economic, political, colonial and electoral aspects. Besides, the modern French scholars are also credited to have put forward three interrelated doctrines, viz. 'Possibilism', 'Genre de Vie' and 'Paysage'.

Regional Geography

Mainly Vidal de la Blache, Gallois and Martonne have produced a large number of regional monographs, particularly on French regions. French regional geography is scientific and is the product of the synthesis of geological and historical schools. Contribution to regional studies has been made through the works like *Geographie Universelle* series, *Bibliographie Geographie Internationale* (founded by Blache and known as *Annals de Geographie*), and Paris Geographical Society (founded in 1921 and the oldest in the world). Blache insisted that the geographical research and training should concentrate on the study of particular areas—small and accessible enough for thorough study in the field. By region he did not mean the clearly bounded area in space as a frame for areal description. Instead he classified it to be an area selected on the basis of some homogenous (similar) characteristics and where there is an inter-dependence of areal phenomena.

Human and Historical Geography

The French have excelled in the study and development of Human Geography. In this, as in regional, they have brought to their geographical work a fine literary style and a firm historical background that is particularly important in this aspect of geography. Here again the work of Blache is pre-eminent. His *Principles de Geographie Humaine*, is regarded by geographers as a classic. Ratzel was used as a springboard in the thinking of social scientists in general. Blache and his colleagues developed from Ratzel's "geographical determinism" what began to be called "geographical Possibilism". This field of human geography as founded by Blache was fully discussed and its limitations were pointed out in an outstanding work by Lucien Febvre (Professor of History in the University of Strasbourg) in a work published in France in 1922, translated into English and published in 1924 as *A Geographical Introduction to History*. This book is a critical reflection of the Vidalian concept of human geography and is an important work in the history of geography. Another early writer in human geography was Jean Bruhnes, a pupil of Vidal de la Blache. Bruhnes published his *Geographie Humaine* in 1910; it was revised and enlarged in 1912, 1925 and 1934. The book was published in America in 1920 as *Human Geography*, being translated and edited by I. Bowman, R.E. Dodye and T.C. Le Comple. This book was the first large-scale attempt in France to examine fully the geographical facts in human economy. He has described and analysed the patterns of man's occupancy of the earth as expressed in housing types in villages and town sites, their shapes and development and changing forms of communication. He has also discussed man's conquest and adaptation of the vegetable and animal kingdoms and man's destructive or robber economies of mineral extraction. All these matters were examined universally as a world picture and in a comparative manner.

Albert Demangeon (1872-1940) and Deffontaines were other outstanding human geographers of France. By the time of his death Demangeon had become the most distinguished exponent of the field. He set himself two aims, first to produce a manual of Human Geography, and secondly, to write an Economic Geography of France. One of Demangeon's greatest interests was Rural Settlement in France. On the question of settlement, he regarded the degree of concentration or dispersion as the vital matter. His work attracted great attention at the International Geographical Congress at Cairo

(1925), Cambridge (1928) and Paris (1931). Demangeon defined Human Geography as the study of human groups and societies in their relationships to the physical environment. He laid emphasis on the work of man in modifying the environment by means of communications, artisan wells, the control of rivers, and the evolution of new plants for human food. If one considers the principle behind his studies, he rejected crude determinism. A further examination of the nature of geography is given by Deffontaines, a collaborator of Bruhnes. One aspect of the study is the struggle of man against his raw environment and the other aspect analyses the results achieved and classifies the types and limits. Thus, on the one side are man's struggles against deserts, mountains, forests, seas, rivers, and on the other side the geography of settlement, transport, and colonization (i.e. winning the harsh environment). Deffontaines differentiated two types of regions of human occupancy; one in which there is an individual pattern of human economy and the other where through intensive and long contact present day life is the result of many complicated influences.

Political and Colonial Geography

Andre Siegfried and Jean Ancel have brilliantly promoted political geography. Siegfried has been most successful in securing recognition of the value of geography among economists and politicians inside and outside France. His lectures on the political and economic geography of the Great Powers are regarded as outstanding in the French academic world. His books on England, the United States, the Suez and Panama Canals and the Mediterranean have been highly successful among geographers and general readers in the world and have had many translations. Seigfried has indeed interpreted to the general reader the significance of the geographical factors. Jean Ancel, a pupil of Demangeon, produced studies on the political geography of the Balkans and Central Europe. Ancel was a staunch opponent of German "geopolitics" as developed by Haushofer.

Among other fields, it is Colonial Geography which French have made their great contributions. The interest in Colonial Geography was stimulated particularly after the Franco-Prussian War. Demangeon best expresses the French conception of Colonial Geography in his book *"L' Empire Britannique"* (1923, with English translation 1925 and German 1926). Here he explains that it is not merely a question of studying the regional geography of the colonies, for that is part of the normal work of the regional geographer. Rather, it is a study of the geographical effects of the contact of two types of peoples associated by the fact of colonization.

Physical Geography

Since Geography first evolved in France from history, the whole emphasis of its trend has been with the humanities. The development of Physical Geography was slower and was effected by geographers taking further training in geology or by geologists becoming interested in surface forms. It is Emm de Martonne who was the first French geographer to concentrate on Physical Geography. He has produced one of the fundamental geographical texts for geographers the world over, namely *'Traite de Geographie Physique'* (1909). This classic work has been translated into many languages. Martonne also brought to completion in 1943 the magnificent *'Atlas de France'*. Among the other French physical geographers Blanchard needs a special mention. Blanchard, apart from studies in Human and Regional Geography, to which reference has already been made, also did lengthy work on the French Alps; and distinguished himself in analyzing the Central Massif, particularly in its several erosion and peneplain surfaces. He also tried to work out the consequences of physical factors upon man's economy.

The Doctrines of 'Possibilism', 'Genre de Vie' and 'Paysage'

With the union of regional and human approaches, Blache was able to put forward the doctrines of 'Possibilism' and 'Genre de Vie'. As a proponent of 'Possibilism', Blache considered that the main task of geographer is to correlate the physical and human conditions on the basis of their spatial inter-relations. He viewed this man-nature relationship from the point of view of man and, thus, opposed the doctrine of 'Environmental Determinism'. The main theme of 'Possibilism' is that the physical surroundings provide a range of possibilities to man and man makes a choice. This choice depends on his needs, aspirations and capacities. Secondly, man establishes relationships with nature not as an individual, but as a group or society. In other words, man views this relationship through the prism of culture. 'Culture' here includes beliefs, institutions, traditions, attitudes, religion, language, diet, habits, customs, skills, technology, etc. In this way, man creates his own habitat, the local modified environment, within which the man lives and acts. This habitat is a distinct landscape created because of man's occupancy and usage. This landscape has been termed as *'pays'* at micro level and *'paysage'* at macro level in French by Blache. Thirdly, with time and cultural advancement the level of interaction of man and environment increases. This brings changes in human life-styles as well. Thus, the possibilistic approach to the study of man-environment relationships demands a three-fold interpretation, i.e. in terms of 'spatial', 'cultural' and 'historical (temporal)' perspectives. In other words, the present ways of life of a society, its cultural, socio-economic and overall development, is the result of the history of its occupation of the physical surroundings. Lucien Febvre, a friend of Blache and also a notable historian of his times, supported the conceptual framework of 'Possibilism'. While Blache recognized geography's relatioship to the sciences, both natural and social, neither was as important to geography's identity as its connection with human activity. He stressed the importance in viewing man in relation to his natural environment. As a regional geographer, he preferred to view man within a physical milieu, or 'pays', or within his cultural milieu, or 'genre de vie'. A 'pays', accordingly, is a natural region with some homogeneous physical characteristics, such as the mountainous Massif Central region in France. A 'genre de vie', on the other hand, is a way of living or local culture. It includes the traditions, institutions, language, habits, foods, etc. of a people. In other words, when ones studies geography regionally, one may identify a region with regard to both, its physical characteristics and its human characteristics. According to determinists, the human response to environment is the same in similar physical circumstances. Contradicting this viewpoint the possibilists believe that even the similar physical circumstances can generate different life-styles (*Genre de Vie*) depending upon the culture, economy and system of occupancy of land. The nature sets limits and offers possibilities for development, but the way man adjusts to natural conditions of the area of his inhabitancy is largely a function of his own tradition and mental structuring. Therefore, the same environment carries different meanings to people with different *Genres de Vie* (ways of life or culture). According to Blache, culture (inherited trait) is the basic factor in determining which of the many possibilities in natural environment should be selected or exploited by a given community. He drew attention to the fact that the complex of institutions, traditions, attitudes and technical skills which make up a *Genre de Vie*, largely determine which of the many possibilities offered by nature in any given area shall be selected for development. Blache was opposed to the concept of dichotomy between 'natural' and 'cultural' aspects of earth's surface; instead, he considered them as the two ends of the same continuum. To him, natural and cultural aspects of earth's surface cannot be separated from each other. It is because, he reasoned, in every inhabited part of the earth's surface, the original landscape is significantly transformed as a result of human habitation. Such changes are greater in the case of culturally advanced societies where, owing to the more developed technology, the degree of man's intervention in nature is more far-reaching (Dikshit, 1997, pp. 85-86, 92-93, 245). It is, therefore, impossible to study landscapes meaningfully without due reference to the interlocking roles of the nature and culture. The relationship of a community to the physical landscape of the area of its inhabitancy is so intimate that it is difficult to think of the one without the other. As such, each segment of the cultural landscape, with the passage of time, acquires a unique personality of its own. Such areas were named as *'pays'*. According to Blache, the study of such regions constitutes the primary task of geography as a professional field. Thus, in his scheme of geography, Blache asserted that human occupancy creates homogeneity even where natural features are not homogeneous. The 'cosmopolitan urban culture' may be cited here as an example.

Blache sought to establish geography as a distinct discipline. Its field of study is the "ensemble of phenomena that occurs in the zone of contact of solid, liquid, and gaseous masses that make up the planet" (Recent German scholars refer to this as the 'geosphere'). These phenomena are studied in relation to 'place, localization and distribution'. Thus, the field of study has a double aspect—Nature and Man. To explain their expression on the earth's surface one needs to borrow from both the natural sciences and the humanities, especially from geology and history. He outlines the distinctive characteristics of geography and claims that study has its goal in the characteristics and groupings of the phenomena of the landscape as the expression of man's presence and works in it. Blache lists the distinctive concepts of the discipline as: (i) Geography is a distinct discipline essentially concerned with the phenomena that occur in the zone of contact of solid, gaseous and liquid materials (geosphere); (ii) The field of geography has double aspects, i.e. nature and man and therefore, it borrows from both the natural sciences and the humanities; (iii) Geography recognizes the unity of the earth phenomena and is concerned with their inter-dependence and inter-connections; (iv) Geography seeks for scientific methods (as human, regional and historical approaches followed by Blache) of defining and classifying terrestrial phenomena; and (v) Geography recognizes the great role of man played in modifying his basic environment.

To conclude, the French School had been in the forefront between the two World Wars. The French geographers were read and appreciated outside the country and in social sciences as well. However, unlike in Germany, there did not exist any dualism between physical and human studies in France. Because, the French believed in only one kind of landscape or 'paysage', the cultural (human) landscape. *Principles de Geographie Humaine*, the *'Annals de Geographie'* (Blache) 1891 onwards; *'Atlas of Geography and History'* in 1894; *'Tableau de la Geographie de la France'* in 1903; *'Geographic Universelle'*; *"Geographie Humaine* (Bruhnes, 1910); and *'Traite de Geographie Physique'* (Emm de Martonne, 1909) are important literary contributions by modern French geographers.

AMERICAN SCHOOL

American Continent as part of the New World was discovered only in the 15th century. Geography as academic discipline was introduced by Europeans in American Institutes. The Harvard, Dartmouth, Mary, Yale, Columbia, Princeton and Pennsylvania were the earliest Universities imparting education in Geography. However, geography remained an unorganized discipline here at least up to mid-19th century. The 'American Geographical Society' was constituted in 1851, and thereafter the discipline started making steady progress. The initial development is credited to the scholars like P. Marsh, M.F. Maury, Samuel Forry, J.H. Coffin, Elias Loomis, William Ferrel, G.K. Gilbert, J.W. Powell, Agassiz, Arnold Guyot, etc. By 1880s, the influence of Darwin had become strong on American Geography. There was a shift from Ritterian teleological tradition to hypothetic-deductive based geography. Darwin's paradigm in conjugation with that of Davis's replaced all the Old traditions till the emergence of Landscape view of Carl Sauer in 20th century. The year 1880 may, therefore, be called 'the watershed' in the history of American Geography. The period thereon marks the beginning of professionalism in the discipline, giving it an independent status. The writings of W.M. Davis played a great role in bringing this professionalization in the continent. As a result more and more scholars were involved in geographical research, converting the subject into a discipline of advance learning and specialization and continued research. Just as W.M. Davis was the leading promoter of geography in the late 19th and early 20th century, so was Carl Sauer the dominant figure of the 2nd quarter of 20th century.

Davis sought to integrate the two elements, viz. Nature and Man. This he did after a period in which a marked dualism prevailed in the discipline, i.e. between exclusively physical geography and the geography of man. Davis found the key in the Darwinian evolutionary approach that regarded the facts of nature and man as existing in a 'harmonious relationship' in a continuous chain of cause and effect. Definite shifts of emphasis took place from the environmental approach in the 1920s. Now, the lead came from men trained in Chicago, notably Carl Sauer and Harlan Barrows. There was an eager search for an alternative to the so-called 'naïve environmental determinism' that tended to fell into a simple routine of thinking and writing; and a search for more effective methods of geographical analysis, once the goals were clearly set. The revised approach to land-man relations assumed various expressions—*possibilism* in France and in the form of *ecology* and *chorology* in America. Barrows claimed geography to be the 'ecology of man'. He shifted the notion of environmental influence to one of environmental adjustment, meaning thereby that man adjusts himself to and utilizes his environment, according to 'attitudes, objectives and technical abilities' of the group. Further, in focusing geography upon the life of man on earth, the physical aspects of the earth would only be investigated in so far as they were relevant to man's occupancy. Barrow's views expressed those of the Chicago School[8] in the 1920s and many geographers from that school were influenced by them. The most active group, however, emerged through contact with Carl Sauer. They rejected the concept of environmental influences as the central purpose of geography, and in its place advocated the study of 'landscape'. This 'landscape' was to be studied with respect to its forms, functions and patterns of the elements and their associations in area—whether as the expression of adjustments to the physical environment or as the cultural deposits of the group occupying the area. These were the views of the self-styled 'chorographers'. They formulated a field for geography based on areal characterization, rather than environmental relationship[9]. Their credo is expressed in the statement: "We conceive of the earth's surface as composed of a 'mosaic of spaces', differing in their surface impressions as exhibited through both natural and cultural features. To delimit these spaces, to describe and interpret their characteristics is the field of geography. By this definition, geography becomes a comparative study of the earth's regions." The physical bias in geographical studies is also clearly evident in the works of American scholars prior to Davis. This bias continued even in the thinking of Davis and his followers, with an inclination towards environmentalism. Since the times of Carl Sauer, a chorological tilt has replaced everything. However, the American writers after Sauer seem to be mainly interested in defining the nature of geography and its development over the years in general. Notable among these were Richard Hartshorne and Thomas Kuhn.

Contribution

The American scholars have ceaselessly contributed to geography since the beginning of 19th century to the end of 20th century. The prominent ones include, for instance, M.F. Maury, Louis Agassiz, Arnold Guyot, Samuel Forry, Elias Loomis, William Ferrel, John Wesley Powell, Grove Karl Gilbert, W.M. Davis, Ellen Churchill Semple, Ellsworth Huntington, Harlan Barrows, Carl Sauer, Richard Hartshorne, Thomas Kuhn, etc.

Of the earliest American scholars, M. F. Maury (1806-1873) truly deserves in terms of greatness and fame. He joined the Navy in 1825 and inspired by his naval profession, he began recording information on the ocean currents and winds. This information was the basis for many of his publications that earned him the title of *"Pathfinder of the Seas"*, Jean Louis Rodolphe Agassiz (1807-1873) earned him the title of *"Pathfinder of the Seas."* was a Swiss-born American zoologist, glaciologist, and geologist, and one of the first world-class American scientists. Louis Agassiz's primary work was on the classification of living and fossil fish—ichthyology. He also promoted the concept of a great "Ice Age" in the recent geological past. Agassiz was the first to scientifically propose that the Earth had been subject to a past Ice Age. On this theme he published the work entitled *Etudes sur les glaciers* ("Study on Glaciers") in which he discussed the movement of glaciers, their moraines, their influence in grooving and rounding the rocks over which they travelled, and in producing the striations seen in Alpine-style landscapes. A friend and contemporary of Agassiz was Arnold Guyot (1807-1884). Guyot's interests were in glaciology, physical geography, meteorology and cartography. His early studies on the flow of ice and the distribution of glacial erratics in Switzerland served to underpin the theory of glaciation that had been advanced and championed by his close associate, Agassiz. His main activities focused on hypsometric measurements of the eastern mountains from New England to North Carolina, on meteorology, and on the reform of geographic teaching in colleges and secondary schools. Guyot made fundamental contributions to the study of geology, glaciology, and meteorology of two continents, namely Europe and America. Samuel Forry (1811-1844) was basically a physician. He worked in the U.S. army as surgeon and also practiced in New York City contributing many articles to the *New York Journal of Medicine*. He is the author of *"The Climate of the United States and its Endemic Influences"* (New York, 1842) and *"Meteorology"* (1843). Although Elias Loomis (1811-1889) was a professor of natural philosophy and not strictly a geologist, his contributions to atmospheric science and geomagnetism are important components in the development of the earth sciences. Loomis encouraged the use of the growing national telegraph system in the 1850s, especially as used by military to record weather conditions around the United States. He was the first to use these synoptic data to map the air pressure differences across parts of the country. On these maps the isobars define high pressure and low-pressure areas. Loomis, in a sense, was the originator of experimental as well as observational meteorology. William Ferrel (1817-1891), a meteorologist, developed theories which explained the mid-latitude atmospheric circulation cell in detail, and it is after him that the Ferrel cell is named. Ferrel demonstrated that it is the tendency of rising warm air, as it rotates due to the Coriolis Effect, to pull in air from more southerly, warmer regions and transport it pole ward. It is this rotation which creates the complex curvatures in the frontal systems separating the cooler Arctic air to the north from the warmer continental tropical air to the south. John Wesley Powell (1834-1902) is an important name in the American and world geomorphology of 19th century. He is well known for exploration of the Great Colorado Canyon (1869), emphasizing river erosion, for the theory of absolute and temporal erosion base, classification and terminology of landforms, drainage basins, genetic types of rivers and structural types of valleys. In 1879 he helped in foundation of Geological Survey and Land Office. Grove Karl Gilbert (1843-1918) was another important geologist of the late 19th century. He is known for his geological study on mechanics of river processes. But he also has merits for many new ideas, e.g. the theory on dynamic equilibrium of river profile and concept of grade, which were crucial in constructing of Davisian geographical cycle. His geomorphologic inventions were also the explanation of many other aspects of geomorphology, as for example, the up warping of the shore-lines of ancient lakes, the study of climatic influence on land forming, the law of uniformity of the slopes, the laws of declivities and structure, the law of divides, the idea of lateral planation and origin of pediments, etc.

All the above scholars who wrote prior to W.M. Davis (1850-1934) covered numerous aspects of physical geography in their studies, ranging from meteorology, oceanography, glaciology and ichthyology. However, the influence of Davis became stronger than anyone else in the late 19th and early 20th century in America. The writings of Davis, in turn, do bear the influence of his predecessors, and more specifically of Powell and Gilbert. William Morris Davis was an eminent geologist and geomorphologist. Davis was very excited about the discipline of geography; and he worked hard to increase its recognition. He is also known as the founder of modern geography in America in the late 19th and early 20th century. Davis is often called "the father of American Geography". He is responsible for enlarging the scope and systematizing the study of geography. Publishing over 500 works, Davis was certainly one of the greatest academic geographers of the century. He is not only responsible for that which he accomplished during his lifetime, but also for the outstanding work done across geography by his disciples. In 1904, he founded the Association of American Geographers (AAG). The geographers like Mark Jefferson, Isaiah Bowman, Ellsworth Huntington, Ellen Semple, etc. were the pupils of Davis. W.M. Davis brought professionalization in geography. The writings of Davis, particularly on physical and general geography, bear an imprint of N.S. Shaler, Agassiz and also Powell. From Shaler, he was directly influenced, as he was his contemporary and also his senior. N.S. Shaler is often regarded as a geologist by training but a geographer by instinct. Shaler transmitted to Davis the vision of the earth as the resource base on which the human habitants are dependent. Moreover, Davis is believed to have acquired the habit of careful observation from N.S. Shaler when he worked with him as an assistant. From the works of Powell and Gilbert, Davis has respectively incorporated the concepts of 'Base Level' and 'Gradation' while postulating his famous theory on '*Cycle of Erosion*'. Davis' most influential concept was the "cycle of erosion". He worked on refining and detailing this concept for most of his professional career. Davis had great impact of Darwin's Organic Evolution Theory. The '*Theory of Cycle of Erosion*' is the most important contribution to Physical Geography by Davis. This revolutionary concept was later applied to the study of evolution of Glacial Landforms, Limestone Region, and Islands bordered by Coral Reefs as well.

A shift in emphasis is evident in the interest of Davis in the later years of his life, i.e. from physical geography to defining the nature of geography in general, by adding the human component. The core of geography, Davis restated, is the study of the relationship between the physical environment (the control) and the human behaviour (the response). This kind of explanation represents the influence of deterministic thinking that that predominated in the first decades of 20th century.

Davis sought to classify the facts of man in reference to the facts of the physical earth, calling them respectively 'Ontography' and 'Physical Geography'. The physical earth required to be studied by physical geographers with respect to origins and processes. The resultant forms of land, water and air were to be examined by the geographers in relation to the 'human responses' and 'controls'. The human facts on the earth were called '*Ontography*' as opposed to the facts of the physical earth (physiography). This was an extension of the evolutionary interpretation of man on the earth in so far as human characteristics show direct results of geographical (physical) influences and controls. Davis believed that such a system would form the framework of an integrated science. This viewpoint very clearly set the pattern for the pursuit of man-land relationships on a deterministic basis. Davis identified '*Ontography*' as the other half of geography, and defined it as 'geographical description of life forms on the earth'. This '*Ontography*' (study of human groups, plants and animals), according to him, demands the appraisal of adjustment to the physical earth. This leads to migration (mobility, movement) and segregations (assemblages, groupings) over space: a view that is close to Ratzel's work.

The end result is: different regions, landscapes. Davis also recognized the importance of regionalization of the phenomena on the surface of earth. He stated that this regionalization of the earth phenomena is the product of three forces, viz.: site-based, migratory and associative (combinations, segregations). Regional Geography, he wrote, 'seeks to describe the geographical elements of a given area in totality as they exist together in their natural combinations and correlations'. When human responses are to be studied, regional approach is more meaningful. Davis envisaged the field of regional geography as the one elucidating the modes of areal associations of phenomena on the surface of earth. He considered regional geography synthetic, so that all the geographical elements and activities occurring, organic and inorganic shall be appreciated in their true spatial relations. There are many methods of geographical presentation, but it still remains true that pure regional geography is the final object of a geographical efforts. Then, the larger regions are the assimilations of smaller regions or landscapes. Further, Davis insisted that regional descriptions must be homologous, i.e. all related aspects—land, climate, vegetation, animals and man—having equal importance or emphasis in the study. Thus, two important trends were noticed in the writing of W.M. Davis, viz. *Geomorphic* and *Geographic*. Davis emerged as one of the greatest geographers of the late 19th and early 20th century. His works had profound influence not only in America, but also in European scientific thought, particularly in Germany and France. He set forth a tradition, which continued to dominate the American geographical scholarship for several decades. His ideas on geomorphology remained largely unchallenged for a long time in America and outside. His '*Ontography*' paradigm, which was subjected to criticism in the earlier period, developed into a major paradigm (i.e. ecological) in the late 1960s, when the emphasis of geographical studies was attached to the ecosystem as a geographical principle and method.

The important paradigm of geographical studies in the early decades of 20th century was 'environmentalism'. A strong physical bias was followed in evaluating man—nature relationships or any study emerging out of this interaction. Davis himself has propounded this kind of approach most lucidly initially. Besides Davis, the contemporary thinking was also dominated by the ideas of Ellen Churchill Semple (1863-1932), Ellsworth Huntington (1876-1947), Mark Jefferson (1863-1949), Harlan Barrows (1877-1960), and Isaiah Bowman (1878-1950). Ellen Churchill Semple was a very intriguing geographer and was the first woman to teach at Clark University. A follower of the German geographer Ratzel, Semple helped to develop the study of anthropogeography and lectured on the subject at the University of Chicago. Semple was an environmental determinist, as proved through her publications on *American History and Its Geographic Conditions* (1903), *Influences of Geographic Environment* (1911) and *The Geography of the Mediterranean Region* (1931). Ellsworth Huntington was a professor of Economics at Yale University in early 20th century. He is known for his studies on climatic determinism, economic growth and economic geography. He was devoted chiefly to climatic studies and anthropogeography. *The Climatic Factor* (1914), *Civilization and Climate* (1915, rev. ed. 1924), and *Climatic Changes* (1922) were among his works. He also wrote *Principles of Human Geography* (1934) and *Mainsprings of Civilization* (1945). The environmentalist approach was taken up in the diligent and far ranging researches of Huntington. He sought to measure the influences of physical environment on human activities and historical development. In 1934, Huntington formulated '*Principles of Human Geography*' on the basis of the study of nature and distribution of the relationships between geographical environment and human activities. He particularly examined the influence of climate and climatic change on men and the course of history. He linked natural and cultural or historical phenomena in causal relations. This book sold 85,000 copies in its fourth edition alone. Harlan Barrows was known for his book '*Geography as Human Ecology*', published in 1923. In American Geography, Barrows presented the concept of geography as human ecology for the first time. He stated that the physical environment does not cause man's adaptation to the conditions of his habitat. This adaptation, in fact, is the function of human choice. Barrows proposed human ecology as the unifying theme in geography[10].

Just as William Morris Davis was the main exponent of geography in the first quarter of 20th century, so was Carl Ortwin Sauer (1889-1975) a dominant figure in its second quarter. And just as the contributions of Davis can be appreciated in the light of prevailing philosophies, so also the appraisal of Sauer's work demands consideration of the background of thought among his contemporaries. Carl Sauer was appointed as the Professor of Geography in the University of Michigan, his first post. Thereafter, he moved to the University of California, at Berkeley and stayed there till his retirement in 1957. He remained a leading geographer and carried on research and teaching for about 40 years in America. He was conferred with the title of 'Professor Emeritus' in 1957. He is famous as a human geographer and a chorographer. During his three decade career at Berkeley, Carl Sauer led the field of cultural geography. Sauer's influence in the field became known as the '*Berkeley School*' of geography. The best-known work of Carl Sauer is '*Morphology of Landscape*' published in 1925. His other notable publications include: '*Ozark Highlands of Missouri*' (1920), '*Aboriginals of N.W.*

Mexico' (1935), *'Colima of Spain in 16th Century'* (1948), *'Agricultural Origins and Dispersals'* (1952), *'Cultural Geography'* (1947), and *'Land and Life'* (1963). Sauer's main interest was in the study of the impact of physical environment on man. Sauer insisted that instead of going into the field with a set of *a priori* principles concerning the effect of physical environment on man, one should seek to observe facts himself and draw conclusions later. In other words, he favoured the use of inductive approach and emphasized on the importance of field studies in acquiring knowledge. Therefore, he advocated for the formulation of *a posteriori* theories, in which the conclusions are drawn on the basis of the facts empirically observed and collected on the field. According to Sauer, geography relies on a purely evidential system, without prepossessions regarding the meaning of its evidence. "Geography is concerned only to establish the connections of phenomena in a visible landscape; and these connections are ones of spatial associations and not of some hidden causality." According to Sauer, the field of geography is based on areal characterization, rather than on environmental relationships, which he considered too general and wider field to be enfolded within the content of geography. This was the main philosophy of the school of chorographers, which Sauer actually belonged to. Sauer's Concept of Landscape served as an important paradigm in the second quarter of 20th century. Roots of the concept lie in German term *'Landschaft'*, which emerged in the writings of Otto Schluter. Its French equivalent is *'Paysage'*. *'Landschaft'* means "a piece of land with certain physical and human attributes". The term *'Landscape'* was accepted by many of Sauer's followers as the basic term in geography. Sauer has provided the detailed exposition of 'Landscape Concept' in his book entitled *"The Morphology of Landscape"* (1925). The views of Carl Sauer, related to this concept may be summarized as: Geography is a comparative study of the landscapes that comprise the earth's surface; Description in an essential part of geography, and is equal in importance to explanation; Man's activities, as such, are not of prime importance to the geographers but rather those material features, resulting from man's activities; Features of the natural environment give character to an area, independent of human and cultural influence; The human occupation and use of the natural area results in various forms, patterns and associations of phenomena; and Cultural forms and patterns are explainable in terms of historical antecedecents, racial characteristics, etc.[11] The real shift in American geography to chorological perspective as the basic paradigm for geographical work came only after Richard Hartshorne (1899-1992), who taught at the University of Minnesota (1924-40) and the University of Wisconsin (1940-1970). In 1939, there emerged the monumental work of Hartshorne, *The Nature of Geography* that synthesized what had been thought and written concerning the nature of geography. This book presented an authoritative account of developments in modern geography, and had soon become essential reading for graduate students in every geography department. From then on, the concept of geography as a chorological science became enshrined as the mainstream concept of the discipline, and it continued to be so until late 1950s[12]. In 1959 he published *Perspective on the Nature of Geography*, which provided the benefit of 20 years' further thought and reflection by the author. In 1989, on the 50th anniversary of the former work, the Association of American Geographers published *Reflections on Richard Hartshorne's*

The Nature of Geography. The American School of Geography was influential even after the Second World War, the time by all the major modern schools of geographical thought had come to a decline. The most notable writer of this phase was Thomas Kuhn (1922-1996) who wrote extensively on the history of science and developed several important notions in the philosophy of science. Kuhn is most famous for his book *The Structure of Scientific Revolutions* (1962) in which he presented the idea that science does not evolve gradually towards truth, but instead undergoes periodic revolutions which he calls "paradigm shifts." The enormous impact of Kuhn's work can be measured in the revolution it brought about even in the vocabulary of the history of science: besides "paradigm shifts," Kuhn raised the word "paradigm" itself from a term used in certain forms of linguistics to its current broader meaning, coined the term "normal science" to refer to the relatively routine, day-to-day work of scientists working within a paradigm, and was largely responsible for the use of the term "scientific revolutions" in the plural, taking place at widely different periods of time and in different disciplines, as opposed to a single "Scientific Revolution" in the late Renaissance.

Thus, looking back, various American scholars, from M.F. Faury to Thomas Kuhn enriched geography in several ways. In general, the writings of scholars in the early 19th century (Faury, Agassiz, Guyot, Forry, Loomis, etc.) showed an inclination towards physical geography. In the late 19th century this physical bias continued, but there also emerged a deterministic tilt, particularly in the works of Davis, Semple and Huntington. Thereafter, i.e. since the beginning of 20th century, the American Geography has been shaped by chorological perspective, a trend in consonance with the German and French Schools of Geographical Thought. The American scholars continued to dominate the geographical scene even after Second World War. More recently, the tilt of their work was seen more on defining the nature of geography as a science and explaining its course of evolution (Hartshorne, Kuhn). The fields of Climatology and Physiography have been immensely contributed by the American scholars. Besides, the theory of '*Cycle of Erosion*' and the '*Concept of Landscape*' also go to their credit.

Climatology

Maury, Forry, Coffin and William Ferrel contributed to climatology in general. M.F. Maury established the tradition of climatic studies in U.S.A. He was the one to produce the first map of the floor of North Atlantic Ocean. He is also credited with his generalized model of atmospheric circulation prepared in 1850. Following Maury, Samuel Forry prepared one of the earliest maps of U.S.A. showing the distribution of temperature by making use of isotherms. J.H. Coffin completed a wind-map of Northern Hemisphere in 1854. In 1875, the map was extended to include the whole world and explained the laws of atmospheric circulation. William Ferrel provided a mathematical explanation of the way moving bodies behave on a rotating sphere. His was the first explanation on the subject.

Physiography

The American school was physically biased since beginning and certain scholars as Powell, Agassiz, Guyot, Gilbert and Davis were very prominent in the field. Powell and Gilbert set forth new paradigms in physical geography. John Wesley Powell, who was also an explorer, developed the concept of 'Base-Level'. G.K. Gilbert, on the other hand, gave the concept of 'Grade'; i.e. equilibrium reached between slope, volume of water, velocity of flow and load of detritus. Davis, in his "*Cycle of Erosion*", incorporated both these concepts. Agassiz and Guyot further enriched physical geography. Both worked towards the development of the concept of Universal glaciation, and studied glaciers and the effects of glacial actions in producing distinctive kinds of landforms.

Geomorphic Cycle of Erosion

According to Davis the cycle begins with the uplift of mountains. Rivers and streams begin to create V-shaped valleys among the mountains (youthful stage). During this first stage, the relief is steepest and most irregular. Over time, the streams are able to carve wider valleys (maturity stage) and then begin to meander, leaving only gently rolling hills (old stage). Finally, all that is left is a flat plain at the lowest elevation possible (the base level). Davis called this plain a "peneplain". Then, "rejuvenation" occurs and there is another uplift of mountains and the cycle continues. Davis' theory was quite revolutionary and outstanding at its time and helped to modernize physical geography and create the field of geomorphology. Although the real world is not quite as orderly as Davis' cycles and certainly erosion occurs during the uplift process, Davis' message was communicated quite well to other scientists through the excellent sketches and illustrations that were included in Davis' publications. Davis claimed that the study of landforms in terms of 'the cycle of landforms' is a 'model', in common parlance, into which an endless variety of forms can be fitted. It directs attention to the 'visible facts of present landscapes, instead of diverting attention to the invisible conditions and processes of past time', which investigations belong, as Davis often emphasized, to the field of geology. Davis pleaded repeatedly for the 'explanatory description of existing landforms'. Past geological conditions, he argued, should be telescoped to a minimum of meaningful explanation, in order to reach the geographical goal which, he argued, is an understanding of landforms, and regarded even this, for the geographer, as a means of memorizing the surface varieties of relief (Dickinson, 1976, p. 196). The German scholars like Penck, Alfred Hettner and Passarge have criticized the work of Davis. But their criticisms are directed to different parts of his theoretical explanation and not the complete theory itself.

Concept of Landscape

Sauer considered that geography must find its entire expression in the 'landscape' on the basis of the significant reality of chorological relations. The facts of geography, he maintains, are place facts. Their associations in area create this 'landscape'. He rejects the terms 'region' and 'area' and adopts 'landscape' to designate such terrestrial associations. Thus, 'landscape' is an area made up of a distinct association of forms (physical and cultural). The geographers should understand its origin, growth, pattern (design) and functions. The design of a landscape includes both (1) the features of natural area; and (2) the forms superimposed over it by the activities of man, converting it into cultural landscape (or '*Kulturlandschaft*' as identified by the Germans). Originally the 'landscape' is a natural landscape, and later it turns into cultural landscape or human habitat. Thus, a landscape at a later stage is the creation and workplace of human beings. Various elements of this cultural landscape are related to its: (1) Habitation (type and grouping); (2) Production (land use and workshops); and (3) Communication (mobility). The relationships of human beings with their habitat are conditioned by their group habits and cultural habits. There is also a succession of these landscapes with a succession of cultures (e.g. Primitive-to-Ancient-to-Medieval-to-Modern). Hence, the idea of landscape may be defined in terms of both: its time relations and its space relations (i.e. in its spatial-temporal framework). It is a continuous process of development or dissolution and replacement. The landscape, therefore, changes continuously. The natural landscape is the assembly of natural features of an area that are relevant to human occupancy. Geography, Sauer claimed, therefore, is 'distinctly anthropocentric'. The study of natural landscape, however defined, is directed to understanding its significance as habitat. The landscape is studied in terms of the process of human activity in time and in area. The landscape elements are studied in terms of their form, function and pattern. These are interpreted in terms of the way in which they are associated in area as the spatial impact of cultural processes. In addition, Sauer considered landscape as an individual unit of area, i.e. geographical unit of study. The various forms, which characterize this landscape, therefore, may be identified, its structure can be recognized and its origin and growth functions can also be understood. Thus, Sauer represented geography as a science that finds its entrifield in the concept of landscape. Geographer systematically observes and records both the natural and cultural features, describes them in their associations and patterns, and seeks to explain the growth and development of the cultural panorama in its natural setting. He does not limit his study to search for relationships between man and earth. Rather, he shifts the emphasis from a study of relationships to a study of the material features, natural and cultural, within an area. The 'characteristics', 'associations' and 'patterns' of these surface forms, together with the bonds of relationships that connect them, are his chief lines of investigation. The region thus studied is viewed as a functioning organism composed of natural and cultural groups. Since, it is the observable material features that give chorographic individuality to any area on our planet, the chorographer does not directly include man's activities among his primary objects of study. When such activities express themselves in material forms, which can be observed and recorded, they (the forms) are then a part of the landscape, and as such open to geographic analysis. If the material features of an area are the 'primary facts' with which a geographer deals, then all of those other facts used to explain and interpret the observable, areal scene may be classed as 'secondary'. Thus, the geographic analysis is both 'descriptive' and 'explanatory' of the man-made landscape. There are no restrictions as to what these secondary facts may be. In spite of its popularity Sauer's Landscape Concept was a short-lived concept. Because, no clear-cut definition was provided for it. It was often confused with that of Regional Concept that continued to exist as an important paradigm in geography later.

To conclude, the American School of Geographical Thought, during the modern phase of its development, was primarily guided and dominated by the thinking of the scholars like W.M. Davis, Ellen Semple, Carl Sauer, Huntington, Barrows and Hartshorne. Four phases of change in the school may clearly be identified, viz. (i) prior to Davis in the third quarter of 19th century, when physical bias dominated the scene; (ii) the time of Davis and his associates towards the close of 19th century and beginning of 20th century when along with physical there emerged a deterministic slant in geographical studies; (iii) the period of Carl Sauer in the second quarter of 20th century when there developed an anthropocentric and chorological shift, a change parallel to other modern schools of thought; and (iv) the period after Second World War when attempts were made to assess the nature of geography in general. A heap of literature supports these developments including, for instance, *Etudes sur les glaciers* ("Study on Glaciers") (Agassiz, 1840); "The Climate of the United States and its Endemic Influences" and "Meteorology" (Samuel Forry, 1842 and 1843); *American History and Its Geographic Conditions*, *Influences of Geographic Environment* and *The Geography of the Mediterranean Region* (Ellen Churchil Semple, 1903, 1911 and 1931); *The Climatic Factor* (1914), *Civilization and Climate* (1915, rev. ed. 1924), *Climatic Changes* (1922), *Principles of Human Geography* (1934) and *Mainsprings of Civilization* (1945) (Ellsworth Huntington); '*Geography as Human Ecology*' (Harlan Barrows, 1923); '*Morphology of Landscape*' (Carl Sauer, 1925); *The Nature of Geography* and *Perspective on the Nature of Geography* (Hartshorne, 1939, 1959); and *The Structure of Scientific Revolutions* (Thomas Kuhn, 1962).

BRITISH SCHOOL

The development of professional geography in Great Britain took place much later as compared to Germany, France and America. Interestingly, the extension of knowledge of earth's surface through explorations and research appeared to have continued to be a major concern of British geographers and British Geographical Societies. A large part of the papers published in British geographical periodicals dealt with reports of exploration of the still relatively unknown parts of the earth. Too much of dependence on reports of voyages, explorations and expeditions in the 19th century prevented individual attempts to give the subject an organized and professional outlook. The British scholars themselves, therefore, could not contribute to geography as a discipline as such in its initial stage of modern period. It was in the early 20th century that geography started taking shape as an academic discipline. Such a beginning was made in the two oldest universities of Great Britain, Oxford and Cambridge. In the meanwhile, the Royal Geographical Society of London (established in 1830) took special initiative in the promotion of geography. Thereafter, geography began getting matured in its new academic discipline status gradually. However, the British school had already come into picture during the second half of 19th century. It was through the writings of Charles Darwin, a naturalist, who achieved lasting fame by proposing the scientific theory of natural selection and evolutionary change in species. This theory was proved not only a cornerstone in biology but also revolutionalized various other related fields, including geography. After Darwin, there emerged numerous scholars in the British school. More recently, the British school is known to have become very instrumental in bringing the quantitative and theoretical revolution in geography.

Contribution

Besides Charles Darwin, the other prominent scholars of British school included, for instance, Herbert Spensor, David Livingstone, Patrick Geddes, Halford J. Mackinder, Andrew J. Herbertson, Griffith Taylor, Sir Dudley Stamp, O.H.K. Spate and David Harvey. Most of these scholars were fellow of the Royal Geographical Society and their interests have been variegated, ranging from urban, social, political, historical, regional and physical geography and also land use planning. Charles Darwin (1809-1882) was born to a wealthy society doctor Robert Darwin in Shropshire, England. In 1825, after spending the summer as an apprentice doctor, helping his father with treating the poor, Darwin went to Edinburgh University to study medicine. However, his revulsion at the brutality of surgery led him to neglect his medical studies. He became active in student societies for naturalists. He became an avid pupil of Robert Edmond Grant[13], who pioneered development of theories concerning evolution by acquired characteristics. Darwin took part in Grant's investigations of the life cycle of marine animals and discovered evidences for homology, the radical theory that all animals have similar organs and differ only in complexity. Channellizing all his thoughts and observations on the subject, Darwin wrote his most famous book entitled *On the Origin of Species by Means of Natural Selection* (abbreviated to *The Origin of Species*). It was published in 1859 and established evolution by common descent as the dominant scientific explanation of diversification in nature. On the basis of this work, Darwin was made a Fellow of the Royal Geographical Society. Darwin's Theory that evolution occurred through natural selection changed the thinking of countless fields of study from biology to anthropology. Evolution by natural selection proved to be a significant blow to notions of divine creation and intelligent design prevalent in 19th century science. The British philosopher and sociologist Herbert Spencer (1820-1903) was another major figure in the intellectual life of 19th century. Spencer was the author of several works, for example, as *Synthetic Philosophy, First Principles* (1862), *The Principles of Biology* (2 vols, 1864-67), *The Principles of Sociology* (3 vols, 1876-96), and *The Principles of Ethics* (2 vols, 1879-93) etc. Spencer was one of the principal proponents of evolutionary theory and his reputation rivalled that of Charles Darwin. He was initially best known for developing and applying evolutionary theory to philosophy, psychology and the study of society—what he called his "synthetic philosophy". Although he wrote mostly about political theory in his works, he is widely described as the father of Social Darwinism[14]. Spencer often analyzed human societies as evolving systems, and coined the term "survival of the fittest". He contributed to a wide range of subjects, including ethics, metaphysics, religion, politics, rhetoric, biology and psychology. The Scottish biologist and botanist, Patrick Geddes (1854-1932) was known as an innovative thinker in the fields of urban planning and education. He was responsible for introducing the concept of "region" to architecture and planning. Geddes shared the belief that social processes and spatial form are related. Therefore, by changing the spatial form it was possible to change the social structure as well. This was particularly important in the late 19th and early 20th century when industrialization was dramatically altering the conditions of life. A junior contemporary of Geddes was Halford J. Mackinder (1861-1947). He wrote on the relationship of history and geography and on the historical geography of the British Empire. Additionally, he was working on the development of a so-called "synthetic geography" which included many sub-disciplines ranging from physical geography to the humanities. In a paper entitled "The Scope and Methods of Geography" that he wrote for the Royal Geographical Society, Mackinder outlined his ideas of a "New Geography". He defined geography as "the science of distributions". It was his conviction that physical and human geography formed one subject, and he consequently drew the conclusion that history and geography can never be studied separately. Mackinder is widely known for his Heartland Theory, in which he suggests that there was a pivotal area "in the closed heart-land of Euro-Asia" which was most likely to become the seat of world power. His theory was a model based on world history and geographical facts. Mackinder defined a "world island" that consisted of the two continents Eurasia and Africa. The British scholar promoting the concept of regional geography was Andrew J. Herbertson (1865-1915). His most famous work was "The Major Natural Regions: An Essay in Systematic Geography" published in *Geographical Journal in 1905.* The central theme in Herbertson's work was the promotion of concept of regional development. Herbertson's views on regional geography, as in essence natural entities, were most famously relayed through his best known paper on Natural Regions. Griffith Taylor (1880-1963) was born with geography in his blood. He was considered to be the authority on the Antarctic and Australia. He retired from the University of Toronto with the rank of Emeritus Professor. But, his retirement from the University does not signal his retirement from the discipline. He continued to be a prolific writer and publisher. Griffith Taylor is believed to have developed the view of 'Neo-Determinism' in 1951.[15]

One of the influential scholars of the first half of 20th century was Sir Dudley Stamp (1898-1966). He wrote a number of papers on the geology of Britain, Belgium and Northeast France. He undertook a number of official and semi-official appointments in various organizations. The contributions from Stamp were more towards geography than geology, his original field of interest. Throughout his career, a succession of awards, national and international, also testified to the growing width and depth of Stamp's influence. A number of Universities (Clarke, Edinbera, Stockholm and Warsaw) awarded him with the degree of 'Honorary Doctorate'. In recognition of his contribution, Dudley Stamp was conferred with the title of 'Sir' and became 'Sir Dudley Stamp'. After First World War (1914-18) it was realized that each unit of land in Britain should be used judiciously. With the purpose of national planning, Stamp committed himself to the task of conducting a systematic survey of the whole country based on a single scheme of land use classification. These Land Use Survey Maps present a picture of actual use of the land of Britain in 1930s. These maps were supplemented by the publication of a complete series of county monographs, each based on the map but drawing on historical, economic, climatic, relief and soil data as aids to the interpretation of the map and understanding of the agricultural geography of the time. The complete series of this literature with land use maps constitutes a quite remarkable addition to the geographical literature of the country. Stamp distilled the essence of the county reports into a book entitled *"The Land of Britain: Its Use and Misuse"* (1947). The book demonstrates the trend of Stamp's thinking, based on the record of how the land was in fact being used, but concerned progressively with how it should be used. This Land Use Survey laid the genesis of Stamp's conviction of the necessity for physical planning. It may be noted that this was the first official recognition that the concept of regional planning is a geographical concept and that the trained skill of the geographer is essential research component if the areal assessments on which plans must be founded are to be adequate. These Land Use Surveys of Stamp served as the basis for the reconstruction of Britain after Second World War. The British scholar, who tried to strengthen the role of geography as a discipline, was O.H.K. Spate (1911-2000). He was a geographer best known for his role in promoting geography as a discipline in various countries. Another geographer and leading social theorist of international standing from British school is David Harvey (*b.* 1935). He is the world's most cited academic geographer and the author of many books and essays that have been prominent in the development of modern geography as a discipline. His work has also contributed to broader social and political debate, particularly in the development of urban studies which, he argues, must draw on geography and the study of space. His career has seen him move through three areas of geographical enquiry, but he is best known for 35 years of commitment to Marxian theory, historical-geographical materialism and critical geography, later. By the mid-1960s he followed trends in the social sciences to employ quantitative methods, contributing to spatial science and positivist theory. David Harvey's *Explanation in Geography* (1969) was a landmark text in the methodology and philosophy of geography, applying principles drawn from the philosophy of science in general to the field of geographical knowledge. The work of David Harvey was historical in nature, emerging from a regional-historical tradition of inquiry widely used at Cambridge and in Britain at that time.

Thus, the the philosophers from British School made notable contributions in the fields of historical, regional and theoretical geography. The theory of Evolution, the Heartland Theory and the concept of Neo-Determinism are the other contributions to geography from the School.

Theory of Evolution by means of Natural Selection

For thousands of years many philosophers had argued that life must have been created by a supernatural being (creator/God) due to the incredible complexity of Nature (in particular, we humans and our minds). It is remarkable that Charles Darwin was able to explain our existence by means of Evolution from Natural Selection—which is very obvious once understood. Darwin's theory of evolution is based on five key observations and inferences drawn from them. The great biologist Ernst Mayr has summarized these observations and inferences as follows: First, species have great fertility. They make more offspring than can grow to adulthood. Second, populations remain roughly the same size, with modest fluctuations. Third, food resources are limited, but are relatively constant most of the time. From these three observations it may be inferred that in such an environment there will be a struggle for survival among individuals. Fourth, in sexually reproducing species, generally no two individuals are identical. Variation is rampant. And fifth, much of this variation is heritable. From this it may be inferred: In a world of stable populations where each individual must struggle to survive, those with the "best" characteristics will be more likely to survive, and those desirable traits will be passed to their offspring. These advantageous characteristics are inherited by following generations, becoming dominant among the population through time. This is *natural selection*. It may be further inferred that natural selection, if carried far enough, makes changes in a population, eventually leading to new species. These observations have been amply demonstrated in biology, and even fossils demonstrate the veracity of these observations. Darwin's Theory of Evolution may be summarized as: (i) *Variation*: there is variation in every population; (ii) *Competition*: organisms compete for limited resources; (iii) *Offspring*: organisms produce more offspring than can survive; (iv) *Genetics*: organisms pass genetic traits on to their offspring; and (v) *Natural Selection*: those organisms with the most beneficial traitsare more likely to survive and reproduce. Darwin imagined that all life is descended from an original species from ancient times. All organic beings that have ever lived on this earth have descended from some primordial form, into which life was first breathed. There is grandeur in this view of life that, whilst this planet has gone cycling on according to the fixed law of gravity, from so simple a beginning endless forms most beautiful and most wonderful have been, and are being evolved.

Mackinder's Heartland Theory

Sir Halford John Mackinder wrote a paper in 1904 called "The Geographical Pivot of History." Mackinder's paper suggested that the control of Eastern Europe was vital to control of the world. Mackinder's Heartland (also known as the Pivot Area) is the core area of Eurasia, and the World-Island is all of Eurasia (both Europe and Asia). The basic notions of Mackinder's doctrine involve considering the geography of the Earth as being divided into two sections, the 'World Island' which comprised Eurasia and Africa, and the 'Periphery', which included the Americas, British Isles, and Oceania. Not only was the Periphery noticeably smaller than the World Island, it necessarily required much sea transport to function at the technological level of the World Island, which contained sufficient natural resources for a developed economy. Also, the industrial centres of the Periphery were necessarily located in widely-separated locations. The World Island could send its navy to destroy each one of them in turn. It could locate its own industries in a region further inland than the Periphery could, so they would have a longer struggle reaching them, and would be facing a well-stocked industrial bastion. This region Mackinder termed the Heartland. It essentially comprised Ukraine, Western Russia, and Mitteleuropa. The Heartland contained the grain reserves of Ukraine, and many other natural resources. Mackinder's notion of geopolitics can be summed up in his saying "Who rules East Europe commands the Heartland. Who rules the Heartland commands the World-Island. Who rules the World-Island commands the world." His doctrine was influential during the World Wars and the Cold War, for Germany and later Russia each made territorial strides toward the Heartland.

Neo-Determinism

Geography has witnessed a long phase of controversies, especially related to the study of man and environment relationship, since the beginning of the modern period. The major of these controversies was determinism and possibilism. These doctrines were raised and supported respectively by the German and French schools of thought. Determinism is a name employed by writers to denote the philosophical theory that holds—in opposition to the doctrine of free will—that all man's actions are invariably determined by nature. Possibilism, on the contrary, suggested that nature does not determine man's actions; instead, it offers a number of possibilities and man is free to make a choice depending on the level of his cultural advancement. This makes man and his activities and not nature as the starting point in geographical studies. Griffith Taylor, on the other hand, maintained that the role of nature was far from crudely deterministic. Instead, he considered the environment as a strong force in human affairs that man can modify but cannot escape. If technologically superior, then man can alter the physical surroundings as per his requirements; but if he is technologically backward, then he is left with no other choice than to make adjustment with the nature. The human agency, through the use of the technology, can modify the force of nature but it cannot escape it. This view gained popularity as Neo-Determinism, i.e. the philosophy of determinism in new form. It is also known as Stop-and-Go Determinism.

SOVIET SCHOOL

The Soviet School of geography is believed to have evolved during the Age of Discovery. But, the major impetus came after the establishment of 'Imperial Geographical Society' in 1845. Thereafter geography made tremendous strides in Russia as in all over the European and American countries in the late 19th and early 20th century. Many faculties, institutions and departments of geography were established during this period. The important events that have shaped the development of geography in Russia are: (i) Establishment of 'Imperial Geographical Society' in 1845; (ii) The Great Russian Revolution of 1917 and the resultant political instabilities; and (iii) The disintegration of U.S.S.R. into several independent nations in 1991. This makes the years 1845, 1917 and 1991 as the important divides in the history of evolution of geographical thought in Soviet Russia. Accordingly, the various phases of development into which the discussion is organised may be identified as: (1) the Classical Phase (< 1845); (2) the Pre-Revolutionary Phase (1845-1917); and (3) the Post-Revolutionary Phase (1917-1991). Although the roots of discipline lie in Classical Phase, the maximum growth took place during Pre-Revolutionary Phase.

Classical Phase (< 1845)

The geographical tradition marks it's beginning in Unified Russia before 1845. The important developments of this phase include: (i) Eastward expansion of the Russian territory, particularly under the regime of Peter the Great, who ruled during 1682-1725; (ii) The official level beginning of map-making activities in 1719 with the technical assistance from French cartographers; the first Cartographic Office being started under the direction of Ivan Kirilov, who was also the first Russian appointed as the head of this office; and (iii) The establishment of Department of Geography in the Russian Academy of Sciences in Moscow in 1758 under the leadership of Lomonosov, who was also one of the founder members of the Moscow State University. The vast expanse of the Russian empire was the most potent factor in development of geography as an institutionalized discipline[16]. Peter the Great appreciated the need for accurate geographical information to facilitate the eastward march of empire. State supported expeditions were sent to the east and the north to explore the vast uninhabited stretches of territory; and generous funding was provided to prepare maps of the explored regions. In 1719, all official Russian map-making activities were placed under the direction of Ivan Kirilov, the first Russian to be appointed as the head of the Cartographic office. He supervised the preparation of an *Atlas of Russia* which was published in 1734. In the preparation of the atlas, he took technical assistance of French cartographers. The Great Russian encyclopaedist M.V. Lomonosov (1711-1765), who was one of the founding members of the Moscow State University, insisted that the exploring parties be asked to make systematic collection of information about the physical character of land, the population and the conditions of the economy. In 1758, Lomonosov was appointed the head of the world's first officially named Department of Geography, which was in the Russian Academy of Sciences. The recognition of geography department in the Russian Academy of sciences gave considerable academic prestige to geography as a useful field of scientific learning. Under the patronage of the Academy, the Department of Geography launched several schemes of regional surveys and mapping of data. Establishment of the Department of Geography in the Russian Academy of Sciences in 1758, much before the birth of the founders of modern classical geography, Humboldt and Ritter, proved that the tradition of geography in an organized way first developed in Russia rather than in Germany.[17] But the impact of the contemporary Russian geography was not as strong as that of German. This was primarily due to the language barrier. However, some contacts did develop between these two Schools. It was through the works of Auton Friedrich Busching (1724-1793), the German scholar, who had a profound effect on the Russian geographic heritage. By the beginning of 19th century Russian geography had already developed two distinguishing characteristics: the first was an emphasis on regions as the basis for organizing geographical work, and the belief that regions are concrete entities that can be objectively defined; Second was the continued use of geography to include a wide variety of fields. This was sharply contrasted to the contemporary trend in Germany where classical geography was torn apart into physical and human, each establishing its separate existence. In Russia, the classical tradition of geography as a field of study dealing with the physical environment of the earth and its human inhabitants had continued. Unlike in Germany where the deaths of Humboldt and Ritter in 1859 were marked by a break in the continuity of scientific geographical study, in Russia the developments continued unabated. For this reason, in the case of Russian geography, it is difficult to pick up any single scholar as the "grand old man of Russian geography".[18] Thus, the beginning of geography in Russia was made with empirical approach. These developments provided a conceptual framework for further growth of our discipline with a positive outlook. And after 1845, when there came up the 'Imperial Geographical Society', the progress was more steady and widespread all over the region.

Pre-Revolutionary Phase (1845-1917)

The earlier part of this phase coincides with the period of Environmental Determinism in Germany; the middle part with that of Davisian School in America; and the later part with that of Rise of Possibilism in France. Although the professional contacts with the Western Schools of Thought had already developed during 19th century, Soviet School was the least influenced. It shaped itself in a different way. While classical geography was undergoing analysis in Germany as each academic discipline sought to establish its separate existence, in Russia the tendency was for scholars with diverse interests to come together as geographers. In 1840 this created need for some kind of institution to provide a forum for the presentation and discussion of different kinds of studies dealing with the physical world and its human habitants. To formalize the structure of geography, the Imperial Geographical Society was founded by Arsenyev in 1845. The great professionalization came in Russian geography after the establishment of Society. This society promoted what has been collectively called as 'geographical sciences' including geography, with its sister disciplines (geology, meteorology, hydrology, anthropology, archaeology, etc.). The contemporary Russian Tsar, Peter the Great, also recognized the vital importance of having accurate geographical information about the virgin land islands of Siberia across the Urals. Under his royal patronage several expeditions to Siberia and Far East were organized. The German scholar Alexander von Humboldt was entrusted by the Russian Tsar to explore the virgin lands across Ural Mountain. He explored the lands of Siberia as far as the borders of China. He visited the shores of Caspian Sea. In the course of his expedition, Humboldt observed that the temperatures varied at the same latitude in accordance with the distance from ocean. The Russian Tsar was so much influenced by Humboldt that he agreed to set up weather stations where related data could be recorded. By 1935 the Russian network of recording stations extended all the way from St. Petersburg to an island off the Alaskan mainland. On the basis of the weather data collected, Humboldt deduced 'the concept of Continentality'. This heritage that Humboldt founded, had a profound effect on the contemporary Russian geography, and empirical studies of areal phenomena with the positivist approach formed an inevitable symbiosis for the conceptual framework.

Most of Russian geographers of the pre-revolutionary period seemed to have had direct or indirect contact with the professional German geographers. Some of them studied in the German universities, and were pupils of Ritter, Richthofen and Ratzel. But the impact of German geographical ideas had produced quite different results in Russia. Russian geographers of the pre-revolutionary period always believed in the mechanical explanation rather than in teleological explanation, where the phenomena and observation were understood as outcomes of prime causes. The Pre-revolutionary Russian geography was largely centred on the works of Petre Petrovich Semenov, Alexander Ivanovitch Voikov, V.V. Dokuchaiev, D.N. Anuchin and Peter Kropotkin. They formed the true heritage of the Russian Geography that continued its existence even after the October Revolution of 1917. Petre Petrovich Semenov (1827-1914) is often regarded as the 'grandfather' of Pre-revolutionary Russian geography who acted as a bridge between the scholars of classical period, such as Lomonosov, Busching and Arsenyev, and the scholars of modern period. Semenov is credited with a distinct humanistic approach in the contemporary Russian geography that emphasized questions of social inequality. He developed a 'welfare paradigm' to social relevance in geography that stressed the quality of life. Alexander Ivanovitch Voikov, V.V. Dokuchaiev and D.N. Anuchin were three outstanding followers of Semenov who left their distinct imprints on the contemporary methodologies. The first two are credited with innovative studies of the climate and soils, and Anuchin tried to establish geography as a major university subject and drew up the curricula for the primary and secondary schools. Besides, as regards the pre-1917 phase in the development of geography in Russia, there was Peter Kropotkin, who became influential because of his concept of geography as social ecology, and his concept of 'mutual aid'. A.I. Voeikov (1842-1916) was essentially a physical geographer. His area of interest covered climatology, agriculture and environmental problems. Voeikov founded the Main Geophysical Observatory (MGO) in 1849 and it is the oldest meteorological research institution in Russia. He studied the earth's heat and water balances and is credited with the expertise in snow-science. His studies of climatology were directed to the improvement of agriculture. He laid down the foundation of systematic agro-climatic geography through elaboration of the comparative method as expounded by Carl Ritter. His concern with the improvement of Russian agriculture led him to compare the farm practices in places similar to those of European Russia. It was because of his suggestions that the Russians first introduced tea-plantations in Georgia, cotton in Turkistan (Turkmania) and wheat in Ukraine. His book *'The Climate of the World'* was published in 1884, and a monograph *'Distribution of Population on the Earth in Relation to Natural Conditions and Human Activities'* was in 1906. Both of these works were later translated into German. V.V. Dokuchaiev (1846-1903) was the first professor of geography in Russia, appointed at St. Petersburg in 1885. He also specialized in physical geography. Dokuchaiev's pioneering work is in the field of soil science, including the study of soil formation processes and soil layers or horizons. His soil studies followed systematic approach. He observed that different kinds of soil could be identified by looking closely at the layers or horizons, which differed because of differences in the soil-forming processes. According to Dokuchaiev, the soil "reflects the extraordinary complex interaction of climate, slope, plants and animals with the parent material derived from the underlying geological formations". A soil that has been exposed to all these conditions for a long time would more closely reflect the complex of climate and vegetation than it would the parent material. His generalizations were quite different from those of German and French soil-scientists who conceived of the soil as a reflection of the underlying geological formations. The generalization of Doduchaiev was essentially based on empirically observed arrangement of climatic zones that could produce observations soil differences over a large tract of plain. Dokuchaiev, like his predecessor, Voikov, also attempted to adopt a holistic approach to explain the very concept of man as a major agent of change on the surface of earth. Undoubtedly, Dokuchaiev was one of the leading professional geographers of Russia and deserved major place among the world's leading geographers of his time because of his contributions to systematic soil studies. Since all of his works were written in Russian, his major ideas could produce little impact on West European geographers. It was only through the translation of his works into German and French after his death that he came to be known beyond the national frontier as one of the founders of applied soil geography. Peter Alexeyevich Kropotkin (1842-1921) was one of Russia's foremost anarchists and one of the first advocates of what he called "anarchist communism". He left behind many books, pamphlets and articles, the most prominent being his works *The Conquest of Bread* and *Fields, Factories and Workshops*, and his principal scientific offering, *Mutual Aid: A Factor of Evolution*. He was also a contributor to 1911 *Encyclopaedia Britannica*. Kropotkin was writing during the period when social Darwinism was in ascendancy, and no intellectual could escape the impact of evolutionary theories. Kropotkin's view of nature was based on three premises, namely nature is organic (i.e. holistic); it is historic; and it is spontaneous. His originality lay in the manner in which he interpreted the organic (holistic) characteristics of nature, especially in relation to man's place in the web of nature. This theme was the key idea and the central focus of his theory of, and his book on *Mutual Aid*. He demonstrated that both cooperation and competition are present in nature simultaneously, as also in the case of human communities. In course of developing his concept of "*Mutual Aid*" among organisms and human groups, Kropotkin laid foundation for a radical theory of *social ecology*. He viewed nature and social groups as organic wholes so that the action of one part affected all the parts. He considered social groups as being subject to many of the same processes as are found in organic nature. Thus, during Pre-Revolutionary Phase geography in Soviet World was propagated as a discipline at academic and administrative levels both. During this phase, geography witnessed a steady progress in an indigenous manner, with its basic branches (Regional, Human, Social, Physical and Economic) being well established. Although, the maximum contribution was made to physical geography, the foundation was also strongly laid for the humanistic and welfare geography during this period. The fields of Cartography and regional geography were the other areas of contribution by the scholars of pre-revolutionary phase. However, this development was interrupted by the rise and spread of communist philosophies after the 'Great Russian Revolution'[19].

Post-Revolutionary Phase (1917-1991)

The communist thinking dominated the Russian Society after the Great Revolution of 1917, following the principles of the Russian and German philosophers like Lenin, Marx and Engels. Vladimir Ilyich, popularly known as Lenin (1870-1924), was a Communist revolutionary of Russia, the leader of the Bolshevik party, the first Premier of the Soviet Union, and the main theorist of what has come to be called *Leninism*. Friedrich Engels (1820-1895) was a 19th century German political philosopher, who developed communist theory alongside his better-known collaborator, Karl Marx. Along with Engels, Marx took part in the political and philosophical struggle of his times, and co-authored the *Communist Manifesto*. Karl Marx (1818-1883) was an immensely influential German philosopher, political economist, and socialist revolutionary. While Marx addressed a wide range of issues, he is most famous for his analysis of history in terms of class struggles. Karl Marx held that society could not be transformed from the capitalist mode of production to the communist mode of production all at once, but required a state transitional period. Socialism, on the other hand, refers to a broad array of doctrines or political movements that envisage a socio-economic system in which property and the distribution of wealth are subject to social control. For Karl Marx, who helped establish and define the modern socialist movement, socialism implied the abolition of markets, capital, and labour as a commodity. Marx and Engels regarded themselves as, "socialists". For Marxists, socialism is viewed as a transitional stage characterized by state ownership of the means of production. They saw this stage in history as a transition between capitalism and communism, the final stage of history. For Marx, a communist society describes the absence of differing social classes and thus the end of class warfare. According to Marx, once private property had been abolished, the state would then "wither away," and humanity would move on to a higher stage of society, communism. Lenin is known as an 'Economic Determinant'. For him geography could serve the necessary foundation on which the design of a new kind of economy had to be based. Under his leadership, applied aspect of geography became very strong in post-revolutionary phase in Russia. According to him, the most important product of geographical study was the identification of rational regions within which the segments of new economy could be constructed. Thus, in the 1920s, the Soviets seemed interested in 'economic regionalization' of the USSR.

As far as the philosophical developments in the first half of the 20th century were concerned, they were parallel to that in France and Germany. But, the Russian ideology was not in favour of the doctrine of Determinism. Instead, the stress was on Human studies, and with an economic bias. In other words, the dichotomy of physical *versus* economic geography was quite strong. And Economic geography emerged as an important branch of study. The continued emphasis on economic regionalization was one of the distinctive aspects of geography in the Soviet Union immediately after 1917. The State Planning Commission (GOSPLAN)[20] was established in 1921 to prepare the regional plans. The GOSPLAN divided Soviet Union in 21 regions and proceeded to a detailed study of each of them. In this the geographers played an important role along with the economists and engineers. Moreover, the geographers also helped in the selection of industrial location and resource development. Various industrial regions were also planned, e.g. Ural-Kuznetsk and Dnieper Basin Industrial Regions. In 1950s onwards, the focus of geographical studies was found specifically on Resource Planning. This increased the importance of physical geography. Because, the study of economic or resource geography was considered baseless without physical geography. This was evident in the following developments:

(1) Soviet geographers developed the sound theoretical principles of forecasting and a typology of climatic phenomena based on dynamic methodology;
(2) They studied the radiation budget and moisture cycle, and their role in the formation of climates;
(3) They worked on the theory of water budget, the relationship between surface water and soil water;
(4) Soviet geographers developed the theory of glacial processes, based on the study of heat and mass exchange in various types of glaciers;
(5) Geomorphologists studied the crustal movements and developed the morpho-structural approach to geomorphology;
(6) Soil-Scientists identified many soil types; and
(7) The students of Bio-geography gave emphasis on ecological approaches.

By 1960s, however, the dichotomy between Economic and Physical Geography got resolved. The Soviet scholars started believing in Unified Geography. Till 1990s, Soviet geographers have been concentrating on the management of natural resources. Their efforts were directed towards solving the related regional problems and the management of environment. The Soviet leaders, like Leonid Brezhnev, were interested in seeking answers to the questions related to the man's interactions with the natural environment and encouraged studies, which contained practical recommendations in this direction. They promoted the research based on the study of development trends in Soviet Society and its productive economic forces. The issues related to environment, society and industrial economy were largely stressed upon the writings of geographical nature. Most recently, i.e. before disintegration in 1990s, the geographical science in the Soviet Union manifested considerable interest in the concept of Territorial Industrial Complexes (TIC). The Territorial Industrial Complex is the concept of planned organization of the large industrial units. Moreover, the Soviet geographers were concentrating on environmental problems. They were trying to find out new methods for consolidating harmonious relationships between society and nature. Another sphere in which the Soviet scientists were doing constructive research was the study of the theoretical foundation of recreational geography and the establishment and development of recreational zones near large towns. Thus, since very beginning

Geography has been put in the academy of sciences in Soviet Union. Geography remained, more or less, a well understood discipline throughout in terms of its content and methodology. The dichotomies remained for a very short period and were soon resolved. The Russians, therefore, provided a very sound base to geographical thinking. Soviet geographers were highly conscious about man-environment relationships. They always emphasized on the applied aspect of geographical studies. The regional and economic geography were the most sought after fields of geographical study in Soviet Russia. Since the Russians adopted a more practical approach, the theoretical and philosophical aspect of geography always remained poor here.

In general, the modern schools of geographical thought gave the discipline of geography a number of viewpoints. For instance, to the Germans and Americans we owe the Philosophy of Determinism. The roots of Possiblistic thinking lie in the French School. For the introduction of the concepts of Landscape, the Cycle of Erosion, again the credit goes to American School; whereas, the British contributed the Theory of Evolution and Heartland Theory and the viewpoint of Neo-Determinism.

To conclude, the Classical Phase of development of geography witnessed the emergence of various robust schools of thought one after the other. The Ancient and Medieval Schools (Greek, Roman, Indian, Chinese and Arab) laid the foundations of the discipline; whereas the Modern Schools (German, French, American, British and Soviet) developed its structure in the form we are familiar today. The studies on the fields, like Philosophy, Astronomy, Astrology, Mathematics, Algebra, Geometry, Cosmology, Physics, Meteorology, Geology, Chemistry, History, Historiography, Cartography, Botany, Political Science, Medicine, Engineering, Literature and Poetry,. lie at the base of geographic development in its initial stage. Later, the discipline has flourished into a form with numerous branches practically covering all the aspects of knowledge, whether related to physical, biotic or human. No discipline covers such a vast arena of knowledge as geography does. However, along with the accumulation and analysis of vast amount of knowledge, the emergence of various dualistic tendencies in geography is also associated with its Classical Phase. One distinct development that was observed commonly among various classical schools of thought (particularly German, French and American) in geography by the mid-20th century, showed a tilt towards the adoption and dominance of chorological (regional) approach in general. However, the developments in certain modern schools, especially American and British, continued even after the Second World War. In fact, the American and British Schools of thought have been very instrumental in the post-modern development of geographical thought. In general, the fields of physical, regional, human, political, colonial, historical and theoretical geography attracted the scholarly concern during modern phase of development. Besides, the philosophy of environmentalism with its three variants, viz. deterministic, possibilistic and ecological, got completely rooted in this period.

DUALISM

The dualistic tendencies have always been part of geographical studies since very beginning. But, they became more conspicuous during the modern period. The major controversies in geography have been related to its subject matter, approach, and method. Respectively, they are: Systematic *versus* Regional; Physical *versus* Human; and Idiographic *versus* Nomothetic.

Systematic versus Regional

The studies in geography have always been conducted in either of the ways, viz. to study all the aspects in one region (chorological); or to study one aspect in detail throughout all the regions (systematic). The chorological or regional approach deals with study of all the elements within an area, specifically demarcated for this purpose. Another parallel approach is systematic or topical; it leads to a series of studies focused on the character, distribution, process and behaviour of one element or phenomenon throughout the world, or the surface of earth, breaking it into different areal units for this purpose. The confusion, as what should be the correct approach to geographical studies, has overshadowed the development in geographical thought for a long time, and particularly up to the middle of 20th century. The 'Systematic *versus* Regional dualism' became evident particularly after the classic work of Bernard Varenius (1622-1650), a German scholar. He divided Geography into two: (i) General Geography and (ii) Specific Geography. According to him 'General Geography' studies the earth in general, describing its various phenomena. Such a study provides the discipline with general laws and theories. 'Special Geography', on the other hand, is the study of individual regions, countries or areas. However, Varenius also clarified that the laws of General Geography can always be used in explaining and understanding Special Geography that provides concrete examples or case studies to the General Geography. On the other hand, the systematic studies may also lead to regionalization, and the study of regions so formed, e.g. agricultural, climatic or industrial regions and so on. But the ideas of Varenius were not conceived the way he intended to. They were misinterpreted, leading to a division into 'Systematic' and 'Regional' geography. The terms like 'General' and 'Specific' were caught by various geographers later and the dichotomy between the two widened to the extent that there had been periods when there existed double dualism in geography, i.e. 'Systematic and Physical' *versus* "Regional and Human". One such time was that of Emmanuel Kant (1724-1804). Kant substituted the word 'physical' for 'general', and classified all the general or systematic studies under 'physical'. It was under his influence that the geographers in the late 18th century concentrated on systematic or general studies at the expense of regional or specific studies. But, during the 19th century the interest of some of geographers, like Ritter and Ratzel, shifted towards human aspects. Then the so-called human geographers emphasized the concept of region. They considered any law or generalization to be empty and unpredictable without its application to regional or specific geography. Therefore, they gave more emphasis on the regional studies, and conceived general studies as only the honourable occupation of the people without some practical or empirical base. The development of geography during the first half of the 20th century has been marked by an increasing interest in regional geography.

The climax was witnessed in the views of Richard Hartshorne (1899-1992) and Alfred Hettner (1859-1942), when the concept of geography as a chorological science was advocated by full force. This led to the development of concept of geography as a science of 'areal differentiation'. The idea of region was further strengthened by its usage in different forms, e.g. *'Paysage'* in France (Blache), *'Landschaft'* in Germany (Otto Schluter) and *'Landscape'* in America (Carl Sauer). Moreover, this was a period when the 'possibilistic' thinking was at its peak. At this climax phase, this argument took the form of the old 'hen and egg story'. One group believed that the core of geography was regional; whereas the others believed that systematic geography was really indispensable. The compromise could reach only by the middle of the 20th century, i.e. in the 'Ecological Approach' to geography. The dualism between systematic and regional approaches came to be accepted less and less with time. Various arguments were put forward for this. Firstly, the earth's shell is filled with a multiplicity of heterogeneous phenomena of varying nature. At the same time the surface is also divided into a number of areas or places distinguished from each other in terms of some measurable aspects (natural or man-made). Thus, the totality of the phenomena at any given place is not a single integration but rather a complex of inter-related segments. The geographer's task is to break down this complex whole into some understandable forms. This can be done in either of ways: systematically (analytically) or synthetically (by aggregating). Secondly, the purpose of learning has two aspects, viz. (i) the basic human interest, i.e. the need to know the facts about the entire world we live in, and (ii) the intellectual concern, i.e. to comprehend or explain these facts, which one really observes around him, and then to understand them in terms of their inter-relationships. In the former case, the field is more widespread and, therefore, the general facts may be known, as there is no scope for knowing about their intense relationships with other elements. In the second case, the field is small and, therefore, a number of inter-related elements may be studied together. Thus, in an essential form, the former (or systematic) approach is the first step preceding the second (or regional) approach. As a norm also, we generally proceed from general to specific or from basic to intellectual or scholarly concern. Thirdly, the structure (anatomy) of the complex of elements of our earth can be compared with the human anatomy, i.e. the structure of human body. Various relationships can be studied in terms of the categories, like bones, muscles, arteries, etc. (the general parts of the body). Or the same elements may also be studied by anatomists with respect to arms, legs, feet, chest, abdomen, etc. (the specific parts). Each of these specific parts has some parts in common or the general parts link all the specific parts of the human body. The same holds true for the earth's surface. Thus, all studies in geography either analyze the areal variations in terms of individual phenomenon at macro level, or they study interconnections of phenomena in integration within a relatively smaller unit of study. As such there is no dichotomy or dualism between the 'systematic' and 'regional' approaches to geographic study, but rather a gradational range along a continuum, i.e. from those that analyze the elementary complexes over the world to those that analyze their complex integrations within smaller areas. The former approach is 'systematic' and the latter is 'regional'. The geographical studies require the use of both approaches of area study. Because the systematic geography attempts to formulate the rules and laws, which may later be applied to regional studies. On the other hand, the regional geography serves as the basis for explanation. It also provides data for analysis to the systematic studies. Thus, it is appropriate to say that the systematic and regional approaches are complementary to each other.

Physical versus Human

The relation between the world of man and the non-human world is of prime concern in geography. Since, the earlier times, the human and the non-human aspects have contributed to the content of geography. The dichotomy is what aspect is more important to study in geography, physical or human. Physical Geography covers the planet, from climate to landforms to hazards and disasters to the distribution of plants and animals, and much more.

The main purpose of Physical Geography is to explain the spatial characteristics of various phenomena associated with the Earth's hydrosphere, biosphere, atmosphere, and lithosphere. Human Geography, on the other hand, is a branch of geography that focuses on the study of patterns and processes that shape human interaction with the environment, with particular reference to the causes and consequences of spatial distribution of human activity on the Earth's surface. It encompasses demographic, political, cultural, social, and economic aspects of man. However, while the major focus of human geography is not the physical landscape of Earth, it is hardly possible to discuss human geography without referring to the physical landscape on which human activities are being played out. Originally, the controversy between man and physical (nature) was external to geography. It appears to have entered into discipline first with the writings of German scholars, and was later supported by the rise of natural sciences and Darwinism. The important sources of problem may, therefore, be identified as: (1) The works of German scholars; (2) Environmentalism (Determinism and Possibilism); and (3) Systematic *versus* Regional Dualism. 'Physical' is the oldest and the most honoured term in geography. It is particularly with the writings of German scholars like Kant and Humboldt that the physical aspect of geographical studies got greater emphasis than human. Kant clearly stated that 'Physical Geography' is the base of all geographical studies and categorically separated it from its other branches. He considered 'Physical Geography' as the only systematic part of the discipline. Humboldt conceived man as the part of nature, dependent upon nature-*minus*-man, which existed before man and which could exist independent of man. He believed that the natural laws govern human beings as well. Since, man was conceived as the product of nature, it was advocated that the study of man-nature relationships should begin with nature. Moreover, the knowledge in the natural sciences was more complete than the social sciences, and the former were on a firm footing comparatively. And in geography, it was physical geography that was considered as the basic to all studies. The German scholars, particularly, have given this kind of understanding. Humboldt, for instance, considered 'physical geography' as 'the science of laws like physics'. His 'Physical Geography' was directly comparable to the 'General Geography' or the 'Systematic Geography' of Varenius. Carl Ritter, a contemporary of Humboldt, on the other hand, had an anthropocentric approach to geographical studies. Ritter conceived Earth as the home of man, as human body is for the soul. According to him, the physical surroundings get meaning only with reference to human beings. They are required to be known only in relation to human beings. Because it is the man who interacts with the physical world. The emphasis of geographical studies shifted from physical to human in the beginning of 20th century and continued to be so up to the Second World War. The French school of thought, led by Paul Vidal de la Blache, dominated geographical thinking of this period.[21] He developed human geography as an entity, incorporating into it the historical and chorological perspective. His "*Principles de Geographie Humaine*" is regarded as a classic by geographers. According to him, the historical evolution of each phenomenon in human geography is followed by a study of its present setting, localization and correlation. He analysed the principles of terrestrial unity, concept of milieu and the significance of environmental factors in relation to man. The dichotomy between 'physical' and 'human' deepened with the coming up of the 'environmentalism', particularly after the publication of the theory of Charles Darwin. After the publication of the Organic Theory of Darwin, the paradigm of environmentalism gained popularity among professional geographers. In this paradigm, the distinction between human and physical (natural) was fundamental. It was because of the split of the concept into two streams, viz. Determinism and Possibilism. The scholars like Ratzel, Ellen Semple and Davis exaggerated the role of nature in human affairs. On the other hand, the writings of Blache, Sauer, Hettner, Huntington and Hartshorne did just the opposite, i.e. overemphasis on the role of man in comparison to nature, considering him the most superior of all the organisms, and as the major decision-making element in the world. The 'physical' *versus* 'human' dichotomy has also been strengthened by the 'systematic' *versus* 'regional dualism', as the systematic approach was considered more compatible with physical studies and the regional with human. Like 'systematic' *versus* 'regional', the 'physical' *versus* 'human' dualism has also been criticized on various grounds. Both man and nature are intrinsic to the study of particular character of areas. Both exist in such a close union that, at times, they cannot be separated. To consider them different from each other is a direct disregard to the reality, which is so complex that nowhere such a distinction occurs. It is just a matter of degree or relative importance that what aspect is more emphasized and what remains in the background. In the study of diverse features in interrelation, the 'invisible' aspects are generally overlooked. These 'invisible' aspects may be physical or human. For instance, in the case of landslides in a hilly region, the human cause (deforestation) may generally be overlooked. Further, for a clear-cut dichotomy between the two a sharp separation is required. But, this cannot be established, as the study of one leads to the other automatically. For example, the study of 'population distribution' ultimately culminates on the factors responsible as terrain, water availability, climate, soil type, etc. Or the study of physical is incomplete without relating it to man, i.e. in terms of his impact (as habitation, deforestation, mining, agriculture, etc.). There is only one landscape, i.e. 'geographical landscape', and the distinction between the 'physical or natural' and 'human or cultural' is purely theoretical, and does not exist in reality. As far as the formulation of generalizations, laws and theories is concerned, they are not the domains of physical sciences only in the present day context; they are equally possible in human studies. Thus, the distinction cannot be created on this basis as well. Since geography, in particular, must examine phenomena in the actual complexes in which they are found, it is impossible for it, in practice, to separate natural and human phenomena. All modern geographers are agreed that geography cannot adapt itself to the conventional division between natural and social studies; not only does geography as a whole fit into neither group, but neither can it be divided into two halves, natural and human. The separation of things natural from things human is possible only in theory, in reality they are interwoven. Thus, it is found that even after the laborious research by many able scholars it has not been possible to separate the features of earth resulting exclusive of man and from that exclusive of nature. Even the minor landforms, micro-climates, soils and wild vegetation—all are affected by the human culture. On the other hand, every material work of man, whether a house, plantation, a farm or a settlement—all represent a composite of natural and cultural elements. Moreover, the meaning of 'physical' has also changed in the modern period. Earlier this meant the 'non-human world', i.e. including organic life (flora and fauna as well). Now, it means purely 'inorganic world'. For the organic world (minus man) the term 'biotic' is used. In fact, the matter of importance in geographical studies is not the emphasis on human or physical aspects but the 'cause and effect relationship' between the two. Sometimes the cause is human and at others, the physical. This earth's surface is a living whole, and it cannot be bisected into something purely dead and something purely alive. Indeed, such a dualism exists only in the minds of people. Moreover, since now we have also abandoned the concept of 'environmentalism', with the dichotomy between 'Determinism' and 'Possibilism' and 'Systematic' and 'Regional' being resolved, the 'Physical *versus* Human Dualism' too does not carry relevance in geographical studies.

Idiographic versus Nomothetic

Of all the problems of current concern in the thinking of geographers, the most disturbing appears to be the question whether geography 'like other sciences' can develop 'the knowledge of principles, laws, and general truths'—and thus lay claim to the name of science—or whether its function is merely to describe innumerable unique areas. The Idiographic *versus* Nomothetic dichotomy is related to which of the above two concerns, unique or general, represents the ultimate purpose of geography[22]. In other words, it concerns itself to the method of obtaining and organizing information in geography. Here the dualism is between that aspect of science that can be expressed in universals and that which is concerned with the individual attribute of object. Nomothetic means 'stating laws', whereas Idiographic means 'pertaining to the study of individual cases or events'. The idiographic method deals with discrete facts or events. As far as geography is concerned, the contrast between that aspect of science that can be expressed in universals and that, which is concerned with the individual phenomenon, received the particular attention at the turn of 19th century. This was in the attempts of two German philosophers, Windelband (1848-1915) and Rickert (1863-1936). 'Nomothetic' and 'idiographic' are terms used by Kantian philosophers Wilhelm Windelband and Heinrich John Rickert to describe two distinct methods of acquiring knowledge, each one corresponding to a different intellectual tendency, and to a different branch of academe. Nomothetic is based on what Kant described as a tendency to generalize, and is typical for the natural sciences. It describes the effort to derive laws that explain objective phenomena in general. Idiographic, on the other hand, is based on what Kant described as a tendency to specify, and is typical for the humanities. It describes the effort to understand the meaning of contingent, unique, and often subjective phenomena. Besides in geography, the two terms have also been used in philosophy, psychology, sociology, anthropology and law with differing meanings. Windleband and Rickert classified the different branches of science on the above basis, distinguishing between "nomothetic" or law-making sciences, and "idiographic" sciences, those concerned with the unique. In a way this seemed to conform to the conventional division between the natural sciences and the social sciences. In the earlier writings of scholars, particularly after 1900, geography has emerged as an 'integrating science', i.e. completely different from other sciences and studying all the physical and human elements in an area. Because of this kind of growth, geography developed a unique methodology, termed as 'idiographic', i.e. studying different areas with their own individual significance. The proponents of this particular methodology argued that every area or region on the surface of the earth is a unique combination of various elements. This unique combination provides that area a unique outlook. Therefore, there is only one Alps, Himalayas, Amazon, Germany, London, New York, Paris, and so on. They all appear to have their own individual significance. To cite another example, Africa and North America both have a system of lakes, but as a system of connected fresh water lakes the Great Lakes of North America have no counterpart in the world. Thus, geography began with the description of the earth's surface and it has always shown a great concern for individual cases in the situation. Then, for a long time, the bond of geography with history (and human culture) dominated geographical thinking. This also strengthened the belief of geographers in the uniqueness of areas or areal combinations. The most important supporter of this kind of thinking was Emmanuel Kant (1724-1804), popularly known as the 'Father of Exceptionalism' in Geography. Kant's philosophy and writings are responsible to a great extent to promote the study of unique cases in geography. Therefore, his philosophy is known by the name of 'Exceptionalism in Geography'. Because, his methodology to the study of geography was not considered scientific and he appears to have profoundly changed the trend of development of geographic thought in his times. According to this view, geography is different from all other sciences or disciplines and is methodologically unique. Therefore, it deserves a name of its own. Kant laid an exceptionalist claim not only for geography but also for history. Some of the later geographers, namely Ritter, Hettner, Hartshorne, etc. made 'idiography' a very important method of geographical studies. Kant wrote that the history and geography are two parts of the whole world of perception of man. Both the disciplines are based on descriptions, one according to the time and the other in terms of space.

History is narrative, i.e. reporting the events which fall one after the other in time. Geography, on the other hand, is the study of events (phenomena), which have taken place side by side in space. The major point taken from the statement made by Kant in the support of the idiographic method is that there is no limit placed in the class or number of objects in geographical studies (as in systematic sciences). If there is no limit to the factual content of geography, then the discipline must be defined and recognized in terms of its distinctive method of study, i.e. idiographic. It is the description of unique or specific areas or of elements in unique or specific combinations in an area, rather with developing generalizations about the elements. Kant argued that the locations are unique. If locations are unique, then the descriptions of what existed at those unique locations could not be accomplished by referring to general laws. Rather, it requires the employment of the idiographic method. The idiographic view was very strong in the earlier part of the 20th century. There were many supporters. The view generated great professional activities and a strong reaction to deterministic thinking. These supporters constructed false similarities between history and geography. For instance, according to Alfred Hettner (1859-1941), the German scholar, as no historical event or period is like each other, so is geography of no two or more phenomena or areas or regions is alike. Thus, both the fields explain the unique. Since 1950s onwards various objections were raised against idiographic method of study in geography. The American scholar, F.K. Schaefer, for instance, pointed out that Kant's views were immature. According to him, this immaturity was the result of the fact that they were expressed before Newton's scientific laws were put forward. The similarity between geography and history as proposed was not acceptable to him. The concept of uniqueness proved too much. For example, though there are not really two stones that are completely alike in all minute details as shape, colour, chemical composition, etc. yet Galileo's law of falling bodies holds equally for both. The idiographic method of study suggests considering everything on the surface of the earth. If this is taken as a practice without knowing the importance or significance of the study, then geographers would be lost under the heap of useless and insignificant facts and details and it will become unfruitful. Thus, the idiographic method had adversely affected the course of geographic thought. Kantian influence, however, was a disaster, since it led to Exceptionalism in geography. The modern trend is to simplify the complexity and predict for the future. Since, there exists a heap of information, one cannot deal with this information without following a general approach. The prime concern of geographers in the present times is to form generalizations and establish laws and theories. The idiographic technique cannot help in this process. On the basis of this argument, the idiographic method has been repeatedly criticized in contemporary period. Then, the essential point is that the uniqueness does not preclude similarity. From these similarities the relationships can be formed to provide the basis for scientific generalizations. To cite an example, although the moon, an apple and a ball are unique in themselves with no formal similarities, yet they share certain common attributes. In a trivial descriptive terms, they each have an inside and outside, and are round; whereas Newton recognized a major functional similarity when he proposed his laws of gravitation. The 'Nomothetic' method emphasizes the task to search for similarities and especially the functional similarities. Different cities, for example, may not be the same or similar in terms of size, pattern, structure, shape, density, culture, etc., but most of are similar, at least, in terms of their non-agriculture function. Such generalizations or laws may not be true and applicable to the entire world (as such an attempt is rarely possible), but they are true and applicable to a large extent and to a majority. The patterns produced by the individual facts and events in geography are spatial. This implies that geographers must describe and explain the manner in which the phenomena combine to fill an area or are distributed over space. Humboldt, the father of scientific geography, stressed that geography must pay attention to the spatial arrangement of the phenomena on the earth's surface, and not to the individual characteristics of the phenomena themselves. This is sufficient to call geography a scientific discipline and not the unique. It is agreed that logically everything is unique. But similarities can be found among unique things. General statements can be elicited from these. These general statements after testing may be expressed as formal laws. In spite of the difficulties involved, geography has been able to develop general principles of the relationships among the variable phenomena. Nomothetic method is, therefore, the key to the study of geography. There cannot be any kind of compromise between the two absolute opposite methods, Idiographic and Nomothetic. The geographers' task is to study the general character of the individuals and not their uniqueness. The geographers' task is to seek for the most complete presentation possible. His method of study should be such that a majority, if not all, aspects should be brought under investigation. In this process of seeking universal truths some individual cases may be left out, which are not explained or described.

Thus, except 'Nomothetic' *versus* 'idiographic' the other two controversies have been proved useless and are responsible for posing obstacles in the development in the development of the discipline. The other two controversies could not generate demands and, therefore, are now discarded and the geographers have freed themselves from such pseudo-issues, which forced them to waste their valuable time and energy in debating over useless issues. At the same time geographers have also come out of the disputative situation regarding the dualism of Idiographic and Nomothetic. The other two controversies, namely 'systematic *versus* regional' and 'physical *versus* human' are just differences in the viewpoints of the scholars and not the dichotomy in the true sense.

NOTES

1. Dikshit, 1997, 65.
2. Dikshit, 1997, 67.
3. Pearcy, 1957, p. 22.
4. Dickinson, 1969, 68.
5. *'Geopolitiks'* is the method of political analysis, popular in Central Europe during the first half of the 20th century, that emphasized the role played by geography in international relations.

6. Dikshit, (1997), 74.
7. Franco-Prussian War or Franco-German War, 1870-71, was the conflict between France and Prussia that signaled the rise of German military power and imperialism.
8. The Chicago School (sometimes described as the Ecological School) refers to the first major body of works emerging during the 1920s and 1930s specialising in *urban sociology*, and the research into the urban environment by combining theory and ethnographic fieldwork in Chicago, now applied elsewhere. While involving scholars at several Chicago area universities, the term is often used interchangeably to refer to the University of Chicago's Sociology Department—one of the oldest and one of the most prestigious.
9. Dickinson, (1976), 324.
10. Dikshit, (1997), 107.
11. Martin, Geoffrey J. (1994), 480-492.
12. Dikshit, (1997), 110.
13. Robert Edmond Grant was a stalwart of the Plinian society for student naturalists which Darwin joined in the autumn of 1826 while starting his second year of medical studies at Edinburgh University. Darwin became Grant's keenest student and assisted him with collecting specimens as well as learning from Grant's knowledge and theories.
14. Social Darwinism in the most basic form is the idea that biological ideas can be extended and applied to the social realm. It has been applied to the claim that Charles Darwin's theory of evolution by natural selection or by "survival of the fittest" can be used to understand the evolution of society: just as competition between individual organisms drives biological evolutionary change (speciation), competition between individuals or groups in human societies drives social evolution.
15. Dikshit, (1997), 246.
16. Dikshit, 1997, 94.
17. Adhikari, (1992), 169.
18. Dikshit, 1997, 95.
19. The Russian Revolution took place in October 1917, affecting the course of world history for many decades. The revolution was widely regarded as a reaction to the strains that had been placed upon Tsarist Russia (headed by an autocratic Emperor or Tsar) as a result of the Great War (First World War). The revolution overthrew the *Russian Provisional Government*, which led to the *Russian Civil War* from 1918-1920, followed by the creation of the *Soviet Union* in 1922.
20. Gosplan **or** State Planning Committee (Russian: Госплан, Russian pronunciation: [g' plan]) was the committee responsible for economic planning in the Soviet Union. The word "*Gosplan*" is an abbreviation for Gosudarstvenniy Komitet po Planirovaniyu (Russian:, Государственный комитет по планированию, State Committee for Planning). One of its main duties was the creation of Five-Year Plans.
21. Adhikari, 1992, 138.
22. Hartshorne, 1959, 148.

REFERENCES

Adhikari, S. (1992) *Fundamentals of Geographical Thought*, Chaitanya Publishing House, Allahabad, India.

Dickinson, Robert E. (1969), *The Makers of Modern Geography*, Routledge and Kegan Paul, London.

Dickinson, R.E. (1976), *Regional Concept: The Anglo American Leaders*, Routledge and Kegan Paul, London.

Dikshit, R.D., (1997), *Geographical Thought: A Contextual History of Ideas*, Prentice Hall of India.

Hartshorne, R. (1959) *Perspective on the Nature of Geography*, Rand McNally & Co., Chicago.

Martin, Geoffrey J. (1994) "Richard Hartshorne, 1899-1992", *Annals of the Association of American Geographers*, Volume 84, Issue 3, pp. 480-492.

Pearcy, E. (1957) *World Political Geography*, Crowell Company, New York.

Rana, L. (2008), *Geographical Thought: A Systematic Record of Evolution*, Concept Publishing Company, New Delhi.

SECTION-III
PARADIGMS

CHAPTER-6

Paradigmatic Shifts

Any discipline always proceeds within a framework of some conceptual matrices, called 'paradigms'. Paradigms focus the philosophical perspectives and methodologies that have been in vogue and which continue to provide the theoretical feedback for new researches. Any model, idea, concept, technique or method that is capable of generating scholarly concern in a particular time period may be called a 'paradigm'. The term 'paradigm' was coined by Thomas S. Kuhn (1922-1996), an American Historian of Science, while postulating a very important theory about the growth and development of science. He wrote extensively on the history of science and developed several important notions in the philosophy of science. Kuhn is famous for his book *"The Structure of Scientific Revolutions"* (1962) in which he presented the idea that science does not evolve gradually towards truth, but instead undergoes periodic revolutions which he calls 'paradigm shifts'. Kuhn became as familiar to students of geography as Hartshorne or Humboldt. In his classic book, *"The Structure of Scientific Revolutions"*, he defined 'paradigm' as 'the entire constellation of beliefs, values, techniques, and so on shared by the members of a given community'. According to Kuhn, science is not a well-regulated activity where each generation automatically builds upon the results achieved by earlier workers. No scientific discipline progresses in a normal steady manner. This growth is a process of varying tension in which peaceful periods of steady growth are separated by periods of crisis. He further defined 'paradigm' as "universally recognized scientific achievements that for a time provide model problems and solutions to a community of practitioners". The paradigm tells researchers what they should be looking for and which methods they should use in this connection. A viewpoint becomes a paradigm if it has a large following and if it has a clearly defined theoretical and methodological base. In order to elucidate this process of development of science, Kuhn prepared a model that he termed as the 'Paradigm of Science'. Kuhn in his postulates advocated that development of science consists of: 'Pre-Paradigm

Phase', 'Professionalization Phase', 'Paradigm Phase-1', 'Crisis Phase with Revolution', 'Paradigm Phase-2', 'Crisis Phase with Revolution', 'Paradigm Phase-3', and so on (Fig. 6.1).

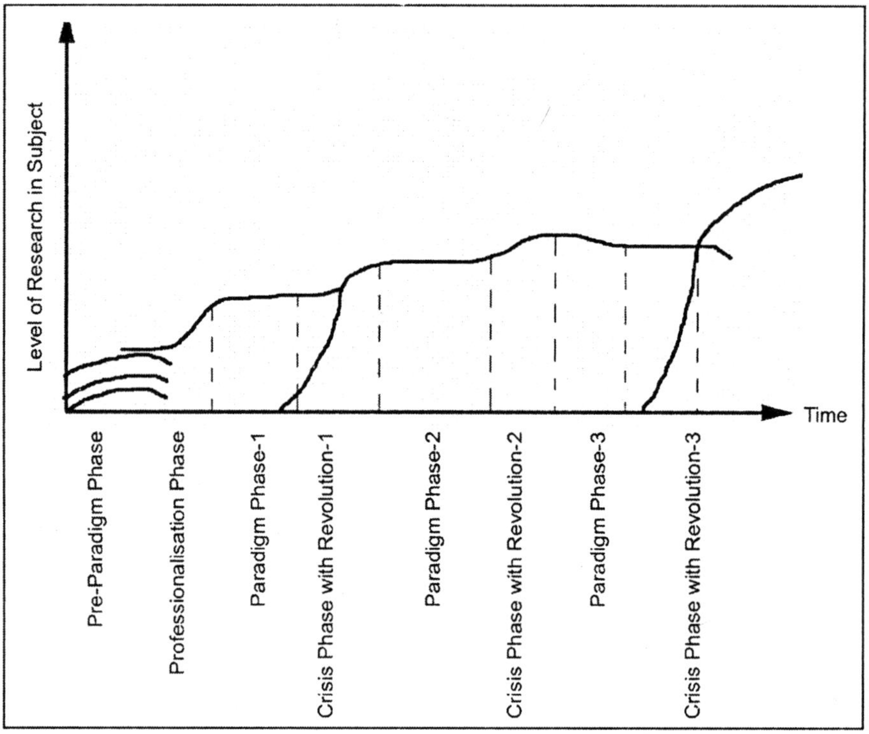

FIGURE 6.1 Paradigm of Science

According to Kuhn, the scientific knowledge progresses and develops like a plateau. There are sudden upheavals and abrupt rise, followed with smooth and slow progress. The first phase, i.e. *the pre-paradigm period*, is marked by conflicts between several distinct schools, which grow around individual scientists. This period is full of communication between various schools of thought. But none of these schools considers itself to be more 'scientific' than the other. There exists a low level of specialization in the discipline. From the pre-paradigm phase the scientific development marches and enters into the *professionalization phase.* Professionalization takes place when one of the conflicting schools of thought begins to dominate the others and thus a clear answer to the question is given. A particular school of thought may become dominant because it develops new methods or puts questions that come to be regarded as more interesting or significant. New researches are thereby attracted and the research makes progress. The third phase is the *paradigm phase-1.* It is characterized by a dominating school of thought, which has, often in quite a short space of time, displaced others. A paradigm is established which leads to concentrated research within a clearly distinguishable problem area, an activity described as *'normal science'*. After 'normal science' phase there occurs stagnation in research, which leads to chaos and turmoil. This period has been termed as the crisis phase (temporary dark phase) in the development of scientific knowledge. This *crisis phase with revolution* is the starting point for the *paradigm phase-2*, which in turn is followed with '*crisis phase*' and to again '*paradigm phase*'. This sequence of crisis, revolution and paradigm continues throughout the history of science and helps in the advancement and decline of disciplines. In this dynamic world, the period of 'normal science' also does not last indefinitely. A period of 'normal science' is sooner or later replaced by a 'crisis phase'. This occurs because more and more problems are accumulated which cannot be solved within the framework of the ruling paradigm. The crisis phase is characterized by a reassessment of former observational data, new theoretical thinking and free speculation. This involves basic philosophical debates and a through-going discussion of methodological questions. The termination point of the crisis phase, owing to the acceptance of a new paradigm, becomes the inaugural point of the next revolutionary phase. This involves a break in the continuity of research, with a through-going reconstruction of the theoretical structure of research field rather than a steady development and accumulation of knowledge. The understanding of the truth itself and the scientist's perception of the world can take on a new diversion. This simple evolutionary framework is, however, disrupted by the possibility that certain disciplines, especially in the social sciences (e.g. geography), to be multiple-paradigm sciences. In a multiple-paradigm science, many paradigms are competing for hegemony in the field.

PARADIGM OF GEOGRAPHY

The history of development of geographical thought is also not a story of continuous progress. The periods of intellectual progress have been interrupted by the periods of stability or regression. In the light of Kuhn's Model one can discuss the evolution of paradigms in geography through various stages in succession. Accordingly, six Paradigmatic Phases interrupted by six Crisis Phases may be identified during the whole course of development in geography till the contemporary times (Figure 6.2). However, the paradigmatic shifts in the discipline have been relatively faster in its modern period than earlier.

Pre-Paradigm Phase: Scholarly Observations of the Eastern and the Western Civilizations (Ancient Period)

In a way, enquiries of geographical nature had begun since the man first originated over this surface of the earth. Man's preliminary observations and reactions towards nature were of geographical type. As such, no other science can claim a longer genealogy than geography. However, as far as the literary evidences are concerned, they date back to about 5th Century B.C. It is during this time that some scholars in the West and the Orient had made observations of the world around. Although, the beginnings of the first paradigmatic phase in geography are associated with the West, the civilizations in the east, particularly those of the Indians and the Chinese, were

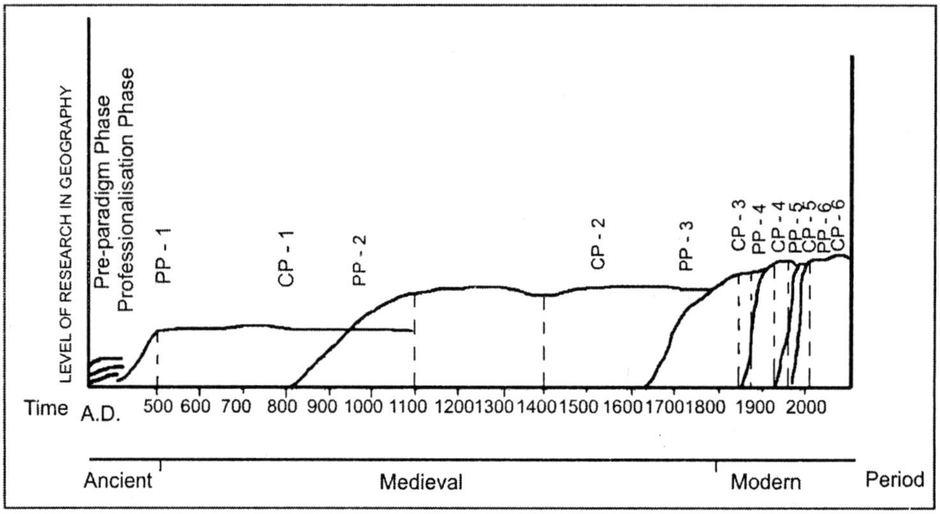

FIGURE 6.2 The Paradigmatic Shifts in Geographic Development

Pre-Paradigm Phase: Scholarly observations of the Eastern and Western Civilizations (Ancient Period)

Professionalization Phase: Concentration of Geographical Research in the West (Ancient Period)

PP-1: Paradigm Phase-1: Period of Greeks and Romans (up to 500 A.D.)
CP-1: Crisis Phase-1: Dark Age in Europe (500-1100 A.D.)
PP-2: Paradigm Phase-2: Rise of Arab School of Thought (800-1450 A.D.)
CP-2: Crisis Phase-2: Age of Discovery (1450-1650)
PP-3: Paradigm Phase-3: Dominance of German School of Thought (1650-1859)
CP-3: Crisis Phase-3: Emergence of Dualistic Tendencies in Geography (late 19th Century)
PP-4: Paradigm Phase-4: Rise of Determinism (late 19th century-1920s)
CP-4: Crisis Phase-4: Criticism to Determinism and Paradigmatic Shift to Human from Physical (Early 20th Century)
PP-5: Paradigm Phase-5: Dominance of French School of Thought and Rise of Possibilism (1930s-W.W.II)
CP-5: Crisis Phase-5: Quantitative Revolution (1940-1970)
PP-6: Paradigm Phase-6: Rise of Behavioural and Humanistic Approaches (1970-2000)
CP-6: Crisis Phase-6: Methodological Revolution (Computer Based) in the 21st century

also making efforts to gain knowledge of the human habitat and of man's spread over the earth. The studies of geographical nature began in India with the dawn of Indian civilization in the ancient times. The Indian contribution to geography came largely from astronomers and astrophysicists like Bhaskaracharya, Varahmihira, Aryabhatta

(A.D. 476), etc. The views of these scholars are available on the study of the Universe, the Earth and the Indian Subcontinent. The ancient Indian literature deals with many problems pertaining to cosmology, cosmogony and cosmography. The theory of Philosophical Origin of the Universe is very close the presently accepted theory of Nebular Origin. In the *Rigveda*, there is a description of various heavenly bodies including the Sun, the Moon, different planets and constellations. Observations about the earth are related to its origin, sphericity, eclipses, size and dimensions, latitudes-longitudes and local time, cardinal points, earthquakes and volcanoes, atmosphere and seasons, and physical divisions of this planet. The estimates of Aryabhatta regarding the earth's volume, circumference and mass are nearly correct and reliable. Geographical knowledge about the Indian Subcontinent is associated with its physical identification, people and culture, and relief and drainage. China was another major centre of learning in the ancient world. There are certain fascinating parallels in the concepts and methods of study with that of Indian and European civilizations. The most important contributions of Chinese scholars go to the fields of mathematical and practical geography, physical geography, human and regional geography, and the accounts of travels and explorations. The Chinese mathematics had developed the Decimal System, which was introduced to the Arab world in about A.D. 800 from the Indians. The Chinese were expert in map-making on the latitudinal-longitudinal grid, and this art is believed to have developed much earlier than in Greece or Rome. Also commendable are their observations on weather and climate, since they are known to have prepared a number of weather reports dating back several centuries before Christ. These reports are now preserved in the National Archives. The discovery of the rest of the world by Chinese travellers is an aspect of the history of geography that is often overlooked in western writings. However, with time the West became the centre of continuously increasing scholarly research.

Professionalization Phase: Concentration of Geographical Research in the West (Ancient Period)

Towards West the earliest developments took place in Western Asia (present Turkey, Lebanon, Syria and Israel), Egypt and Europe, and various cultural groups, namely Sumerians, Assyrians, Babylonians, Phoenicians, Egyptians, Greeks and Romans were involved. The geographers credit the beginning of geographical writings to Homer, a Greek scholar, but the earliest known map was made by the Pre-Greek civilizations (particularly Sumerians) about 2700 B.C. The scholars of Babylon had already collected a large number of data on the motions of the stars and planets, and they had developed the concept that the position of the celestial bodies had a fundamental effect on human activities.

Paradigm Phase-1: Period of Greeks and Romans (up to 500 A.D.)

Geography, as a discipline, has grown out of three related activities, viz. exploration, mapping and charting and speculation of the collected information. The first two activities were common among all the ancient civilizations, but the Greeks were the first to scientifically examine the geographical information collected up to their period. Thus, the first amazing period of intellectual ferment that became the part of the written tradition of the Western World was that of Greeks and the Romans. They are credited to have begun, both, the 'mathematical' and 'literary' traditions in geography in true sense. Besides 'geography', a number of other terms, like 'cartography', 'chorography', 'topography', 'frigid', 'torrid', 'temperate', 'parallels', 'meridians', 'delta', etc., which are part of present geographical dictionary, were given by the Greek and Roman scholars.

Crisis Phase with Revolution-1: Dark Ages in Europe (500-1100 A.D.)

There was a long period of decline during the Early Middle Ages, when geographic horizons closed in due to theocratic slant in human thinking and writings. The wide-ranging Greek and Phoenician explorers and the Greek geographical concepts were largely forgotten, except among the Arabic scholars. Observations piled up in Christian monasteries, but the intellectual climate was not favourable for the formation of new interpretations. In Europe, this period is known as Dark Ages. During this period, geography made a little progress as the organized body of knowledge. The quest for scientific learning stopped and there began a phase of regression. The approach to the science, including geography, became theocratic. The image of the world was moulded to fit the Biblical references. This kind of thinking had an adverse impact on the art of map-making, which got distorted. The tradition of making fairly accurate maps, as started by the Greeks and Romans, got lost. Instead, there emerged maps, which were purely fancy and unscientific. The ideal map of this period was *'orbis-terrarum'* ('T-in-O') Maps (Fig. 4.1). This type of cartography was in practice for a long time. These maps were made beautiful and extremely decorative, but they were very far from reality. The earlier concepts of sphericity and the nearly correct distances of places, latitudes and longitudes were ignored. When Europe was plunged into Dark Ages, the Arab School of Thought was rising to prominence in medieval times.

Paradigm Phase-2: Rise of Arab School of Thought (800-1450 A.D.)

There was rise of Islam in Arab in 7th century and the new zest of learning in the medieval period spread through the world of Islam. The vast expanse of the Muslim Empire facilitated the coming together of intellectual and scientific traditions of various civilizations, particularly Greek and Indian. The Arabian conquest of Palestine, Syria, Persia and Bactria, where Greek culture still flourished, gave to the Muslims the rich fruit of Greek learning. Their contacts with the Indians helped to diffuse the system of numerals from India to the West, and this system was called *'Hindsa'* in Arabic language. In the Muslim universities from Persia to Spain, the scholars studied the Greek and the Indian heritage again. Baghdad was the centre of intellectual world, and the other important centres of learning spreading throughout the Arabian Empire included Jerusalem, Cairo, Toledo, Istanbul, Damascus and Shiraj. Climatology, mathematical and practical geography, and regional and historical geography became the important areas of Arab contribution. Thus, the renaissance movement had begun in the times of Arabs in the medieval period. Starting in the 8th century the renewed study of Greeks revived much of the forgotten learning. But the classical spirit of inquiry was not revived. The renaissance remained primarily a religious movement, more concerned with preserving the knowledge of the past than expanding it through scientific investigation. Nevertheless, the Arabs did keep the Greek learning alive, and it was from their hand that the Europe received it, as the Dark Ages drew to a close. The Arab scholars and travellers presented a new outlook of the world, which later led to the onset of the Age of Discovery. The Arab civilization, in fact, became the main link between the ancient civilizations and the modern world. Arab Empire started disintegrating after A.D. 1400 and the Arab period culminated in the Age of Discoveries and Explorations.

Crisis Phase with Revolution-2: Age of Discovery (1450-1750 A.D.)

Largely inspired by the writings of Arab scholars, the Great Age of Discovery began in the late 15th century, and the geographic horizons were pushed back. The discovery of new lands, improvements in cartography and correction of old maps, and various scientific and technological inventions characterized this age. Later, various geographical societies were constituted to finance explorations and promote research in the discipline. With the improvements in the art of navigation and the use of magnetic compass the travels became very popular. Great explorers emerged in the scene and discovered the world that was unexplored earlier. Geography was enriched by an immense collection of facts and information about the world. Ptolemy's maps were restudied and the missing details were filled in and the discrepancies removed. For the first time, there appeared the complete and accurate world maps and globes. Mercator's projection was devised (1569) and the new maps were prepared on it. As a result of the developments like the establishment of the heliocentric concept of the universe; numerous astronomical observations by Nicholus Copernicus (a Polish scholar) about the planets, moons and stars; the presentation of Laws of Gravitation by Isaac Newton; and the invention of Printing Machine in Europe, there came a scientific revolution not only in materialistic terms but also in the ideas and thinking of man. All the disciplines, along with geography, were benefited. The dependence of geography increased on the other disciplines. The ground became ripe so that geography could move from consolidation to splitting and specialization. This brought in two fundamental innovations to shake the world of scholarship. First, the battle to establish the principles of what we call 'academic freedom' began; and Second was the 'separation' of the academic world into distinct fields, each devoted to the study of a specific group of related processes, and regulated its own paradigm based on its own theoretical structure. There emerged lot of scope for scientific investigations and generalizations began in geography. The renaissance period marks the end of recession phase in geographic development and a return to more scholarly research in the discipline. Post-Arab period was not only an age of discoveries; it was also a transitional phase from the medieval to the modern, i.e. from the old to the new. The subsequent period saw the emergence of first two modern schools of thought in geography, viz., German and French, one after the other.

Paradigm Phase-3: Dominance of German School of Thought (1650-1859)

The knowledge of geography was not organized in the late medieval and the early modern period. In the absence of an appropriate methodology, it was no more than encyclopaedic compilations of places and their descriptions. The principles behind spatial arrangement of phenomena, their causal connections and inter-relationships that could give geography a true scientific status were yet to be established. The credit for this goes to the German scholars. The Greeks laid the foundations and the Germans built the structure of geography. For the next 200 years after the Age of Discovery geography was purely a German science. Lot of research was carried on towards defining geography and deciding upon its methodology (approach, scope, content, method of study). Ultimately, the place of geography was seen established among the scientific disciplines. The initial impetus to the German School of Thought came from the works of Bernard Varenius (1622-1650) who brought together contemporary knowledge of astronomy and cartography and subjected the different theories of his day to sound critical analysis. He gave the first modern literature on geography in the form of *'Geographia Generalis'* published in 1650 itself. Through this he tried to provide a scientific methodology to geography by advocating two approaches for its study, viz. General (systematic) and Special (regional). Since, Varenius died a premature death, Humboldt and Ritter, popularly known as the fathers of modern geography, ultimately handled the task, respectively in the form of a natural scientist and a human geographer, in the years to come. In the meanwhile, writings of Emmanuel Kant (1724-1794) guided the research in Germany. Emmanuel Kant, an armchair geographer, was an outstanding example of a professional philosopher of the Western Thought. He considerably modified the scientific structure of geography as proposed by Bernard Varenius. He laid an exceptionalist claim not only for geography, but also for history. Therefore, he has been labelled as 'the father of exceptionalism' in geography by F.K. Schaefer, a 20th century American scholar. Schaefer considered Kant a poor and immature geographer. However, in spite of all the criticisms put forward against Kantian philosophy, the basic tradition of the field of geography has persisted in his writings, i.e. 'geography is a description of the whole earth, where the earth is understood to mean the surface of the earth and the dwelling place of the man'. The exposition of geography as 'chorology' by Alfred Hettner and Richard Hartshorne in the late 19th and early 20th century had closely followed the Kantian philosophy. Hartshorne's *"Nature of Geography"* (1939) also followed the line of thought first formalized by Kant. Through his book, it became the dominant concept of geography as a field of learning in the mid 20th century. On the other hand, Alexander von Humboldt and Carl Ritter revived the tradition of Bernard Varenius. In his best-known work, *Cosmos*, Humboldt regarded man as an integral and constituent part of the nature. Through this work, particularly, Humboldt was able to make geography "an original science" rather than a collection of facts from the physical and biological sciences. Two main philosophies, viz. the Natural Philosophy and the Moral Philosophy, dominated the minds of the scholars in the first-half of the 19th century. Natural philosophy was related to the physical world while the latter to man's inner world and opposed the former. But Humboldt regarded both of these in harmony and shared the concept that envisaged the earth as a whole, an organic whole, of all parts of which being mutually interdependent—an idea which runs right through his Cosmos. Humboldt followed systematic approach and his scientific procedure was empirical and inductive—contrasting to theoretical and deductive (as that of Kant). Humboldt included man and his works in the concept of nature and natural areas, but he did not consider man as a primary determinant. Humboldt recognized geography in the form of a scientific discipline. Carl Ritter's work was not only contemporary but also complementary to that of Humboldt. He exercised profound influence on geography by adopting a different methodology. Unlike Humboldt, Ritter was an anthropogeographer who had an inclination towards human and regional geography. Since Ritter was deeply a religious man, he gave teleological interpretations to man-nature relationships. He thought that divine has created the earth as a school of man and looked upon it as home of man. Ritter did identify geography as an empirical and descriptive science, but he himself was and armchair scholar. According to him, geography deals with local conditions and embraces the attributes of a place in the form of topographical, formal and material characteristics. These three attributes constitute the basic content of his most important work, *Die-Erdkunde,* based on regional descriptions of the earth covering largely Asia, Africa and Europe. Like *Cosmos* of Humboldt, Ritter's *Die-Erdkunde* is also a professional work—a product of his lifetime reading, selection, interpretation and organization of a mass of available data. This work was written with a distinct theme, i.e. comparative and inter-relationships. On the basis of this he proved that geography also has a right to be distinctively considered as a science with the same dignity as the other sciences. Ritter claimed that the central principle of geography is 'the relation of all phenomena and forms of nature to the human race, examined and organized within the framework of the unique geographical associations of land and man on the earth's surface'. Through his *Erdkunde,* Ritter made a rhetorical claim for geography as the science of the earth. He believed 'the earth to be an organism made, even in its smallest details, with divine intent, to fit the needs of the man to perfection; as the body is made for the soul, so is the physical globe made for mankind.' One has to overlook teleological approach of Ritter in order to discover a treasure of geographic insights in his works. His plan of study, based on the concept of regional associations of terrestrial phenomena, became the model for geographical presentations and had far-reaching influences. The classical period of modern geography terminated with the death of both Humboldt and Ritter in the year 1859.

Crisis Phase-3: Emergence of Dualistic Tendencies in Geography (late 19th Century)

The followers of Humboldt and Ritter exaggerated certain aspects of views of each of the founders. The geographers in the last decades of the 19th century clang to a dualistic concept of geography, i.e. the study of the earth as a natural body on the one hand, and as the dwelling place of man on the other. This gave birth to two dichotomies, viz. 'systematic *versus* regional' and 'physical *versus* human'. And, very soon the discipline entered a phase of double-dualism, i.e. 'systematic-physical' *versus* 'regional-human'. Thus, geography bifurcated in two main directions, and its position as a branch of knowledge was thereby brought into serious question. And, the geographers all-Around the world remained busy, at least up to Second World War, in seeking to establish the status of geography as an independent discipline, distinct in its concepts and procedures from other disciplines. Because, the foundations of geography as secured by Humboldt and Ritter did not provide a clearly unified field and, therefore, differences in the opinions and viewpoints persisted for a long time in the discipline. The emphasis of research shifted fast from one viewpoint to the other. Ritter's followers tended to emphasize the 'historical and human' aspects even more than he himself did. However, soon interest was deflected from Erdkunde and Ritterian philosophy was criticized. It was focused on systematic studies in a manner reminiscent of Humboldt, though there was no revival of Humboldt's broad interest in all aspects neither of physical geography nor of his idealistic stressing of the unity of nature. The increased interest was on physical or natural geography and the systematic studies in their new guise were highly specialized, and prosecuted by scientists who tended to be geomorphologists and climatologists. The systematic studies happily accepted in geography were, particularly, in the fields of geomorphology (Peschel, Frobel, Davis, and Penck), climatology (Buchan, Loomes, Koeppen, and Thornwait), plant geography (Haberlandt, Grisebach, and Warming) and oceanography. This was a time when 'systematic' and 'physical' became synonymous and the study of landforms emerged as the dominant and most attractive part of geography. This attracted the researchers not only from Germany but also from outside, especially U.S.A. The development that started after 1859 culminated in the works of W.M. Davis (1850-1934). In comparison, Human Geography lagged behind. Because, some geographers wanted to make geography a preserve of pure science. Since human behaviour did not appear to be subject to general laws, they proposed to exclude man entirely from the field of geography. This had a tremendous influence on geographic thought and it led to a long phase of controversies. These dichotomies overshadowed the discipline for about hundred years. The important doctrines emerging during this period were those of Determinism and Possibilism.

Paradigm Phase-4: Rise of Determinism (late 19th Century–1920s)

For one period of time, i.e. late 19th and early 20th century, physical geography dominated the scene. The maximum contribution came from geologists (Gilbert, Powell,

Agassiz and Davis). The emphasis was on morphological research and landform studies. At this point of time Ratzel corrected the prevailing tendency to overstress the physical aspects of geography and established a more balanced viewpoint. Indeed his work served to fix interest still more firmly on systematic studies, and regional geography continued to receive scant attention. The year 1859 marked not only the death of Humboldt and Ritter but also the publication of Charles Darwin's *"On the Origin of Species"*. Darwin's ideas on adaptation to the environment and on evolution stimulated social scientists to reappraise old concepts. New observations in the biological sphere seemed to provide the key to scientific understandings of differences in cultural levels and economic activities. Among geographers it was, above all, Freidrich Ratzel who explored the influences of physical environment on mankind. In *Anthropogeographie* (Volumes I and II, 1882 and 1891) and *Political Geography* (1897), his major contributions, Ratzel treated human geography systematically and from Darwinian's point of view. He stressed "man is a creature of his environment in much the same way as Darwin had demonstrated adaptation and survival of the fittest in the animal world." Accordingly, he attempted to show how the distribution of man on the earth had been more or less controlled by natural forces depending on the level of his cultural advancement. However, his second volume of *Anthropogeographie* showed a different spirit, where the emphasis was on distribution and density of population, forms of settlements, migrations, and diffusion of cultural traits. For the explanation of these features Ratzel did not merely resort to environmental influences but equally, or more so, historical-cultural factors. Ratzel saw man as the end product of evolution, in which the mainspring was the natural selection of types according to their capacity to adjust themselves to the physical environment. A deterministic tint thus coloured most of his writings. Ratzel's impact on American geographers was quite strong, owing to the interpretation of his ideas by Ellen Churchill Semple, his American pupil. Through his Political Geography, the other contribution, Ratzel reshaped the old topic along the lines conformable with the principles he had enunciated in his *Anthropogeographie*. In this work, he tried to make a comparative investigation of the relation between 'the political state' and 'the earth's surface'. The revolutionary biology led Ratzel to adopt the organic theory of state and society, i.e. the 'Concept of state as an Organism', a piece of humanity and a piece of earth.

Crisis Phase with Revolution-4: Criticism to Determinism and Paradigmatic Shift to Human from Physical (Early 20th Century)

By 1900, the philosophy of determinism had become particularly widespread in Germany. The thinking also percolated in U.S.A. through the works of Allen Churchill Semple, a pupil of Ratzel and a staunch supporter of determinism. But, from the beginning of the 20th century this doctrine came under attack. The chief fault in the assumption of the extreme determinists was found in their over-simplication of the man-environment relationships. Their generalizations were accused of being 'too-sweeping', without being backed by careful investigations and cause-and-relationships. The determinists over-emphasized the role of physical environment in human affairs.

Man was seen only as a passive element of the environment. Such works of man were overlooked for which environmental forces alone give no satisfactory explanation. Spate, for example, stressed that "environment taken by itself and without man is a meaningless phrase. Equally important is his indication of the need to consider the psycho-physiological influence of the geographical environment *via* social structure. Geographical environment is only one of the factors of territorial differentiation and it acts through society and cultural tradition." Hartshorne put similar argument forward. He rejected determinism purely on the grounds that the latter separated nature and man, and this he found disruptive to the fundamental unity of the field, contradicting the concept of geography as an integrating science. However, the school of determinism did not die out completely. The central theme survived and continued to persist in the form of 'Neo-Determinism' or 'Stop-Go-Determinism'. This philosophy grew stronger particularly around Second World War period. Supporters of this new school of determinism basically believed in determinism but not as rigidly as the earlier determinists. Where some of the early determinants were apt to make unsound generalizations, the neo determinants were more careful, to seek and record, wherever possible, a true sequence from cause to effect. They were aware that man might be subjected to many surrounding influences that are not purely physical ones. Where the neo-determinists differed most radically from the 'possibilists', they did not allow that man has complete freedom to choose, but rather that he narrows down alternative courses of action in his mind, and, in selecting one alternative rather than another, is, influenced (consciously or subconsciously) by his general environment. They were more careful and believed in theoretical and empirical examination of facts.

Paradigm Phase-5: Dominance of French School of Thought and Rise of Possibilism (1930s–Second World War)

Second quarter of the 20th century was devoted to research in human and regional geography. This stream of geography was reinstalled and over-emphasized particularly after the First World War, when a sociological trend was put forward through the writings of Vidal de la Blache, Lucien Febvre and Jeans Bruhnes in France and Isaiah Bowman and Carl Sauer in United States. Blache developed the concept that environment contains a number of possibilities, and their utilization is dependent almost entirely on human selection. This concept gave rise to the school of Possibilism. According to possibilists it is man who is the primary architect of his environment. It depends on the choices he makes among the numerous possibilities offered by the environment. In order to use these possibilities he makes two types of adjustments with the environment, i.e. in the form of adaptation and modification. For the possibilists, the works of man, not the earth, are the starting point and the most important is the freedom of man to choose. There are no necessities, but everywhere possibilities and man as a master of these possibilities is the judge of their use. In spite of the fact that man has numerous possibilities in a given physical setting, he cannot go against the directions laid by the physical environment. On this ground, the possibilistic approach was also criticised, and the possibilism, too, did not hold for many years. Thus, the man-environment debate continued in geography for about a century after the death of Humboldt and Ritter. The controversy could resolve only with a compromise being reached in the Ecological School of Thought followed by Quantitative Revolution. The ecological approach no longer accepted the sharp distinction between man and the external biophysical surroundings. It considered man not as the same creature everywhere and at all times, but diverse in outlook and capability, depending on the society to which he belongs. And each society perceives and interprets its biophysical setting through the selective prism of its culture. Thus, nature and culture are interwoven and complementary parts of a whole rather than opposing forces.

Crisis Phase with Revolution-5: Quantitative Revolution (1940-1970)

In the first half of the 20th century the ideas of Hartshorne dominated geographic thinking and were practiced by various scholars. In 1950s they came under attack. New generation of geographers felt that geography was not truly scientific in its approach. Laws and generalizations were largely restricted to natural sciences (physical geography) and human geography was generally devoid of such practice. The most severe statement of the view that geographers should seek for laws and theories came from F.K. Schaefer, who attacked the views of Hartshorne in a paper entitled "Exceptionalism in Geography: a methodological examination" (1953). Schaefer's paper brought a theoretical and methodological revolution in geography. And geography underwent a radical transformation of spirit and purpose, best described as 'Quantitative Revolution'. The mechanistic approach of environmentalism, which dominated much of the 19th century, was soon replaced by a more nomothetic and scientific approach. By 1970s, the revolution was over and a wide range of geographers, including Hartshorne, accepted that quantitative techniques could be usefully applied to the geographical problems. Thereafter, the best use of such techniques was made in the study of uniformities in locational patterns in geography. The earliest signs of this change were brought by British Geography, particularly through the works of Gregory, Haggett and Chorley.

Paradigm Phase-6: Rise of Behavioural and Humanistic Approaches (1970-2000)

During 1970s, when the need for precision in research through increased use of statistical-mathematical techniques had been commonly accepted, and the controversy regarding systematic *versus* regional geography resolved, with the scale tilting in favour of the former, but both being accepted as independent and equally useful components of the discipline, the euphoria about quantitative revolution had also greatly receded. Now, many workers had begun to feel disillusioned with the quantitative models being developed and used by human geographers of the positivist orientation, since they failed to relate to real-life situations. Consequently, geography again witnessed a paradigmatic shift, as many geographers began to feel that a deeper understanding of man-environment interaction could be achieved if geographers paid greater attention to the study of the mental images that the decision-makers have of the environment to which their decisions relate since it is the perceived reality that leads to action. The realization that decisions may be based on subjective or perceived reality than on objective reality as it exists in nature led to attempts towards derivation of theories of spatial behaviour; theories that concerned more with why certain activities take place rather than what patterns they produce. This shift in focus inevitably led to a shift to the behavioural sciences. The objective of behaviouralism was to replace the simplistic and mechanistic conceptions of man-environment relationships with a new theory that views human actors in space as thinking beings whose actions are mediated by cognitive processes (i.e. the mental processes, by means of which people acquire, organize and use knowledge). The focus was on studying how people come to terms with their physical and social milieu. The behaviouralist rejected the assumption that decisions are always based upon a sequence of reasoned thoughts involving systematic evaluation of all available options. The behaviouralist was concerned with the full spectrum from the sub-conscious impulse to the self-conscious deliberation in the process of decision-making.

Crisis Phase-6: Methodological Revolution (Computer Based) in 21st Century

The 21st century in geography is again marked with another revolutionary phase. With the widespread introduction of computer technology, geography has entered a new methodological revolution. In 1950s, such a change has been brought by the Quantitative Revolution, and now it is computer revolution, along with the use of the techniques of remote sensing and GIS. Today, we know geography as 'the science of spatial distribution'. It is not merely an academic subject but one dealing with numerous practical applications. Though the word 'Geography' is originally derived from the ancient Greek '*geographikos*', meaning a description of the earth. The modern geographer's scope extends far beyond describing the earth. Geography as a field of learning is related to significance of locations and spatial relations of things and events. This implies study of systems of areally inter-connected and inter-dependent parts of diverse origin. A geographer, therefore, is a person who asks questions about the significance of locations, distance, direction, spread and spatial succession. He deals with problems of accessibility, innovation, diffusion, density and other derivations of relative location. The geographer's ambit includes spatial analysis, the creation of theoretical models involving extensive use of mathematics, as an essential component. A geographer is interested in the earth, its produce and in people and their activities. He has mathematical ability and an ability to understand concepts besides the basic drawing skills. A geographer's work calls for patience and meticulousness to handle detailed tasks and analyze data. The revolutionary advances have been made in data collection and data interpretation with the use of computers, remote sensing and GIS techniques. Use of these techniques has brought a methodological revolution in the sciences and geography is not unaffected. Computers, remote sensing, GIS and related fields are opening up new avenues for geographers. It is, in fact, the field of the future. With the use of these modern techniques there may be collected and interpreted a variety of information about the earth's surface. Since geographers are directly concerned with analyzing the variable data on the earth's surface, this is going to open up vast areas for analysis and research. Modern geography is a spatial science increasingly focusing on the study of spatial aspects of social and economic phenomena, and thus being pre-eminently humanistic (anthropocentric). The humanistic tradition in geography is now reflected in the scholarly texts, e.g. by David Stoddart (1986), David Livingstone (1992), Derek Gregory (1994), Richard Peet (1998), and Marcus Doet (1999).

To conclude, formal definitions of the field of geography have been attempted at each stage of its growth, since ancient to contemporary times. For example, some geographers like Hartshorne, projected geography as a discipline concerned to provide accurate, orderly and rational description of the earth's surface. Others have laid greater emphasis on the inter-relations of man and environment. Then there are scholars who see the stress on a more limited range of problems concerning locational and spatial arrangements. However, none of the above definitions has really satisfied all the geographers, but these definitions have some common ground for explanation. We may, at least, identify few important themes through the various attempts at definition: (1) Geography shares with the earth sciences a common base, i.e. the earth' surface, which is not abstract space; (2) At the same time, geography is concerned with the earth as the environment of man, an environment that not only influences man and his activities, but also is influenced or modified by man; (3) Geography focuses on man's spatial organization and his ecological (balanced) relationships to his environment, through the use of its various resources and space; (4) Geography is sensitive to the changing nature of various phenomena of the earth; and (5) Geography is a dynamic discipline and not the static one. These basic themes have shaped and profoundly influenced the nature, scope and content of geography. Largely based on these themes various such viewpoints in geography may be defined which have persisted through times till today, although changing their forms from time to time. Such basic themes or viewpoints that continued to survive and shape the growth, nature, content and scope of any discipline, have been called as 'paradigms'. The persisting paradigms in geography with their exemplars are summarized as follows: (1) Paradigm of Area-Study (Regional): Areal-differentiation, landscape/landschaft/ paysage, chorological viewpoint, regional concept; (2) Paradigm of Man-Environment Relationship: Determinism, possibilism, ecological, behavioural approaches, radical and humanistic approaches; and (3) Paradigm of Space: Spatial distribution, spatial organization.

REFERENCES

Husain, M. (1988), *Evolution of Geographical Thought*, Second Revised Edition, Rawat Publications, Jaipur.

Rana, L. (2008), *Geographical Thought: A Systematic Record of Evolution*, Concept Publishing Company Pvt. Ltd., New Delhi.

CHAPTER-7

Major Paradigms

Beginning with the Greek and Roman period, through the definition of geography being as the description of the earth's surface, Ritter's regional concept, Humboldt's systematic geography, Darwin's environmentalism to the present nomothetic, quantitative and more recent environmental perception approaches applied to geography—we find three major themes (paradigms) have been continuously applied in geographic research. They are Environmentalism, Regional and Spatial.

ENVIRONMENTALISM

People, place and time may be thought of three basic elements out of which geography is fashioned. The relationship of man with nature and the resultant pattern or description over space in a particular time is the core of all geographical studies. Human beings have claimed their contribution using natural resources. With the help of technology, human beings moved from the stage of necessity to a stage of freedom. They have put their imprints everywhere and created new possibilites in collaboration with nature. Thus, we now find naturalized human beings and humanized nature; and the environmentalism studies this interactive relationship. This relationship, in the history of geographical thought, has been viewed in three ways, viz. Determinism; Possibilism; and Ecological Viewpoint. The proponents of 'Determinism' regarded environment to be superior than man; the believers of 'Possibilism' gave man all powers to choose anything better for his living, at his own will and apparent freedom; whereas the 'Ecological Viewpoint' gave equal weightage to both man and environment, one affecting the other in a reciprocal manner. Much of the geographical work since the mid 19th century, however, has either explicitly or implicitly taken its inspiration from biology, and in particular from Charles.[1] Darwin established a sphere of scientific enquiry free from *a priori* theological ideas, and freed natural science from the argument of natural theology (science of God). And finally, and in this he was alone, Darwin established man's place in nature, and in doing so he made man a fit object for the scientific study. Modern geography is inconceivable without these general advances. Darwinism ultimately paved the way for determinism.

Determinism

Determinism is one of the oldest schools of thought. The philosophy of environmental determinism rose to prominence in the late 19th and early 20th century when it was taken up as a central theme by the discipline of geography. The view persisted up to the Second World War in one form or the other. Supporters of the determinism favour the view that environment controls the patterns of man's behaviour, so that all his actions, his settlements, his modes of life are controlled, directly or indirectly, by his surroundings. They believed that nature is superior to man. According to the determinists, man is a passive agent on which the physical factors are acting and determining his attitudes and the process of his decision-making. The essence of the deterministic school of thought is that the history, culture, living style and the stage of development of a social group are exclusively or largely governed by the physical factors of the environment. The origin of deterministic thinking goes back to antiquity, when the Greeks and Roman scholars made first attempts to explain the physical features and character traits of various people and their culture with reference to the influence of natural conditions. Hippocrates (born in 460 B.C.), for instance, discussing his work "on Airs, Waters and Places", commented on the easy going life of Asians endowed by nature with very favourable physical environment, thus making them inactive, relaxed and slackened. In comparing life of Asians with those of Europeans, who live in relatively harsh physical environment, he found the latter as hard working. Aristotle repeated the same generalization about the national character and added the political element. Similarly, Strabo, in his geographical accounts established a close relationship between the shape, relief, climate and space and the rise and strength of Roman Empire. The geographical determinism continued to reflect the writings of Arab Scholars in the medieval period as well. The Arab Scholars divided the habitable world into 7 'terrestrial zones' (*Kishwars*) and highlighted the physical and cultural characteristics of races and nations living in these zones. Al-Battani, Al-Masudi, Al-Idrisi, Ibn-Khaldun, etc. attempted to correlate the environment with human activity and their mode of life. Al- Masudi, for example, asserted that the land where water is abundant, the people are gay and humorous, while people of dry and arid lands are short tempered. The nomads who live in open air are marked by strength and resolution, wisdom and physical fitness. Determinism was also strongly observed in the work of Emmanuel Kant (1724-1794), who stated that the people of New Holland (the Dutch/Northern Brazil) have a half closed eyes and cannot see to any distance without bending their heads back until they touch their backs. This is due to innumerable flies that are always flying in the eyes. Kant further stressed that all the inhabitants of hot lands are exceptionally lazy and timid. Timidity engenders superstition and in lands ruled by kings leads to slavery. In support of his hypothesis of influence of climate, he stated that animals and men that migrate to other countries are gradually affected by their environment. For example, the brown squirrels that migrate to Siberia turn gray and the colour of white cows in winters turns greyish. Humboldt and Ritter, on the other hand, looked on this 'man-environment relationship' more scientifically, i.e. 'cause and effect relationship'. Therefore, the development of astronomy in Arab, they attributed to the physical environment—open skies and desert lands where astronomical observations are easier to be made. The philosophy of determinism, in turn, was greatly influenced by the high profile of evolutionary biology of Charles Darwin (1809-1882), the English natural scientist. After the publication of the theory of Darwin[2], a change was seen in the development of deterministic viewpoint as well. Because, the pre-Darwinian views on Determinism were found relatively more modest in comparison to that in post-Darwinian period, when they attained more extreme form. And the scholars started talking about 'environmental control' instead of 'environmental influence'. Darwin achieved lasting fame by convincing the scientific community of the occurrence of evolution and proposing the theory that this could be explained through natural selection. This theory is now considered the central explanatory paradigm in biology. Darwin compiled evidence for evolution by natural selection for about 20 years and published his theory in 1859 in the form of *On the Origin of Species by Means of Natural Selection,* or *The Preservation of Favoured Races in the Struggle for Life* (usually abbreviated to *The Origin of Species*). The maximum influence of deterministic thinking was exercised after the popularity of Darwin's Theory. The scientific developments in geography that followed Darwin were deeply influenced by the role of nature on organisms, including man. This role of nature was deeply studied in human affairs. In other words, it was tried to prove that man's houses, communications, agriculture, his physiographic characteristics, nature, attitudes, capacities, general way of life and beliefs all are determined by the nature or environment (particularly including climate, relief, soil and vegetation). Gradually, the term 'environmental influence' was replaced by the 'environmental control', i.e. the environment does not simply influence but determines. The observations of various scholars caused them to conclude that 'man's immediate surroundings have a decisive (deterministic) effect not only on his occupations, dwellings and the materials he uses in everyday life, but also on his personal traits, his general character, and even his religious beliefs.' They found environment to be superior to man and mastering all of his actions. Such generalizations were called 'extreme' and not as the result of 'cause and effect' relationship between man and environment. Thus, there emerged extreme form of determinism in the late 19th century. Such assumptions or explanations were generally based on incomplete evidence or research. The scientific milieu in the latter half of the 19th and the early 20th centuries was dominated by Darwin's idea, deductive approaches and an acceptance of the Newtonian cause and effect relationships. Fitting well into this intellectual environment, the theme of environmental determinism was prevailing view not only in Germany but also in American geography. Ratzel supplemented 'classical' geographical determinism with the elements of 'Social Darwinism'[3] and developed a 'theory of the state as an organism' that owed its life to the earth, which was ever striving to seize more and more territory. In the opinion of

Ratzel "similar locations lead to similar mode of life". He cited the example of British Isles and Japan and asserted that both these countries have insular locations, which provide natural defence against the invaders. Consequently, the people of these countries are making rapid progress. Ratzel, a follower of Darwin, believed in 'the survival of the fittest' and saw 'man' as the end product of the evolution, an evolution in which the mainspring was the natural selection of types according to their capacity to adjust themselves to the physical environment. He was convinced that the physical features and location of a place in relation to mountains and plains closely influence the course of history, the mode of life of the people and the stage of development. In his deterministic approach he gave more weightage to location in relation to topographic features. At the beginning of the 20th century 'environmentalism' became particularly widespread in the United States, where its leading proponents were Ellen Churchill Semple (1863-1932) and Elsworth Huntington (1876-1947). Based on the philosophy of determinism she wrote two books, namely *"American History and its Geographic Conditions"* in 1903 and *"Influences of Geographic Environment"* in 1911. According to Semple, there are four ways in which the environment can influence the man and they are: (1) *Physiological Influences* (On the features, stature and physical appearance of the human beings); (2) *Psychological Influences* (On the level of thinking, beliefs, knowledge of the human beings); (3) *Influences on the Economy and Society* (On the occupation, culture, traditions of the people); and (4) *Influences on the Human Mobility* (On the mode of living: sedentary or practicing transhumance). Elsworth Huntington wrote the monumental book *"The Principles of Human Geography"* in 1945, was a protagonist of Environmental Determinism. He explained more in terms of 'Climatic Determinism'. Huntington assessed the role of climate on man at three levels, viz. Global, Regional and Personal. The basic philosophy of Huntington was that the supreme achievements of civilization in any region were always bound up with a particular type of climate and variation in climate led to 'pulsations' in the history of culture. He associated with the climatic cycles the 'Golden Age' in ancient Greece, the Renaissance in Western Europe, and cyclical fluctuations in iron production or the price of shares. Huntington divided the world in the 'mild' and 'harsh' climatic zones and established that the ancient civilizations (Egyptian, Mesopotamian, Chinese and Indus) flourished in the fertile river valleys of mild climates. He also established the hypothesis of invasion and tribal warfare. The great outpouring of nomadic people from Central Asia which led to Mongol Conquest of Turan, Turkistan, China and India and the raids in Eastern Europe in the 13th century could be explained by the dying of pastures on which the nomads were dependent. According to Huntington, the religion and racial character are the products of climate. A temperature of about 20°C with variable atmospheric conditions is good for the high mental and physical efficiencies. Such a climatic condition is found in the countries of North-West Europe. Huntington has attributed the advancement of Europeans in the field of science and technology to climatic conditions. The under-development of Tropics, he explained, is owing to the humid, hot, oppressive weather that makes the people lethargic, lazy, inefficient and timid. The subsequent geographers like H.J. Mackinder, Chislomn, Davies, Bowman, Robert Mill, Geddes, Herbertson, Taylor, etc. interpreted the progress of societies with a deterministic approach. Many of the scholars made it vividly clear that the climate influences the physical properties of soil that ultimately determines the cropping patterns, dietary habits, physique and attitudes of human societies. In the 1920s and 1930s, environmental determinism came under attack as its claims were found to be hopelessly inadequate at best, and often destructively wrong. The chief faults in the assumptions of the extreme environmentalists lay in over-simplification in the approach dealing with the issues related to the study of man-environment relationship. Man is undoubtedly profoundly affected by physical conditions. Environment does strongly affect the nomad, and has almost certainly contributed to the nature of religious beliefs of such peoples. But assertions of the kind quoted above are too sweeping, and were not, in the main, backed by careful investigations that modern environmentalists employ in their search for a sequence from cause and effect. The examples above represent the more extreme opinions of a school of determinism which has few followers today, and which went beyond the ideas of the early exponents of environmental control. From the perspective of the late 20th century, it appeared that the determinists over-emphasized the role of environment in human affairs. This does not imply that environmental influence is inconsequential or that the human geographer should not study such an influence. Rather, it suggests that the physical environment is only one of the many forces affecting human culture and rarely the sole determinant of human behaviour and beliefs. The crude determinism was a one-sided approach, as there was a lack of cause-and-effect relationships. As a result, Determinism started fading away in the beginning of the 20th century. Particularly after Second World War, the philosophy of determinism was attacked vigorously. Many geographers in the USA, France, Britain, Canada and other countries were drawing attention to the one-sided approach adopted by the environmentalists in their interpretation of historical reality to their exaggeration of nature's active role and the fact that they only acknowledge man as capable of passive attempts at adaptation. Works of man reveal many facts for which environmental forces alone give no satisfactory explanation. The natural environments may be modified by the technological improvements in order to make the inhospitable and harsh physical situations worthy of human occupancy. O.H.K. Spate (1952) criticized the fanatic approach of environmental determinists. He, for example, stressed "environment taken by itself is a meaningless phrase. Without man environment does not exist. Equally important is his indication of the need to consider the psycho-physiological influence of the geographical environment via the social structure. In the final analysis Spate concluded that geographical environment is only one of the factors of territorial differentiation and it acts through society. Cultural tradition has a certain autonomous influence." Hartshorne also put similar arguments forward. He rejected environmentalism purely on the grounds that it separated nature and man and thus was 'disruptive of fundamental unity of the field'; and contradicted the concept of geography as an integrated science. As a consequence of the above criticism of the doctrine of crude Determinism faded away in the second quarter of the 20th century, but after few decades it revived in a new form known as 'Neo-Determinism'. In the meanwhile there emerged the doctrine of "Possibilism" in France as an antithesis to "Determinism".

Neo-Determinism

In spite of its vigorous criticism the School of Determinism did not die away completely, its central theme continued to survive in the form of Neo-Determinism. Where some of the early environments were apt to make unsound generalizations, the modern determinists were more careful, to seek and record, wherever possible, a true sequence from cause to effect. They were aware that man may be subjected to many surrounding influences that are not purely geographical ones. Sometimes, geographical influences are not strong, nor their effects obvious, but to the environmentalists these geographical influences are always present as part of the general environment which determines man's course of action. The modern determinists differed radically from the 'possibilists' in their view on the nature of human 'choice'. They did not allow that man has complete freedom to choose, but rather, that he narrows down alternative courses of action in his mind, and, in selecting one alternative rather than another is, he is consciously or subconsciously influenced by some part of his general environment. The supporters of this new school of determinism basically believed in determinism but not to the extent as earlier determinists did. They did not use unsound generalizations; they believed in cause and effect relationships between man and nature; they were more careful, and applied theoretical and empirical examination of the facts with the help of modern techniques. Griffith Taylor used the term 'Stop-Go-Determinism' for Neo-Determinism[4]. Taylor maintained that it may be that the well-endowed parts of the world offer a number of different possibilities for making a living; but in most of the earth's land area, nature speaks out clearly—"this land is too dry, or too cold, or too wet, or too rugged". He wrote that the settlers who fail to heed this nature-given limitation must face disaster. According to Taylor, the role of nature was far from crudely deterministic, but the environment was nevertheless a powerful force in human action which man could ameliorate but not escape. It is like the case of the flow of traffic on a busy road. The traffic policeman cannot wish away the traffic that must necessarily flow. All that he does is to regulate the traffic through the temporary stop-go-method. Similarly, the human agency through the use of technology can modify the force of nature but it cannot escape it. The role of human agency is similar to that of the traffic regulator[5].

Possibilism

After the First World War the sociologist trend, put forward by French School of Thought to explain the stage of development of different societies and their history, was termed as 'Possibilism'. The possibilistic thinking grew stronger in the early 20th century due to technological and scientific innovations. Several developments proved the man's conquest over nature. The geographers, particularly those who entered the field after training in history, instead of natural science, have tended to stress this freedom of man to choose. For them the pattern of human activity on the earth's surface is the result of the initiative and mobility of man operating within a frame of natural forces. Without denying the limits every environment sets to man's ambition, they emphasize the scope of man's action rather than these limits.[6] The philosophy of Possibilism is closely linked with the writings of Lucien Febvre, Vidal de la Blache and Jean Bruhnes in France, and Isaiah Bowman and Carl Sauer in U.S.A.The essence of the possibilistic approach to the problem of relationship between man and his environment can be summed up as: "Nature does not drive man along a particular road, but it offers a number of opportunities from which man is free to select". For the possibilists "the work of man, not the earth and its influence are the starting points, the most important being the freedom of man to choose". There are no necessities, but everywhere possibilities; and man as master of these possibilities is the judge of their use. This by reversal that it involves put man in the first place, man and no longer the earth, nor the influence of climate, not the determinant conditions of localities. However, the Possibilities have never claimed that man can entirely free himself from all environmental influences. The debate between determinism and Possibilism continued for more than half a century. According to the determinists 'similar surroundings should generate similar life styles or cultural groups'. But this is not the reality. Because the nature of man's response to the environment depends to a large extent on the particular state of cultural advancement and the extent to which he is mentally and technically equipped. For man, the natural surroundings hold inherent possibilities, out of which some may be chosen by man to be developed and the others may be ignored. Thus, human choice plays an important role in the study of 'man-nature' relationships. This is the basic theme of Possibilism. Possibilism does not deny the influence of environment, but studies man-environment relationship from human point of view. Man has a selective power. He is free to make choices. The possibilists do agree that this selective power has greater scope of choices if the environment is favourable. But when the natural conditions are not favourable or man is not aware of technological developments, human choices are also limited. For example, for the mountain fold of Borneo (or anywhere else), the choice may be between an effort to overcome local hardships or to migrate. The belief is that every region possesses a combination of environmental features that offer a number of opportunities, of which man may or may not take the advantage. It is usually the lack of 'know-how' that prevents people from fully exploiting the possibilities of any region. For example, the soils of the Northern Territories of Ghana (Africa) have been little utilized as yet though many areas are basically fertile, and even though there is over-population in Ashanti, to the south. It has taken the modern disc plough and the advice of soil scientists to open up the possibilities of considerable settlement for the native peoples, and allow them a livelihood that can be based upon crops grown locally on these soils. Yet in Denmark, the Government deliberately altered the form of land use, turning a grain producing country, naturally suited to the production of wheat, into a land of highly organized dairy production; their option, prompted by the competition of cheap grain from the New World, apparently indicating that they had chosen to adopt one of the several possible ways of exploiting the land. According to possibilists it is man who is the primary architect of his culture. It depends on the choices he makes among the possibilities offered by the environment. In order to make use of the possibilities of nature man makes two kinds of adjustments, viz. Adaptation and Modification. The basic concept of Possibilism does not exclude the nature totally from its definition; rather it gives more importance to human choice. The human choice, in turn, depends on his *needs, tastes* and *capacities*. That nature is the base of human activities is well admitted even by the possibilists. Vidal de la Blache, through his important works as "*Geographie Universelle*" and "*Regional Geography of France*" formally put forward the concept of 'Possibilism'. By this he meant that in a given environment, in which a group of people lives, there is no necessity that the people should be influenced in a certain way, there is only possibility that they may react in a number of different ways, depending upon their cultural heritage. Vidal de la Blache is also known as the founder of modern French Geography. He consistently expressed outright opposition to environmental determinism. According to him the nature does not directly dictate man's behaviour. It only offers opportunities and human society makes a choice. This choice is not free and arbitrary one, but it is guided and restrained by the mental and social patterns of the group and their level of technology and cultural advancement. In other words, man conceives the nature of his habitat through the prism of his culture. Central to Vidal's work were the 'life-styles' (*genres de vie*) that develop in the different geographical environments. In his opinion, the life-styles are the products and reflection of a civilization; representing the integrated result of physical, historical and social influences surrounding man's relation to milieu in particular place. He tried to explain the differences among groups in the same or similar environments, and stressed that these differences are not due to the dictates of the physical environment but owing to the variations in attitudes, values, habits, etc. Variations in attitudes and habits create numerous possibilities for the human communities.

The possibilists emphasize that it is impossible to explain the difference in human society and the history of that society without reference to the influence of environment; and they hold that man himself brings his influence to bear on that environment and changes it. After Vidal, possibilism continued to grow and spread on both sides of the Atlantic. In France, Jean Bruhnes, the pupil of Vidal de la Blache himself, was a strong supporter of possibilism. Outside France as well, the possibilistic ideas were accepted by a large number of geographers and anthropologists. Carl Sauer, from the American School, asserted that geographer's role is to investigate and understand the nature of the transition from the natural to the cultural landscape. From such an exercise the geographer would identify the major changes that have occurred in an area as a result of occupancy by succession of human groups. Its importance is often greater in regions where it has been acclimatized than in those where it originated and domesticated. For example, wheat does not have the largest yields in regions where it was first domesticated (Southwest Asia). Cultivation of rice is now done largely in USA, Canada, Australia and India, the places where it diffused later. Isaiah Bowman, another American Scholar, also held the similar views. According to him, men can never entirely rid themselves whatever they do, of the hold their environment has on them. Taking this into consideration they utilize their geographical circumstances more or less completely of their geographical possibilities. But here or elsewhere there is no action of necessity. The limits set by nature by man's action vary from one historical period to another. In marginal environments, such as the hot and cold deserts, and at low stages of culture man's choices may be extremely restricted. In the more favourable areas of the warm and cool temperate zones, and in periods when man's techniques are highly developed, the possibilities are more numerous. But however many skills man acquires, he can never free himself entirely from nature's control. This is emphasized over and over again by Possibilists. Bowman asserted: "While the physical laws to which mankind responds are available in their application and degree of effect, yet this is also true that all men everywhere are affected to some degree by physical conditions."

In spite of the fact that man has numerous possibilities in a given setting, he cannot go against the directions laid by the physical environment. On this ground, the possibilists' approach has been criticized by many of the contemporary thinkers. Therefore, like Determinism, the Possibilism also did not hold for many years. Griffith Taylor (1951), while criticizing Possibilism stressed that society as a whole should make the choices, and since an advisory role is assigned to geographer, his function "is not that of interpreting Nature's plan". Taylor was largely right when he wrote that the task of geography is to study the natural environment and its effect on man, and not all problems is connected to man or the 'cultural landscape'. Moreover, Possibilism does not encourage study of geographical environment and it promotes over-anthropocentricism in geography. Geographical determinism, at least, obliges the geographer to turn attention to nature; and if the question is asked as to who is setting out to destroy geography, then the blame should be placed above all at the possibilists' door.

Ecological Viewpoint

Man's relation to the natural environment is one of intricate and perpetual association historically, culturally and biologically. The very conception of the natural environment is moulded by human history and culture. The natural environment is not to be understood as a causal factor but rather as an all-pervading and all-enveloping condition of human life. The history of the world with all its complexity and the culture of the world with all its constituent varieties belong only in the environment of this world and its regions—as living things, including man, are conditioned by their natural environment. The natural environment can be understood by only in terms of human life, history, and culture; and, conversely, the human life, history, and culture can be understood fully only in the natural environment. Separate studies of natural environment may have analytical value, but only as abstractions made in full view of the indivisible totality. Human life and environment are intimately interwoven in every aspect of their being, biologically and culturally, from the beginning of life on earth. Both nature and man are intrinsic to the particular character of areas, and indeed in such intimate union that they cannot be separated from each other. It is also in this relationship that geography has faced two of its most difficult methodological problems, of the dualism between man and environment, and that between human and physical geography.[7] The ecological viewpoint has been used to overcome these problems. It provides an alternative approach to a central theme in geographical enquiry, that of the relationship of man and environment in area. In man and environment studies, the ecological approach is the latest approach cutting through the themes of both 'determinism' and 'possibilism'. Neither of these two concepts could provide the discipline an analytical tool of sufficient power to lead to new insights. Only in the 1960s did the geographic profession become alive to the methodological potentialities of the ecology and the ecosystem concept.[8] According to the ecological approach the man and environment are equally important. The ecological approach is the most scientific approach and at the same time most commonly accepted by human geographers. With the adoption of ecological viewpoint, geographers got rid themselves of naïve determinism and misinterpretation in both physical and human geography. The ecological concepts provided a research method that geography so badly lacked.[9] The concept of 'Ecosystem' lies at the base of the philosophy of Ecological Viewpoint. Ecology is the scientific study of the relations that living organisms have with respect to each other and their natural environment. For the study of human geography, it only suggests the man and environment relations, and does not allow anyone to be supreme power over the other to exert its influence. It describes the cause and effect interplay between cultural and physical environment. Man and nature are acting and reacting together. As a human science, there are many practical applications of ecology in conservation biology, wetland management, natural resource management (agriculture, forestry, fisheries), city planning (urban ecology), community health, economics, basic and applied science and social interaction (human ecology).

Ecosystems sustain every life-supporting function on the planet, including climate regulation, water filtration, soil formation (pedogenesis), food, fibers, medicines, erosion control and many other natural features of scientific, historical or spiritual value. The scope of ecology covers a wide array of interacting levels of organization spanning micro-level (e.g., cells) to planetary scale (e.g., ecosphere) phenomena. The term *ecosystem* was formally proposed by Arthur George Tansley, the plant ecologist, in 1935. Tansley was an English botanist who was a pioneer in the science of ecology. It was proposed as a general term for both the biome ('the whole complex of organisms—both animals and plants—naturally living together as a sociological unit) and its habitat.[10] All the parts of such an ecosystem—organic or inorganic, biome and habitat—are regarded as interacting factors that are in equilibrium. It is through their interactions that the whole system is maintained. Tansley's ecosystem concept effectively broadens the scope of ecology, which is no longer purely biological in content. An ecosystem is a functioning interacting system composed of one or more living organisms and their effective environment, both physical and biological. It is the self-regulating association of living plants and animals and their non-ling physical environment.[11] The ecosystem system concept has four properties that recommend it in geographical investigation. They are: (1) The ecosystem is monistic, i.e. it brings together environment, man, plant and animal worlds within a single framework, within which the interaction between the components can be analyzed. The ecosystem concept being functional in design, offers a sound principle for geographical analysis of man-environmetal interaction in specific areas and regions. Since the ecosystem concept combines both physical and the biological environment in a single interacting system, it helps to resolve the age-old problem of dualism in geography—that between physical geography (focused on the study of the physical environment) and human geography (concerned with the study of human organization and its spatial consequences); (2) The ecosystem is structured in a more or less orderly, rational and comprehensible way. The essential fact here, for geography, is that once the structure is recognized, it may be investigated and studied, in sharp contrast to the transcendental properties of the earth, and its regions as organisms or organic wholes; (3) The ecosystem is a functioning system involving continuous throughput of matter and energy. In geographic terms, the system involves not only the framework of the communication net, but also the goods and people flowing through it. Once the framework has been defined, it may be possible to quantify the interactions and interchanges between component parts, and at least in simple ecosystems the whole complex may be quantitatively defined; and (4) The ecosystems are a type of 'General System', and the ecosystem processes the attributes of the general system. In general system terms, the ecosystem is an open system tending towards a steady state and obeying the laws of open-system thermodynamics (the science of the relations between heat and other forms of energy). Many of the properties of such systems have been implicitly recognized. For example, the idea of climax in vegetation, of maturity in soils, and of grade in geomorphology. Ecosystems may be conceptualized at different levels of complexity; e.g. from a single farm unit to the national system of agriculture in any country. Accordingly, with the adoption of the concept of ecosystem, geographical systems may be examined at a series of levels or scales, beginning with the framework level (such as settlement hierarchies or a transportation network), to simple information systems focused on the analysis of the mechanisms of supply and demand, to the still more complex levels of social organization. Ecosystems are hierarchical systems that are organized into a graded series of regularly interacting and semi-independent parts (e.g., species) that aggregate into higher orders of complex integrated wholes (e.g., communities). In the last few years it has began to be applied by geographers both as a research tool and as a methodological instrument offering an alternative to that of Kant and Hettner. It links geography with the mainstream of modern scientific thought, in systems analysis and related disciplines, and opens up as yet unexplored possibilities in the application to geography of the whole field of information theory and communication techniques. In the ecosystem concept, the ecology makes its most profound and powerful contribution to geography. In the period immediately following Second World War, many changes took place in geography and it was largely through the American School. The single, overriding change that characterized this phase was the shift away from the physically biased geography to a more human-oriented focus. This shift became a very significant one in trying to to balance both the physical and human aspects of the discipline. This reorientation in geography involved a few main ideas. The first of these was the idea of geography as 'human ecology'. The idea of human ecology in geography was proposed in 1923 by Harlan Barrows, a leading member of the faculty of the University of Chicago. Human ecology is the interdisciplinary investigation into the ecology of our species. "Human ecology may be defined: (1) from a bio-ecological standpoint as the study of man as the ecological dominant in plant and animal communities and systems; (2) from a bio-ecological standpoint as simply another animal affecting and being affected by his physical environment; and (3) as a human being, somehow different from animal life in general, interacting with physical and modified environments in a distinctive and creative way. A truly interdisciplinary human ecology will most likely address itself to all three."[12] Human ecology is the sub-discipline of ecology that focuses on humans. More broadly, it is an interdisciplinary and transdisciplinary study of the relationship between humans and their natural, social, and built environments. Ecology is not just biological, but a human science as well. Herbert Spencer has an important place in the history of human ecology. Spencer was influenced by and reciprocated his influence onto the works of Charles Darwin. Herbert Spencer coined the phrase *survival of the fittest*. He was an early founder of sociology where he developed the idea of society as an organism, and he created an early precedent for the socio-ecological approach that was the subsequent aim and link between sociology and human ecology. Human ecology is the discipline that enquires the patterns and process of interaction of humans with their environments. Human values, wealth, life-styles, resource use, and waste, etc. must affect and be affected by the physical and biotic environments along urban-rural gradients. The nature of these interactions is a legitimate ecological research topic and one of increasing importance.

REGIONAL CONCEPT

Like 'environmentalism' the 'region' is also one of the important paradigms in geography (Whittlesey, 1954). The aim of geography is to study the world, in which we live. Clearly, the world is so complex that is cannot be studied meaningfully at once and as a whole together. This is true of the human as well as the physical world. So vast is the range of human activities—economic, political, social—and so immense the variety of distribution and patterns of man and man's works over the globe that areal sub-division is a matter of necessity. The areal sub-divisions can best be achieved by dividing the world into regions.

Definition

A number of definitions are available for 'region'. A region, broadly speaking, is a part of the earth's surface which is distinguished in some defined ways from the surrounding area. The word 'region', means an uninterrupted area possessing some kind of homogeneity within its more or less well-defined limits. A region, simply stated, is an area of any size throughout which accordant areal relationships between phenomena exists. But it is different from area, which is universally used to mean a geometric portion of earth's surface, with no implications of homogeneity and cohesion.

The common characteristics of a region and an area are that both possess some scale (dimensions), contents (features), location and more or less clear boundaries. The region has two more characteristics in addition and they are homogeneity and organization.

Concepts of a Region

Region as a Part of Reality or Mental Construct

Among the many ideas about a region, the two most difficult ones are: (i) whether the regions are the facts of the earth's surface, or (ii) whether they are mental constructs. If the regions are considered as the facts of the earth's surface, then they really exist. A region, then, becomes a concrete unit of study. It actually exists. To discover the regions and to study them becomes the ultimate goal of the regional geographers. In other words, the discovery and description of regions become the end purpose of the regional geography. On the other hand, if the region is taken as a mental construct, then it is not a reality. It is simply an idea, a mental construct—to study the world distribution. Then it becomes just a means to come to an end (and not the end itself), a means to serve some purpose of study. Any segment or the portion of the earth's surface is a region, if it is homogeneous in terms of such an areal grouping. Its homogeneity is determined by the criteria, formulated for the purpose of sorting from the whole range of earth phenomena, the items required to express or a particular grouping, areally cohesive. So defined, a region is not an object, either self-determined or nature-given. It is an intellectual concept, an entity for the purpose of thought, created by the selection of certain features that are relevant to the areal interest, and by the disregard of all features that are considered to be irrelevant. As far as the origin of these viewpoints is considered, the roots of the first concept lie in Germany, and for the second in France. The concept of region as a concrete unit of study, as an organism, date back to Ritter. Ritter and his followers regarded the earth as a whole, an organism, and described its constituents as individuals or organs. Till the early 20th century, this concept dominated the geographical thought. The region, accordingly, has been referred to as a definite individual unit that has form, shape and structure like an organism or concrete unit of study. The face of the earth was considered to be the mosaic of individual regions, in this manner. The geographer's task, therefore, was to discover, identify, define and describe them. But, in early 20th century, there emerged another group of scholars who did not favour this idea. Major criticism to the first notion came from the French School. According to this school, although areal differentiation is a naively given fact, it is not true for the regional division of the earth. Because they are not static units. Rather, a region is more or less an arbitrarily chosen fragment of the earth. Its form, limits and characteristics would change with the change in criteria and purpose. Therefore, there cannot be a common or uniform method of regionalization. Regions do not exist in reality but are mental constructions that are later outlined as end products over the surface of the earth. One can regionalize the world depending on the purpose of study—physical, cultural, social, economic, political and so on. Similarly, the analogy of region with an organism is misleading and ambiguous. Because, in an organic growth, all the individual parts develop from a common origin (point); they are nourished from a common food supply; and have a common directive agency. External elements introduced to any part are either expelled or immediately recognized as foreign bodies. But we do not find anything comparable to this when an area of the earth is designated as a region. The area identified as a region is homogeneous only with respect to the criterion adopted. For instance, a nation (political unit) is homogeneous in terms of the type of political control over its territory and not jurisdiction or in terms of its physical, climatic or society point of view. Moreover, the final and ultimate division of earth into various regions is also impossible. Because regions are not static, not even the physical divisions. One time regionalization is valid only from a particular point of view and also to serve a particular theme or purpose of study. Thus, regions are just the mental constructs and arbitrary divisions of the earth's surface, and regionalization is not the objective reality.

Region as a Classificatory Device

All individuals have a number of properties. In classifying, one property that is possessed by all the individuals is selected for the grouping process and this property is called 'the differentiating characteristic'. On the basis of this characteristic, individuals can be grouped into a number of 'classes'. These classes, all being on the same level, are called 'a category'. If this initial grouping is unrewarding, the classes of the first order may themselves be grouped into a second set of classes on the basis of second differentiating characteristic. The process may again be repeated and so 'hierarchy of classes' is formed. Thus, there are two levels of classification. One classifies single region into various individuals, and secondly, where various regions are the individual themselves to become a part of still larger system—according to their size and importance—the hierarchy. In classification we begin with a number of individuals and group them together on the basis of similarity. The reverse procedure is the logical division of entire whole into smaller and smaller constituent parts. This division may be carried further to form the hierarchy of reverse order—categories, classes, and individuals. The former process of classification in this case is the 'Synthetic Regionalization' and latter one is 'Analytical Regionalization'. The synthetic regionalization is the method of aggregation and the analytical regionalisation is that of sub-division. Unstead, for instance, has described his regions by synthetic approach, while Herbertson's method was analytical. In 'Synthetic Regionalization' one begins with a number of individuals, and groups them together on the basis of similarity. On the other hand, 'Analytical Regionalization', as a reverse procedure, is the logical division of entire universe into smaller and smaller constituent units. It is a method of sub-division, as followed by Ritter, Herbertson, Koeppen, Thornthwait, etc. Synthetic regionalization is normally restricted to relatively small areas, while conversely world classifications are invariably as a result of sub-division. This leads to a system of hierarchy in either of the ways. A region is, thus, an individual unit and at the same time is also a classificatory device. For instance, a Nation, a State, a District, a Tehsil and a Village—all are individual units in themselves. It is also possible at the same time, to divide each of these administrative regions into various sub-parts, like a State into Districts, a District into Tehsils, a Tehsil into Villages, and a Village into houses.

Region as an Individual or a Hierarchical Unit

Besides being an individual unit of study a region also shares some characteristics with a greater whole. Though, there is a tendency to keep the regions similar in size at any given level, the region may be of any size according to the intensity of the study. The ideas of size, form and function can be combined to give ranks of regions that change steadily in all these respects. At each higher rank the region is larger and more complicated. The notion that the ranks of the region increase in size and complexity, first came from Passarge and others in Germany. The *Gegend* was taken as the smallest unit, usually with uniform relief—which was combined with others to form *Landschafsted*. In turn several of these units were combined to make up the larger region—*Landschaft*. At this stage the climate differences were sufficiently marked to be taken into consideration. Next in order was *Landsteil*—which, by combination with others, ultimately formed the most complex region—the Land (may be assumed as the whole country or a major division of it). Still another system, as devised by Unstead, begins with *site*, leading to *stow*, *tract*, *section*, *province* and ultimately the *continent*.[13] The basic idea common to these systems is that the amount of generalization increases with every higher order of the region. Whether one starts from a lower limit working upwards, or from a larger unit dissecting downwards, the ultimate product is the region of varying ranks. Further, the ranks of the regions are not fixed. They change according to the change in their form, function and size. Every region is a part of some system of hierarchy. Either of the approaches (Synthetic or Analytical) one follows in classifying regions, the end product is some form of hierarchy of regions. If one considers region as a hierarchical unit, then the inter-links can be better understood and explained. The concept of 'Macro', 'Meso' and 'Micro' region can also be explained, otherwise it is vague; it can be understood only in relative terms. At each level of the above example, the political unit is an individual unit, and at the same time, it is also the part of some hierarchical order. Depending on its hierarchy, a region can be labelled as 'Micro', 'Meso' or 'Macro'. Thus, besides being an individual unit of study, the region also shares some characteristics with a greater whole (linkages). However, the larger the region, the more complex is it in terms of its features and characteristics. The rank of the region increases with its increasing size and complexity. Further, the ranks of the regions are not fixed. They change according to the change in their form, function and size. This change is much more rapid in the functional regions than in the formal regions.

Region as a Paradigm, Model, or Generalization

It means, the regions once formed may be used to explain the other areas with similar content features; or the criteria used once to delimit a region may be tested in other areas possessing more or less the similar characteristic features. In other words, the regions once established may repeat themselves in different parts of the world. The study of one well-demarcated region is applicable to other areas. A model, simply argued, can be a theory, or a law, or a hypothesis, a paradigm or a well structured idea. There are four ways in which the regions or regional systems can be regarded as models or paradigms: (1) The regions are the best abstractions or the best approximations of the reality, i.e. the regions represent reality in simplified form; (2) The regions are 'isomorphic' systems'; (3) The regions are controlled models; and (4) Analytical regionalization is a form of model building in general. The reality is so complex that it is impossible to reproduce all its features, all its functional relationships, or the entire web of interdependences. To represent the reality in words, or symbols, or statistics one always simplifies or generalizes. In this process of simplifying or generalization, the measure of success lies in the significance of the properties, which one chooses to represent the reality. For example, a map for the tourist would be significant if it shows roads, towns, and relief—the things most relevant to tourists' purpose—other details of the regions are irrelevant in this context though things exist in reality. Hence, considerations along these lines suggest that all regional systems are models, because they are based on selection of general, common, relevant and significant features, and not on all the possible properties of the objects regionalized. The regions, therefore, are the best approximations of the reality—climatic regions, agricultural regions, etc. However, this reality is not totality. The regions show similarity among diverse features or forms. Whatever limits are drawn for a region they may be called 'isomorphic lines'—within which uniformity exists among features in terms of criteria. Thus, the law of the known field could be applied to an unknown field. In this manner, it is argued that all regions are models. To understand the various process of the complex whole, it is necessary to look at only one at one time, and ignore the others. For example, to explain the climatic differentiation of the world, the significant process is the general circulation of the atmosphere. But the simplicity of this process is disturbed by certain irregularly distributed mountains, shape of continents, etc. Thus, to demonstrate how climatic regions result it is a common practice to construct a hypothetical continent, lacking all other features. The climatic regions marked on it—are the hypothetical regions or the model regions. The deviations from the model regions may be explained with the help of real world patterns in the same context. Such regional models have been devised to show their impact on other processes, like the impact of climate on the development of vegetation regions or soil regions, or on human beings (as determinists believed), and so on. The construction of such regional models has been developed much more elaborately in the field of economic geography. The most popular of such models is that constructed by von Thunen (though the original model has been modified by a number of later workers—on the basis of changed conditions of the region). His model is based largely on a number of assumptions or hypotheses, like uniform climate, soil characteristics, wages and means of transport. The only variable is the distance from the market (that lies in the centre of the region). Another example is that of Christaller's Central Place Theory. Thus, these regional models are generated by considering only a limited number of processes and ignoring the others. In this way, the regions are controlled models.

Earlier the distinction is made between two methods of arriving at regional systems—synthetic regionalization and analytical regionalization. The latter process is considered here for further analysis. In the process of arriving at Analytical Regions—i.e. by dividing the complex into its sub-parts—a typology of the phenomena is created. A typology is an assumed or hypothetical classification or theoretical classification—i.e. not based on observation or experience (as in the case of synthetic regions). It is yet to be tested in the real field. The regions, thus formed serve as models (or basis) of explanation in the real field. Having created such types or classes, it is assumed that any part of the area would fit into one of these classes or types. Such a process is not followed in synthetic regionalization. Admittedly the properties have to be selected before the regions are established, or there is reasoning in advance is involved in the process. Hence, it is justified that analytical regionalization is a form of model building.

Attributes of a Region

Certain important attributes of regions have been mentioned by Derwent Whittlesey, which include Criteria, Categories, Core and Periphery, Boundary and Regional Consciousness.[14]

Criteria

The region is an area in which accordant relations produce some form of cohesion. It is defined by specific criteria and is homogeneous only in terms of these criteria. Criterion is the basis on which any region is delimited. The uniformity, homogeneity, organization or interrelationships that are talked about for any region are actually in relation to this criterion or basis on which it has been identified. An effective regionalization can be undertaken only on the basis of selection of meaningful criteria. Such criteria may be formulated *a priori* or *a posteriori* of observation. The criteria by which a regions are identified not only determine the outer limits (boundaries of the region), they also determine the amount of variation within a region (i.e. in the characteristic features e.g. population density regions). The criteria are so selected that they represent the reality in the best possible manner. The region, which has some form of cohesion or interrelationships among features, is defined by specific criteria and is homogeneous only in terms of these criteria. One can really have infinite regional patterns brought forth by the application of different criteria. But these patterns must be meaningful and significant. Effective regional study is founded on the selection of meaningful criteria. Such criteria are sometimes formulated in advance of actual field observations and sometimes after the field study. The criteria previously used in other areas or related to other problems should be introduced only after testing.

Categories

There are many different categories of regions. The regions range from relatively simple delineations of single features, such as slope categories, to highly complex areas embracing the entire content of the human occupance of earth-space. Regions also differ according to the nature of their internal cohesion and structure. In general, the various categories of a region may be identified on two bases, viz.: (A) In terms of Number of Features; and (B) In terms of Internal Cohesion and Structure.

***In terms of Number of Features*:** In considering the classes of regions, it is useful to think of them as ranged, in three basic types: (i) those defined in terms of single features; (ii) those defined in terms of multiple features; and (iii) those defined in terms which approach the totality of the human occupancy of areas (compages). The *single feature regions* delineate an individual phenomenon that is examined in relation to other phenomena in the search for accordant relationships. The geographers who prefer the topical emphasis in their approach to regional study favour the construction of regions of this type. For example, the regions like Population Density Region, Slope region, or Wheat Producing Belt of a Country are the examples of single feature region delineated on the basis of one single phenomenon. Such regions are difficult to be divisible further, as they have small range of variations permitted by the criterion. The *Multiple Featured Regions*, on the other hand, are delineated on the basis of a combination or association of two or more features. Sometimes they may be constructed through matching single feature regions; or they may be sufficiently distinctive and cohesive to be observed and mapped directly in the field. Such regions fall into three sub-types: (1) Association of intimately connected features which are highly cohesive and produced by one kind of process, e.g. Climatic Region, defined as interplay of temperature, moisture, winds, etc.; or types of agricultural land use, defined by the association of crop and livestock; or Soil region, defined in terms of slope, soil and drainage; (2) Association of features less intimately connected and produced by different kinds of processes, e.g. an Economic Region, defined by the mode of handling a particular association of crops and livestock; and (3) Association of features only very loosely connected, e.g. a Cultural region, formed as a result of various economic, social, political and religious criteria. Numerous geographers postulate the existence of a region that is assumed to approach totality of the combined physical, biotic and societal content of area. This kind of region has been termed as '*compage*'. In its literal meaning the term 'compage' is an adaptation from the Latin *compage*, formed by '*com*' meaning 'together' plus '*page*' meaning 'to fasten or to fix' and the word as such means 'joining together structure', etc. Thus, in its strictest form it is a system of many parts united; a whole formed by the juncture of parts; a framework of joined parts, a complex structure, a solid or firm structure. This type of region is differentiated in terms of the entire content of human occupance of an area. Such a region is an association of inter-related natural and societal features. Like all regions of one category, the compage is uniquely located. It embraces a combination of elements found nowhere else. Strictly constructed, therefore, it has no counterparts. Hence, no standard list of criteria or of items to be observed is adopted as universally applicable for the purpose of differentiating compages. The selection of phenomena to be observed, classified, and mapped should be made on the basis of relevance to the particular compage under investigation. The common method used so often in German regional studies, e.g., has been to start with geologic past of the region, progressing through the physical and biotic features to the societal aspects of the area. But sometimes even this approach fails to illustrate the unique character of a region. To simplify the problem, a list is made of various possible criteria for the study of compages and it includes: All material features of the landscape; All methods of making a living; and Other aspects of social life. The features covered are: structure, relief, drainage, climate, soil, vegetation, wild fauna, gathering, hunting, fishing, mining, forestry, manufacturing or industry, transport, settlement, population, religion, politics, social organization and social life. Then there is practical problem of procedure as yet unsolved, i.e. how to observe, record and analyse all these features of a compage. It, therefore, involves the use of teams of scholars, each member responsible for the treatment of a rather narrow range of topics. Such teams also require cooperation of scholars from other disciplines than geography. But this is very rarely possible. Theoretically, the compage should represent the entire content of the area, i.e. all the features should be taken care by the criteria. But, this is a task hard to be accomplished. Therefore, compage is a theoretical concept not practical, as a single criterion cannot represent all the content features of an area. Thus, apparently, the idea of a 'total' region appears simple, but in practice it is really very difficult to find such a structure. Therefore, many modern geographers and others have attacked this omnivorous study of spatial totality, and have stated it as undiscriminating, futile and even dangerous.

***In terms of Internal Cohesion and Structure*:** The Regional analysis draws the attention to the contrast between the region as an area throughout which there is substantial unity or uniformity of characteristics and the kind of region which has no such uniformity, rather it constitutes the sphere of activity of a certain group of people, e.g. urban field. Regions that are defined by uniformity of characteristics or homogeneity of content are termed as *formal*, and those that are defined by coherence (interrelationships, connections) of parts as *functional*. The *Formal Region* is also known as a 'uniform' or 'homogeneous' region. It is homogeneous within the limits set by criteria and in terms of association of features as defined by the criteria, as is the case with, for example, a Cultural Region, Political Region or Agricultural Region. Various characteristics of a Formal Region include: (1) A formal region is a part of clear-cut or well-defined hierarchy; (2) It occupies a fixed position in the hierarchy; (3) There is no overlapping of boundaries; and each successive higher order region consists of aggregation of regions at immediate lower order. The uniform regions are homogeneous within the limits set by criteria and in terms of the association of features. But there is always a certain range of characteristics permitted by the criteria and there are irrelevant differences that are disregarded. For formal region, whatever stated about one part of it is true of any other part; it is the largest area over which a generalization remains valid. This is simple enough with single featured regions—like limestone region, farming region or a grazing land and so on. But with multifeature regions it is very rare that each feature remains completely uniform throughout. Thus, the uniform region is homogeneous in terms of the features contained, but it also includes a certain range of intensity permitted by the criteria. The *Functional Region*, on the other hand is also called a 'Nodal' region. Such a region is heterogeneous in terms of their content features, e.g. the metropolitan region and an economic region. The uniting or binding factors are inter-connections, lines of movement and functions. It is always known by a pattern of circulation, expressed by movement of people, goods, ideas, telephone lines, newspaper circulation, etc. The tie with the node or nucleus weakens as one moves away from it. The functional region is homogeneous with respect to its internal organization and not the content features. This structure includes a focus or foci, and lines of circulation (movement) tie the surrounding area to this. For example, an area of newspaper circulation is a nodal or functional region. Nodal regions of like character may lie adjacent to each other or nodal regions of different character may surround them. Internally the nodal or functional regions are heterogeneous, i.e. marked by the diversity of functions and features. Circulation, including the movement of people and goods, is the primary attribute. The interdependence of various parts or features is more important than their similarity or homogeneity. Marked diversity of functions and features is the major characteristic. Unlike uniform regions, the functional regions are of different order (levels) and may overlap. The theoretical regions in Central Place Theory of Christaller, or the various functional regions identified in the National Capital Region (NCR) Plan of Delhi may be cited as examples here. The focus or foci of nodal region serve as the centre of organization. Moreover, the same centre may serve as focus of two or more functional regions. For example, New Delhi in the National Capital Region is the national capital, a transport centre, a point of convergence at a point across the river, besides being administrative centre it also serves as a commercial centre. Thus, for a student of functional regions, the essential unit of study is not the homogeneous area, but it is the complex of different areas which function together as a whole. The functional region, therefore, is essentially diverse. It is also a classic example of unity in diversity. The uniform features of a homogeneous region may just be together by chance, but the parts of a functional region work together and are to a degree dependent on each other.

Core and Periphery

Every type of region has a core and a periphery. The core and periphery are the most dynamic parts of a region. Core is an area of the region where its characteristics find their most intense expression and their clearest manifestation. Although the whole region is homogeneous in terms of the criteria by which it is defined, the peripheral parts of it are distinguished from the core by an increased intermingling of other characteristics. Hence, the selection and presentation of the regional character is best accomplished by exposition of the core area. This core may or may not be the geographical centre of the region. In uniform regions the core is the place of an ideal expression of the criteria. For example, in a single featured linguistic (uniform) region, the core is that part where a single mother tongue is found. The peripheral areas show an increasing proportion of intermingling with other languages until the limits of the region, as defined by the criteria is reached. In nodal regions, on the other hand, the core is the most representative portion of the entire area and is the part most closely tied to the focus. Thus, it is easily detectable here. However, here also it should be noted that the core and the focus are not synonymous, even though the focus ordinarily lies wihin the core. The cores of adjacent nodal regions may closely resemble one another, because the distinguishing feature is nodality and not uniformity. Thus, the core is the epitome of the regional character, whether uniform or nodal. The marginal area of a region is referred to as its 'periphery'. The regional boundary is the part of this periphery.

Boundary

Boundary defines the limit of a region and separates it from the other. There are three types of boundaries identified in the context of a region, and they are: Lines of Discontinuity; Lines of Continuity; and area of Discontinuous Distribution. The *Lines of Discontinuity* separate regions that differ in kind, e.g. linguistic regions, political regions, cultural regions, etc. They are sharp and also easily observable in some cases, as the boundary between different vegetation, relief or soil types (formal regions). Generally, the boundary lines of discontinuity are abrupt. In nodal or functional regions, such boundaries are drawn where the attraction of the adjacent foci becomes nil or the minimum. Boundary lines drawn in continuity are of very different character. The *Lines of Continuity* do not represent divisions between different kinds of features, as the same kind of feature exists on both the sides of such a boundary. The difference is in degree or intensity of the component features of the regions. Generally, such boundaries are the isarithm or isopleths of some feature, e.g. contours (determining regions of varying heights leading into mountains, plateaus, plain areas); or demographic regions based on density of population. The boundaries formed by *areas of discontinuous distribution*, on the other hand, are not simple lines. Instead, they are zones (as no man's land, buffer zones). They may be devoid of any particular type of human activity.

Regional Consciousness

The term 'consciousness' means 'a sense of awareness or belongingness towards something'. Accordingly, the regional consciousness is a kind of awareness towards the region one is associated with (spatially or mentally). It should not be taken synonymous with 'regionalism' as the latter is the extreme form of it, overlaid with special meanings serving the non-geographic interests (James and Jones, 1954, 51). In general, the regional consciousness rises by sharing a common space; whereas the regionalism takes shape by sharing common feelings, culture, religion, traits, etc. The regional consciousness usually applies to uninterrupted space. The most obvious occurrence of regional consciousness is in political, cultural or urban regions. Taken in a healthy sense, the regional consciousness leads to the stability and prosperity of the region concerned.

Regional Classification

Whether, the regions are concrete unit of study or not, it is justified that regional division is necessary to understand this complex world of reality. There may be many different bases for regional divisions, each more or less suitable for different geographical purposes. But, there is no universally valid division, which does justice to all phenomena; one can only secure a division with great possible advantages, and the least possible disadvantages. In the process of regionalization, one geographical fact may be observed at a series of places, or various geographical facts at the same place. The former leads to the study of spatial distribution, and latter to the study of locational inventories. The former involves studies of spatial variations, and the latter implies the areal differentiation in its holistic sense. The former method leads to functional regions and the latter to formal generally. A regional system, thus, may be a series of hierarchical places with similar characteristics, or an area with functionally related diverse features. Then, the regional classification may be arrived at inductively (aggregation) or deductively (sub-division). To achieve the goal of regional presentation the method (chosen from any of the above) requires a well-defined purpose, recognition of various attitudes and conceptions to choose the area and the skill in the use of the techniques. The ultimate aim of regional geography is to study the entire world. However, in most respects, the world is much too large and complicated to be readily comprehended all at once. Some distinction, some delimitation is necessary. The world can be classified into the Natural, Cultural, Political and Economic Regions in general. Each one of these regions is discussed in detail with an emphasis on its various attributes, particularly criteria, categories, boundaries, etc. Here the world level regionalization is followed, in which the nation (country) forms the lowest order unit.

Natural Regions

A "natural" region in its simplest form means a considerably large portion of earth, homogeneous in its natural aspects, i.e. similar at least in relief, climate, and natural vegetation. Within a number of similar natural regions the kind of human activities developed are also believed to be the similar. The natural region, as defined, varies in its size and the smaller regions may be aggregated to form one larger natural region. The concept of natural regions was developed around 1900 with reference to the division of the world into systems of major entities by A.J. Herbertson in Britain in 1905.[15] This division is based on a system of generic regions aiming at a worldwide classification of homogeneous natural entities. Herbertson used the term "natural region" to describe physical environments that were "inherent and not arbitrarily imposed". He defined the region strictly on a natural (or physical basis) as a framework for the evaluation of human occupancy. According to Herbertson, the facts of configuration and of climate are of first importance, but the distribution of vegetation, and even of man, may also be examined. As such, the various criteria to distinguish natural regions include: (i) configuration; (ii) climate; (iii) vegetation; and (iv) density of population. On the basis of the above criteria, Herbertson divided the world into six natural regions, viz. Polar Regions, Cool Temperate Regions, Warm Temperate Regions, Tropical or Hot Regions, Lofty Tropical or Sub-Tropical Mountains and Equatorial Lowlands. Herbertson's ideas about natural regions spread with rapidity and became an important part of teaching. The prime divisions were based on the thermal belts of the world—maintaining the Aristotle's astronomical terms; and these are defined as: (1) *Polar*: with no month with a temperature over 10° C (50° F); (2) *Cool Temperate*: roughly between latitudes 40° and Polar circles north and south; (3) *Warm Temperate:* approximately between 30° and 40° north and south; and (4) *Hot Belt:* with temperatures over 20° C (68° F). The sub-divisions of each of these belts are based mainly on rainfall and partly on relief and vegetation. Herbertson's regions are almost identical with the later Classifications of climates with V.C. Finch and G. Trewartha in America (that in turn are based on Koeppen's classification), and of Siegfried Passarge in Germany. The concept of Natural Regions is one of the fundamental concepts of Regional Geography. The Natural Region is an area delineated on the basis of the physical attributes of land and within which the homogeneity is maintained with respect to these attributes. According to Herbertson, differentiation in terms of natural phenomena is inherent in the nature, and geographer's task is to identify such natural differences on the basis of which are devised natural regions. Generic or Systematic natural regions as developed by A.J. Herbertson are single feature regions—like relief region, vegetation region and so on—at the lower levels of hierarchy; while higher order Natural Region—like Polar, Cool Temperate, Warm Temperate, Tropical or Equatorial—are multi-featured, or the result of the combination of various elements serving as criteria. Natural regions are generally the formal regions. However, the various attributes of Natural Regions may be identified as: (1) The most significant criteria used in the delimitation of natural regions are configuration and climate and these, in turn, are substantiated by soil, vegetation—and human and economic activities; (2) Both single-featured and multi-featured natural regions may be distinguished; for instance, a slope-region at the lowest order is a single feature region, while a Mediterranean or Equatorial region may be multi-featured; (3) In natural regions we seldom talk about core areas, unless it is a lake region (as of North America), or a River Basin. But, it has more or less clear-cut boundaries—which in the case of natural regions may be dissociating oceans, coastlines, mountain chains, inhabitable deserts or river-lines. In most of the cases these boundaries are lines of discontinuity (e.g. coastlines) or zones of discontinuous distribution (as mountain ranges); (4) Like other regions, Natural regions also develop well-marked hierarchical orders. For instance, in their ascending order they form—slope regions—stows or chores-tracts—realms—Biomes; (5) The systematic natural regions (as of Herbertson) serve as Models or Paradigms of analysis. Because these generic regions are devised in such a way that they repeat themselves or have counter parts with similar characteristics, like Congo Basin and North Brazil as Equatorial regions, or Thar and Sahara as Tropical deserts, or Tibet-Pamirs and Plateaus of Bolivia and Peninsular South America as Lofty Tropical Highlands or Plateaus. The generalizations found for one are also true for the other in the same category.

Cultural Regions

About seven billion people inhabit the limited space on this earth, and they live their lives in thousands of different ways. Their racial sources and ancestors vary. Their religious beliefs are not the same. Their languages are mutually unintelligible. They build their houses in different ways. In their technologies, their systems of education and in countless other ways the societies of the world reveal their contrasts and differences. Societies, large and small, all occupy a part of the earth's surface—and all have, in some special way, exploited their particular physical resources and environments. They have in the process, developed different kinds of spatial organization. In cultural regional concept we focus on these spatial aspects of human cultures. A culture, then, is the way of life of a population, all their ways of doing things, all kinds of behaviour they have learnt and transmitted to successive generation.[16] Language, religion, architecture, music—even food preferences and taboos form part of a culture. Culture consists of people's beliefs and values (religion, politics); institutions (legal, educational, government) and technology (skills, equipment). It is expressed in the ways people communicate, in the way they perceive and exploit their resources; and most important for geographers is the way they organize that part of the earth's surface that is theirs. The composite of human imprints on the surface is called the 'cultural landscape', a term that came into general use in geography in 1920s. Carl Sauer defined it as—"the forms superimposed on the physical landscape by the activities of man". He stressed that such forms result from cultural processes prevailing over a long time period, so that successive generations contribute to their development. Sometimes, these successive groups are not of the same race or culture. The study of culture landscapes and culture regions involves so many factors and criteria that it is best carried on a manageable scale. The larger the area, the greater is the complexity of detail, and the more difficult do the inevitably necessary generalizations become. Still there can be no question about it; the world as a whole remains divided into a number of huge Cultural Realms (12 in total), even today in our age of mobility and interaction. Each one of these is a formal cultural region in which there is clearly a homogeneous cultural identity on a wide variety of levels. They are: South-West Asian and North African Realm (Islamic/Arab world); Europe; Indian Subcontinent; Chinese Realm (oldest of the civilization, communist society); South-East Asia (Buddhist faith predominates); Black Africa (a primitive culture); Middle and South America (also called Latin America); North America (most urbanized culture); Australia (an isolated realm); Soviet Union (now disintegrated politically, but cultural characteristics remain more or less the same); Japanese Realm (a Buddhist land away from the mainland); and The Pacific World (including thousands of islands between Australia and Americas, a complex culture). The major criterion used in the delimitation of a cultural region is the life style (*genre de vie*) of the resident population. The cultural regions are multifeatured. The boundaries of a cultural region are more or less sharp, and are lines of discontinuity. The core of a cultural region is not easily detectable in comparison to political or economic regions. The concept of hierarchy carries less meaning in the case of cultural regions. However, the attribute that is most significant in the case of cultural regions is regional consciousness.

Political Regions

Among the criteria used to distinguish homogeneous areas on the earth, the one most widely used is that of political authority or power. Political authority—organized in a system of independent, sovereign states, each with unique characteristics—is a dominating force in determining how the people live and make use of their resources. The regional concept examines the political area as a geographic or spatial entity.[17] The foundations of such a study were laid at the end of the 19th century (1897) through the work—*Politische Geographie*—by Freidrich Ratzel, the German geographer. A political region may be defined as a portion of the earth's surface, throughout which a common type or types of political behaviour takes place. The behaviour most frequently cited in the delimitation of political region is that of 'political power' over territory by a particular government. In terms of political control the land surface of the world is divided into many political regions, ranging in level of organization from national units down to countries, townships and villages. States are part of a hierarchy of politically organized areas, and are unquestionably the most important level in this hierarchy. Above them are international organizations. The political area may take two forms—the formal political organization and the functional political organization—one at the same time. However, the regional concept strictly concerns itself to the spatial aspects of political groupings formally established. The State or Nation serves as the best example of an individual political region. Political uniformity implies that with respect to the operation of all government functions all parts of the region are alike. The present day world political pattern is the result of 4 basic trends, viz. (i) The break-up of empires and the establishment of independent states; (ii) The expansion of Communist Control; (iii) The Union or Division of existing political units; and (iv) The political partitioning of previously unclaimed areas. A State (*Nation*) is the fundamental unit as a political region and has at least six essential elements to be considered in detail viz. (i) A clearly defined Territory; (ii) A substantial Population; (iii) Certain Organization; (iv) A measure of Power; (v) Frontier and Boundary; and (vi) The core and capital. The world political pattern has been a complex and constantly changing phenomena—which reacts to a great variety of forces, and which in turn influences other forces and phenomena, both political and non-political in nature. Geographers have long recognized the need for studies of organized political areas at ranks above that of the Individual Sovereign State. The most common of such associations are, for instance, an *Empire* composed of one independent state and its dependent areas, like Russian Empire or Chinese Empire; the *Commonwealth of Nations*; a variety of *Leagues* of independent states, each organized for specific and limited purposes; as, for example, Arab League, NATO (The North Atlantic Treaty Organization), Council of Europe, SEATO (South East Asian Treaty Organization), CENTO, Benelux, etc., and the *League of Nations*—which began in 1919 and has been resumed since 1945 as United Nations. Apart from all these forms of international organizations is the one related to American Foreign Policy with respect to the Communist States in Europe and Asia. Accordingly, the present day political states can be grouped under three political forces in the world today: The Western Bloc, The Communist Bloc, and The Neutral Bloc. Various attributes of a political region, at national or international level may be discussed. The major criterion used in the delimitation of the political region is the type of political control and the resultant political organization. Political regions are multi-featured, and may be considered either formal or functional depending on the level of hierarchy. For instance, they are formal at state or national level, as they are homogeneous in terms of control or power, while at an international level (*above*) or district level (*below*), they may be considered functional. The boundaries of a political region are very difficult to draw, as it is not easy to come to a common agreement. But wherever they have been drawn they are more stable than in the case of natural or economic or cultural regions. Either type of boundary is the characteristic feature of the political regions, like lines of discontinuity (among Western and Neutral states); lines of continuity (among Communist states); or zones of discontinuous distribution (as a number of buffer zones and no-man-lands among nations). Every political region has developed around a nucleus, and possesses at least one or more core areas depending on its organization—unitary or federal. The political regions form a clear-cut hierarchy at least theoretically. In its ascending order, for instance, political regions may be identified as—locality, district, province, nation or country, leagues or groups of nations (organizations) and ultimately UNO (United Nations Organization). The regional consciousness is the most evident in political regions. The synthetic regionalization (by aggregation) is the more popular approach to arrive at the political regions. The individual state or nation is the best expression of a political region. Though, the nation may be further subdivided into its sub-regions, or various individual nations may be grouped to form a larger region, it is only at nation-level we find the best expression of the various attributes of a political region—like homogeneity, criteria, boundaries, cores, regional consciousness, etc. Finally, the political regions, once demarcated, are more concrete and at the same time all inclusive and heterogeneous. Therefore, they may also be compared with 'compages'—the total regions. The only uniting force among the multitude of features—economic, physical, social and cultural—is the political power. Therefore, political regions are more or less unique in their nature. The analysis of political regions—either at state (nation) level or at international level forms an essential part in the study of regions as such. The political regions consider human aspects more than the physical (though the latter are also important). The discovery and description of the political regions is not the end purpose, rather it is a means to assess the present and future problems related to territory, people, organization and power of the political region.

Economic Regions

The world of economic reality consists of a mosaic of 'regions'—each of which is the location of a complex set of inter-related processes—all of which are geared to a single type of production. The idea of economic region is useful in the identification of the areas that are devoted to particular types of economic activities. Seen in this perspective, an economic region may be defined as 'a spatial entity demarcated on the basis of dominant economic activity or activities in relation to the resources available and is homogeneous in respect to that'.[18] Then, there is uniform and functional view of economic regions. Uniform economic regions are concerned with 'what-is-where' (static view), whereas the functional view is concerned with the organizational regions, i.e. 'what-connects-to-what'. *Uniform economic regionalization* is based on one major mode of production, such as agriculture, manufacturing or potential consumption, which is usually referred to as the level of living of a society. Uniform associations may be based on a single factor or a variety of interrelated multiple factors. A single factor association has a single criterion, such as the production of wheat. A multiple factor association means the co-variance of a number of inter-related single factors—like agricultural system, industrial system, and so on. *Organized or functional economic regions*, on the other hand, embrace both market-oriented activities, such as retail or all production or productive wholesale activities; and productive-oriented activities, such as fresh milk for nearby market. The functional regions are concerned with market areas served through central places. Among the several regions with which the man has close economic relationships 'the state' is the most important. States, as such are mainly concerned with promoting their economies and protecting them from disruption. The state may, therefore, be viewed as an administrative (political) as well as economic region. The two major characteristics of the state may be viewed in terms of *external (or spatial) elements* of location, size, shape and boundaries; and the *internal elements* including the cultural and physical qualities of the inhabitants. The various elements that make up the state are common enough, but they vary in their quality and combination from country to country. Depending on the quality and combination of all these external and internal elements, a country assumes a particular economic character—which represents in average the total national character. Therefore, a nation forms an ideal unit of economic regionalization. Just like the political regionalization, an economic regionalization may be achieved at a level higher than a nation, by using the four criteria: (1) The Per Capita Income; (2) The Occupational Distribution of the Working Population; (3) The Age-Structure of the Population; and (4) The Geographic Distribution of Population. On the basis of these criteria, it is possible to make a four-fold division of the type of economies: (1) *Highly Developed Economies*: or predominantly industrial-commercial economies, supporting only 8 per cent of the world population; or the economic aristocrats of the world with a large industrial-commercial base. (2) *Semi-Developed Economies*: or the mixed industrial-agricultural economies supporting about 12 per cent of the world population; still developing. (3) *Less-Developed Economies*: or predominantly agricultural economies. These economies are very widespread and support about 45 per cent of the world population. (4) *Planned Economies*: which are essentially of either type '2' oe '3', but have major characteristics of '1'. In addition, these have other special features of their own that they are best considered as a separate category. The planned economies at present include about 30 per cent of the world population. As far as various *Attributes* of economic regions are concerned, the *criteria* used in the delineation of world economic regions are per capita income, occupational structure, age-structure and the distribution of population. Along with these may be considered a whole host of other elements of economic significance to study the economic identity of individual states—like location, size and shape or territory, boundaries, numbers and cultural attributes of population. The *single-featured* economic regions are very difficult to construct, as the economy is the result of the combination of a number of features. Thus, economic regions are always *multi-featured* and heterogeneous. Homogeneity is super-imposed. They are more complex than as suggested by the regional framework. Consequently, the homogeneity of economic regions is *functional* and rarely formal. Even the so-called uniform regions—agricultural or industrial—are basically functional, through the lines of communication, exchange and flow of money, people and products. In simple words, economic regions are the best expressions of functional or nodal character of regions. The *boundaries* of the Economic regions coincide more or less with political regions, but *core* areas may distinctively be identified. As stated, the economic core areas are of two orders. **First** order core areas are the primary or extensive productive areas—as agricultural fields, mining regions, forested areas, etc. The **second** order core areas, on the other hand, are small but densely occupied administrative, political or urban centres with specialized and secondary economic activities as manufacturing. One can say about political consciousness, but it is really absurd concept for the economic regions. Because, for instance, the dwellers of under-developed regions are always consciously or sub-consciously tied to more-developed regions and aspire for the same. Two levels of *hierarchy* are well marked in economic regions—national level and world level. At lower levels the economic regions become highly ineffective, as it is an integrated system on multitude of elements. Economic regions are highly dynamic witnessing a rapid change through time with the change in the economic status of individual nations that form them. Most of the present theories and models of economic development are based on more developed regions of the world. Thus, they to some extent, serve as models to the under-developed world. Alternative is provided by the Planned Economies. But, recently, with the intensive studies of the under-developed or developing nations, we find a break from this trend. The latter now believe more on indigenous theories based on their own historical economic development and combination of physical and human resources within their own limits, and the socio-cultural set-up of their nations.

SPATIAL ANALYSIS

The concept of geography as a science of space has been exceedingly important in the history of geographic thought. Space and spatial organization have been important concerns of scholars in ancient times. This space in geography is territorial and not abstract or astronomical. The concept of territorial space is the basic organizing concept of geography. Certainly, a central concern for location and distribution is a hallmark of geographical studies. The Spatial paradigm belongs to the oldest tradition in geography that emphasizes the need to describe and explain the distribution of phenomena on the earth's surface. The fundamental question in geography is 'where?'. The terms like site, situation, location, position, distribution, patterns, arrangements, etc.—all pertain to space and emerge out of this fundamental question. Geography is essentially a spatial science. A heavy dependence of the discipline on the concepts of geometry and mathematics strongly proves this fact.

The usage and meaning of 'space' have changed, improved and expanded with time. The descriptions pertaining to physical space have been one of man's important concerns for a long time. But the evidence of this kind of geographic thought is first evident in Greek and Chinese literature. Eratosthenes was the first scholar to devise crude locative grid system with latitudes and longitudes to provide framework for map construction and to answer where questions with great accuracy. Besides this, he was the first to measure accurately the size of the earth. Later advancement in the same direction helped in dividing the circle into 360 degrees. Ptolemy applied this division to the terrestrial globe; which enabled describing locations and preparing maps with two-dimensional representation—latitude and longitude. With minor modifications we still use Ptolemaic locative system and some of the map-projections he invented. The need and practice of mapping increased further due to increased mobility and travels during the expansion of the Roman Empire. Towards the end of the Greek period or the beginning of the Roman period, the geographers began to ask: 'what is where' instead of 'where'. In the Far East, the Chinese geography was developing into a sound spatial science in the ancient period. By 2nd and 3rd centuries, the Chinese geographers had started using rectangular grid system and graduated scale to construct their maps, which were extremely accurate. After the death of Ptolemy (i.e. about 155 A.D.) a decline was observed in the accuracy of locations in the world maps until 14th century in Medieval Europe. The causes were the collapse of Roman Empire and the increased faith in Christainity. Thus, in the European Middle Ages-Theology (the science of religion) replaced the scientific and mathematical methods of ordering the world. The collapse of Roman Empire was followed by a decrease in mobility, which adversely affected the developments in geography as well. The answers sought to 'where' questions, were by modern standards—the most erroneous, as they were based on 'T-in-O' maps. As Europeans became increasingly immobile, the geographical theory and practice were not tested against experience. Answers sought were more theological than scientific. Therefore, not much progress was made in this direction. But, in Medieval Muslim World (8th–14th century) need for accurate knowledge about locations was continuously felt by the Arabs. The reasons were three fold: (1) Vast expanse of Muslim Empire; (2) Promotion of travels due to religious requirement (pilgrimage); and (3) Increased commercial links with the rest of the world. The great extent of Muslim empire in the 7th and 8th centuries was the prime reason for the need for accurate knowledge about locations and places. All of Ptolemy's spatial works were translated into Arabic. Besides, commerce and interest in other places promoted travels.

The religious requirement of *Hajj* (or Pilgrimage to Mecca) also anticipated travelling at least once during lifetime. Al-Biruni and Ibn-Batuta emerged as two famous travellers during this period and their writings contained valuable record of information about places. But, the spatial viewpoint in Muslim geography was not more than encyclopaedic description of places, or travel accounts (itineraries). Thus, the objective standard of Medieval Muslim geography was no better than that of Medieval Europe. Fixing locations of individual places was the first order geographical business in the ancient and medieval times. In other words, queries related to absolute location of places were more prominent before the Age of Discovery. But, afterwards, the concentration directed towards understanding and study of the relative importance of locations as well. The reasons were: (1) Completion of world map and no more queries to be made towards absolute space; and (2) The Age of Discovery was also the age of commerce; therefore, the questions of minimizing time and cost in travelling were important. The rediscovery and translation of Ptolemaic literature, advances in maritime technology and voyages of explorations, and the subsequent broadening of the geographical horizons combined to produce a Golden Age (i.e. 1450-1800A.D.). This revived interest in locations that again became fundamental to human activity. With the discovery of new places, the existing maps and atlases were revised, especially in 16th and 17th centuries. As the Age of Exploration was also the Age of Commerce, the question of relative importance of locations also emerged. This was to minimize time, cost and distance in travelling. Thus, by the end of the 18th century the problems of absolute locations became less important as the world maps had been filled completely and geographers began to explore the questions pertaining to relative space. Now, the questions pertaining to absolute location of places became completely outmoded. Geography up to 1800 was primarily a locative enterprise. Fixing locations of individual places was the first order business and this had been finished largely by the time of Humboldt and Ritter. Alexander von Humboldt (1769-1859) and Carl Ritter (1779-1859) were the ones who firmly established the methodology of geography and its place among the modern sciences. Between 1800 and 1950, classifying places on the basis of their contents and characteristics and dealing with groups of places became the major concern of the discipline. Thus, the interest shifted from individual locations to regionalization and inter-relationships among places. Imposing order on the chaos of all possible places by segregating them into sets has become a practice. And in the more recent times the greater emphasis is laid on the geographical questions which evoke hypothesis, laws and theories. Thus, considerable changes took place towards the spatial view in the modern period. The major ones included: (1) The scholars started dealing with groups of places (or regions) instead of individual locations or places. In other words, interest shifted from individual locations to regionalization and inter-relationships among places. There began attempts to classify places on the basis of their contents and characteristics; (2) The enquiries started revolving around the questions like "Why what is where?"; and (3) A need was felt to formulate laws and theories in geography to explain various distributional patterns.

The shift to relative spatial context is still in progress presently and is the most fundamental change in the history of geography. The measures of distance and locations have also changed accordingly. Up to the middle of the 20th century such measures have been primarily the unchanging (or absolute) units of miles, kilometres, and so forth. But now the geographers talk about new kinds of spaces, which are stretchable or shrinkable; for instance, a distance to be measured in terms of travel-time (as 'two-hour journey' or 'ten-minutes-drive'). It has opened up an almost infinite number of new worlds to explore and map. In fact, it has become very difficult to list all the possible questions one can ask in the context of relative space today. The combinations of, e.g. where, when, what and why questions—which can be devised—numerous; and to which they can be applied are almost infinite. The answers to 'where' and 'what is where' are not the end in themselves; rather, they constitute the preliminary enquiries towards explanations one requires by answering the 'why' and 'how' questions. At the same time a change in subject matter of the discipline is also witnessed. Increasingly, geographers view themselves as social rather than physical scientists. In the present context, 'what is where' is more related to socio-cultural aspects than to the physical environment. It is why the problems of data collection are more complex today than in the past.

Concepts of Space

The concept of space in spatial viewpoint has several manifestations in geography, expressed in terms of: Location and Place; Absolute and Relative Location; Site and Situation; Distribution; and Pattern. 'Where', as a point of reference is fundamental to geographic thought. The terms like 'Location'. 'Position', 'Situation', 'site', 'distribution', 'arrangement', etc.—all is frequently used by geographers in reference to the placing of things on the earth's surface. But before going on further discussion, a clear distinction may be sought at least among space, location and place. The *space* is the basic organizing concept of the geographer, and it refers to the surface of the earth. It should not be confused with the astronomical space as in ISRO or NASA (Indian Space Research Organization or National Aeronautical and Space Administration). The *location* is a particular position on the surface of the earth. However, unless it is identified with some phenomenon, feature, object or place on the surface of the earth, the meaning of location remains abstract like space. The location identified with some occurrence would become a *place*. Thus, place is an area or space confined to an identifiable location on which we load certain values. For example, the position of "20°59′ North and 86°56′ East" is just a location in the graticule-until it is recognized with the position of 'Mount Everest'—the highest point on the earth's surface. The information pertaining to location may be either physical or human or both. The location, in turn, may be of two types, viz. absolute and relative. The *absolute location* is a position in relation to a conventional grid system designed solely for locative purposes. It is distinct, physical, real, empirical and generally immobile. The latitudes and longitudes are the most common means to describe absolute locations. The absolute location is fixed with respect to an abstract network of X and Y coordinates, but identified with some particular places on the space. Such locations are absolute in the sense that once a locational description of this kind has been adopted it does not change over time, and there is no overlapping. The relative location, on the other hand, is a position with respect to other locations over space. Relative location can also be expressed in values other than usual distance units. We could describe the distances between Delhi and Mumbai in terms of train fare or the cost of a plane-ticket. There are large number of ways of describing distance and location in a relative context, but in the absolute context we are restricted to customary and unchanging units such as miles, kilometres, or degrees of latitudes and longitudes to measure distance. The relative location of two places may change radically, even though absolute locations remain constant. An example would be Delhi and Mumbai again, which were about more than 2 months apart little more than a century ago, and which are now only about 2 hours apart by an aircraft. Today geographers are primarily concerned with relative locations of the phenomena. The relative distance is the basis of relative space, since spaces are defined by distances along the dimensions. By choosing different distance measures, we can change space between two points. The shortest distance between two points may be either a straight line or a curved line or the like. Attempts to explain human spatial behaviour are an increasingly important concern of geographers, and such attempts will be more successful if they are studied in relative spatial contexts. People shipping goods or taking trips between towns A, B and C, for instance, are not as much concerned with absolute distance as they are with accessibility, cost and travel time in covering that distance. The absolute and relative locations when identified with some physical occurrence or characteristics of the surface of the earth may also be called *'site'* and *'situation'* respectively. To be more precise, *site* is the location of a given place with its local internal features or resources. For example, the site of Delhi is across the river Yamuna. But, the *situation* is the site in relation or interaction with other sites. Taking the same example, the situation of Delhi is at the confluence of mountains, plains and plateaus.

The features like houses, hotels, hospitals, trees, factories, shops, offices, fields, roads, railways, etc. all are elements of location or the individual occurrences over space. Each of these occurrences is an identified phenomenon of a specified magnitude (i.e. having shape, size and form). For example, a house is an occurrence that can be identified by location and with a magnitude indicated by its floor area, number of bedrooms, or value. Several houses of similar magnitude at different locations constitute a distribution. Thus, the *'distribution'* would mean the frequency with which each of these elements occurs over the space in relation to each other. In other words, distribution is the assembly of various occurrences related to each other. For example, a house is an occurrence, and an assemblage of several houses makes a settlement, i.e. distribution of houses. Research in geography features the study of distributions rather than individual occurrences. Explanations based upon a single occurrence are unreliable because they do not reveal the causal process. A wide range of factors could influence a single occurrence, whereas the less important factors tend to cancel each other out when many occurrences, i.e. distribution, are studied. The nature of distribution suggests which processes might be causally associated and what research strategy might be appropriate. Distributions are of three types, viz. Discrete, Continuous, and Contingent. Discrete distributions consist of just an assemblage of different occurrences without seeking any kind of interrelationships. For instance, houses, factories, gas stations, parks, roads, etc. when represented separately for an area constitute discrete distribution. The continuous distribution exists when occurrences are dependent. For instance, temperatures are continuous over an area because they are dependent upon air circulation. However, a contingent distribution occurs when the 'magnitude' of distribution is expressed in terms of either area or time. The distribution of agricultural production areas in terms of yields per acre; or population distribution by density or sex ratio, or distance travelled per hour are examples of contingent distributions. In the case of discrete and continuous distributions, the nature of elements or occurrences is more important than their magnitude, volume, quantity or value. Further, the contingent distributions are more dynamic and change very fast in comparison to the other two types.

Spatial indicates that an occurrence occupies a portion of the earth's surface. When geographers talk about distributions, however, they are usually interested in the frequency with which things occur in terrestrial space. Often the two relevant variables are latitude and longitude. Just as the X and Y axes determine locations in a scatter diagram by giving each location a numerical space, latitude and longitude, i.e. define terrestrial space. In the same way as the distribution of points on the scatter diagram is a starting point for further analysis, a distribution of phenomena in terrestrial space is the point to begin with for geographical analysis. The distributions that geographers study are primarily 'two-dimensional'. But, they can also be observed in three-dimensional terrestrial space, e.g. 'latitude, longitude and elevation' produce a distribution in three-dimensional space. To cite an example, due to the vertical growth of cities in the last several decades, human geographers have also become increasingly aware of the third dimension. Areal variations, in terms of spatial differences in occurrence and density, are characteristic of almost all distributions in terrestrial space. Or the distributions vary in pattern and intensity from place to place. There would be no geography if all things were truly ubiquitous, i.e. same or available everywhere over the surface of the earth. Spatial distributions and the processes that generate them are observable at several scales, some of which are within the geographers purview and some of which are not. In size, the geographical scale of analysis and observation is bound on the lower end by the space immediately around observer, and on the upper end by the size of the earth. Geographers as such are neither immediately concerned with very small spaces (or micro-spaces), e.g. of atomic physics; nor do they study distribution in the very extensive spaces, as of the 'astronomers' (the macro spaces). In practice they avoid both the extremes and concern mainly with distributions observable at the local and terrestrial scales.

The pattern, on the other hand, simply means 'a design'. Once the nature and magnitude of distributions have been established, the emphasis shifts to the pattern of distribution. *'Pattern'* is the recurring association of occurrences that may indicate causal processes. Four types of patterns have been recognized, viz. Static, Dynamic, Network, and Normative. The Static Pattern is the pattern of distribution represented at a particular time, for example, the distribution of shopping centres or hospitals in a metropolitan area. Measures of location arrangement and magnitude can be derived for specific time points. In the case of *Dynamic Pattern*, the distribution displays change occurring at different time periods, for example, the population distribution patterns, population growth patterns (as defined by the Demographic Transition Theory). In *Network Pattern*, the nodes and the links between nodes define the pattern. In a transportation system, for instance, the terms as circular, hexagonal, stellar and linear are used to describe transportation network patterns. The *Normative Patterns*, on the other hand, are theoretical patterns and they may or may not be the part of reality. Such patterns are based on certain norms. The spatial patterns formed and studied on the basis of the theories as propounded by Christaller, Von Thunen, Burgess, Hoyt, Ullman and Harris, etc. may serve as examples here. However, the reality may be different from that which is represented by such models. Or it can be obtained only with the imposition of certain norms (assumptions or constraints).

Concepts of Spatial Organization

Geography is the study of the location and arrangement of phenomena on the surface of the earth and the processes that generate these distributions. Traditionally, both physical and human processes were studied. But the work of Fielding (1974) emphasizes human processes; he considers physical elements only in terms of man's perception and use of these elements. Man's behavioural processes also generate spatial distributions. Attempts to analyse spatial distribution in terms of behavioural processes necessitates the separate discussion of each process. The regional descriptions dominated geography until the mid-20th century. Contemporary geography emphasizes the regularity in spatial organization and attempts to develop models that can explain how human behaviour may account for these distribution patterns. The regularities in spatial distribution result from man's attempts to use the land efficiently. The interrelated concepts of agglomeration, distance and accessibility, utility, interaction and satisfaction help to understand these regularities in spatial organisation[19]. These concepts seldom operate independently.

The Concept of Agglomeration is the concept of Clustering. In the present world, the people, industries and various other human activities tend to cluster or agglomerate at certain locations over the surface of the earth. This is to take the advantage of co-existence. The tendency to cluster similar activities is apparent in agricultural industrial, and population distribution. In the United States, for instance, 70 per cent of the population resides on less than 2 per cent of the land area. At the world scale, the polar zones and some of the deserts are virtually uninhabited, whereas most of the population is clustered along the margins and in the fertile valleys of the continents. The agglomeration of people in cities, the clustering of homes in rural townships, and the association of shops in shopping centres improves economic efficiency and social satisfaction. There are numerous examples of the spatial distributions that result from agglomeration of various activities. To cite few more, steel-making plants and iron ore and coal fields are the common foci for agglomeration. Agglomeration is also illustrated by consumer behaviour. For specialized goods and services the customers like to visit central places of the city, while for their day requirement a local shopping centre or market is sufficient.

The Concept of Distance and its corollary Accessibility are fundamental to the understanding of spatial organization of physical and human elements. Evidence of their effect is apparent in economic, social and psychological behaviour. Since many human activities involve men coming together to exchange goods and services, distance and accessibility affect behaviour in numerous ways. Where distance impedes interaction, man devises communication system to minimize its effect. Roads, rails and telecommunication facilities are manifestations of this process. The locations of various production activities—farm, factory, mill, and mine—are primarily determined by the costs of obtaining inputs (material and labour) and marketing outputs (products).

In retail and service activities, for instance, the minimization of distance helps to maximize patronage. Consumers prefer locations where several activities are agglomerated, and operators prefer locations that maximize accessibility to customers. But not all the enterprises are willing to pay the highest price for the most accessible locations. Similarly, the consumers are not always willing to travel long distances for their every requirement. In various models, explaining the distribution of urban areas, agricultural land uses, industries, or uses within a city (Christaller, Von Thunen, Weber) etc., the concept of distance and accessibility has been used. In urban areas, for instance, most accessible locations are also the areas of very high land rent. It is why, the land values in metropolitan areas decline with distance away from the city centre. The CBD is the most accessible location. Therefore, all the specialized functions and services are located here. Even in rural landscape the association between land values and accessibility is apparent; agricultural land adjacent to the city, e.g., is used more intensively. Ideally greater the distance poor the accessibility. But in the contemporary world, the reverse may hold true. Because, today, and particularly after the Automobile Revolution in the 20th century, the accessibility is measured not in terms of absolute distance, but in terms of time, money and efforts involved in covering that distance. The perception of distance is another aspect related to this concept. Geographical distance is not the same as the man's perception of distance, which is referred to as Psychological distance. Psychological distance also affects man's use of earth surface. Man arranges cognitive space (mental space) differently from physical space. The mental map carried by individuals affects the way in which they react to proposed changes, whether these be proposed environmental changes or personal migrations.

The Concept of Utility is related to the value or usefulness of goods and services. It is applicable to both tangible and intangible goods, like farm products, education, or affection; and may be measured in monetary or non-monetary units. Utility is what farmers seek to maximize through choosing a particular crop and livestock association; or what industrialists seek to maximize by choosing low-cost locations for factories, or what the home owner seeks to maximize when he chooses his residence in a certain neighbourhood at an acceptable distance from the location of employment, schools or relatives. Thus, the utility of a good, service or location may differ between people, and identical amount of the same good, service or location may be valued differently.

The Concept of Interaction is related to exchange, interactions, etc. The spatial organizations of various phenomena are also based on the concept of interactions or exchange of interactions. This concept is evident in every level of spatial organization, micro to macro. It explains why the core areas of a city or any other settlement are more developed than others, i.e. of greatest interaction with activities, people, goods and ideas. In fact, this concept of organization has been important since beginning. Therefore, the human civilizations were the river valley civilizations. Or even earlier, the organic (or human) life originated in the contact zones of air-land-water (biosphere). The concept of interaction is evident at every level of spatial organization, ranging from the behaviour of individual or among large political entities. Interaction creates social bonds between individuals. It is through interaction in the form of verbal and non-verbal messages, exchange of goods, favours, and friendships—that the objectives of nations, cultures, families and communities are established and developed. Within these groups, shared objectives, values, and attitudes provide norms for individual behaviour. They restrict the individual choices to those that are socially acceptable and thereby influence the patterns and arrangements that result from human behaviour.

Lastly, there is the Concept of Satisfaction. Man is usually satisfied by a reasonable level of achievement rather than by the optimum or maximum, as there is no limit to aspirations. Satisfaction is a psychological concept because it refers not only to man's perception of the opportunities and resources provided by his technology, but also to the level of their utilization. A location cannot satisfy every aspect of human requirement. Man has to make choice out of many alternatives. For example, in searching for a location for a house—man has to make choices out of school-facility, market, employment, relatives, etc. Any location can give a reasonable or optimum level of satisfaction, and not a complete satisfaction.

NOTES

1. Davies, 1972, pp. 52-68.
2. *Darwin's Theory of Evolution:* Scientists at the beginning of the 1800s knew of some kinds of fossils, and they were very aware of homologous and vestigial structures. Many scientists suspected that some kind of evolution had given rise to living things around them. However, they had no unifying theory to explain how evolution might have occurred. Two scientists, namely Jean Lamarck and Charles Darwin, led the way in the search for a mechanism of evolution. Jean Baptiste de Lamarck (1774-1829), the French scientist, was the one to make the first systematic presentation of evolution in 1809. Lamarck described a mechanism by which he believed evolution could occur. This mechanism was known as "the inheritance of acquired characteristics." Since Lamarck presented no experimental evidence or observation and his theory fell out of scientific favour. The next significant idea came from the British scientist Charles Darwin. From a young age Darwin disliked school and preferred observing birds and collecting insects to study. He was sent to medical school in Scotland when he was 16. Young Darwin found medicine "intolerably dull." He was much more interested in attending natural history lectures. Seeing that Darwin lacked enthusiasm for becoming a doctor, his father suggested him to study for the clergy. Darwin was agreeable to the idea and enrolled in the university at Cambridge, England, in 1827. Here, Darwin found that his friendship with John S. Henslow, Professor of Botany, made life in Cambridge extremely worthwhile. Henslow encouraged Darwin in his studies of natural history. In 1831 Henslow recommended that Darwin be chosen for the position of naturalist on the ship the HMS Beagle. Darwin's job as ship naturalist was to collect specimens, make observations, and keep careful records of anything he observed that he thought significant. During the Beagle's five-year trip, Darwin trekked hundreds of miles, observed thousands of species of organisms and collected many different types of fossils. On the long sea voyages he used his time to catalog his specimens and write his notes. When Darwin returned to England in October 1836, he sent many specimens to experts for study. A bird specialist,

(ornithologist) studied Darwin's bird collections from the Galapagos Islands, located about 1,000 km west of South America. He reported that Darwin had collected 13 similar but separate species of finches (a kind of small bird). The similarities of the Galapagos finches led Darwin to infer that the finches shared a common ancestor. The similarities between the fossil mammals Darwin collected and modern mammals led him to believe that species change over time. In 1837 Darwin began his first notebook on evolution. For several years Darwin filled his notebooks with facts that could be used to support the theory of evolution. Comparing homologous structures, vestigial organs, and embryological development of living species gave him additional evidence of evolution.

3. 'Social Darwinism', term coined in the late 19th century to describe the idea that humans, like animals and plants, compete in a struggle for existence in which 'natural selection' results in "survival of the fittest." Social Darwinists base their beliefs on theories of evolution developed by British naturalist Charles Darwin. The term social Darwinist is applied loosely to anyone who interprets human society primarily in terms of biology, struggle, competition, or 'natural law'.
4. Taylor, (1951).
5. Dikshit, 1997, 246.
6. Tatham, 1951.
7. Hartshorne, 1959, 65-80.
8. Dikshit, 1997, 238.
9. Eyre, 1964; Stoddart, 1972.
10. Tansley, 1946, 206.
11. Christopher, 1994, 586.
12. Foster, J. (2003). 63-74.
13. Unstead, 1933.
14. James and Jones, 1954, 19-68.
15. Herbertson, 1905.
16. De Blij and Muller (1971).
17. Pounds, 1972.
18. Roepke, 1967.
19. Fielding, 1974, 3-51.

REFERENCES

Christopher, R.W. (1994) *Geosystems: An Introduction of Physical Geography*, Macmillan College Pub. Co., Englewood Cliffs, New Jersey.

Davies, W.K.D. (1972) *Conceptual Revolution in Geography*, Rowman and Littlefield, New Jersey.

De Blij, H.J. and P.O. Muller (1971) *Geography: Regions and Concepts*, John Wiley and Sons Inc., New York, pp. 1-18.

Dikshit, R.D., (1997), *Geographical Thought: A Contextual History of Ideas*, Prentice Hall of India.

Eyre, S.R. (1964), "Determinism and Ecological Approach to Geography", *Geography*, Vol. I, No. 49, pp. 369-376.

Fielding, G.J. (1974), *Geography as a Social Science*, Oxford University Press, Harper and Row Publishers, New York.

Foster, J. (2003). "Between economics and ecology: Some historical and philosophical considerations for modelers of natural capital". *Environmental Monitoring and Assessment*, 86.

Hartshorne, R., (1959) *Perspective on the Nature of Geography*, Rand McNally and Co., Chicago.

Herbertson, A.J. (1905), "The Major Natural Regions: An Essay in Systematic Geography", *Geography*, pp. 300-310.

James, P.E. and C.F. Jones (1954), "The Regional Concept and the Regional Method", *American Geography: Inventory and Prospects*, Syracuse University Press, Syracuse, pp. 19-68.

Johnston, R.J. (2004), *Geography and Geographers: Anglo-American Human Geography Since 1945*, 6th Ed., Edward Arnold, London.

Pounds, N.J.G. (1972) *Political Geography*, Ed. 2, McGraw Hill, New York.

Rana, L. (2008), *Geographical Thought: A Systematic Record of Evolution*, Concept Publishing Company, New Delhi.

Roepke, H.G. (1967) *Readings in Economic Geography*, Wiley, New York, pp. 17-37.

Stoddart, D.R. (1972) "Darwin's Impact on Geography" and "Geography and the Ecological Approach" in W.K.D. Davies, *op. cit.*, pp. 52-68 and 301-311.

Tansley, A.G. (1946), *Introduction to Plant Ecology*, Allen and Unwin, London.

Tatham, G. (1951) "Environmentalism and Possibilism", in Taylor, G. (Ed.) *Geography in the Twentieth Century*, Methuen, London, pp. 128-162.

Taylor, G. (1951), *Geography in the Twentieth Century*, Butlar & Tanner Ltd., London.

Unstead, J.F. *et al.* (1933), "Classification of Regions of the World", *Geography*, Vol. 22, pp. 253-282.

Whittlesey, D. (1954), "The regional concept and the regional method", in P.E. James & C.F. Jones (Eds.), *op. cit.*, pp. 19-69.

SECTION-IV
CONTEMPORARY TRADITION

CHAPTER-8

Quantitative Revolution

In the history of geography, the Quantitative Revolution was one of the four major turning-points. The quantitative revolution occurred during the 1950s and 1960s and marked a rapid change in the method behind geographical research, from regional geography into a spatial science. The main claim for the quantitative revolution is that it led to a shift from a descriptive (idiographic) geography to an empirical law-making (nomothetic) geography. The quantitative revolution had occurred earlier in economics and psychology and contemporaneously in political science and other social sciences and to a lesser extent in history. The greatest impact of the quantitative revolution was not the revolution itself but the effects that came afterwards in a form of the spread of positivist (post-positivist) thinking and counter-positivist responses.[1]

SCHAEFER–HARTSHORNE DEBATE

The philosophy and methodology of geography were apparently debated most earnestly in the writings of two American scholars, namely Richard Hartshorne (1899-1992) and Fred K. Schaefer (1904-1953) during the mid-20th century.[2] Schaefer was originally an economist, and later he joined the group of geographers at the University of Iowa. Schaefer is considered as one of the pioneers of quantitative revolution. Hartshorne was another American scholar, at the University of Wisconsin. He is known for his two important publications, viz. *Nature of Geography* and *Perspectives on the Nature of Geography* published in the years 1939 and 1959 respectively. In *Nature of Geography,* he strongly advocated for chorological (regional) approach in geography stating that: "Geography is an integrative discipline to which society has assigned responsibility for the study of areas. It is expected to satisfy human curiosity about how much of what is where, and why it is there, in an organized manner that will facilitate comprehension and retention. The discipline deals with an enormous range of phenomena and must provide a congenial home for many different kinds of practitioners. Systematic geography generates theories to facilitate an understanding of regions, and regional geography is the proving ground where theories are tested empirically. The idea of the region provides the essential unifying theme that integrates the diverse sub-disciplines of geography. The highest form of the geographer's art is the production of evocative descriptions that facilitate an understanding and an appreciation of regions." But, Schaefer, well-known for his article "Exceptionalism in geography: A Methodological Examination", was in great disagreement with the views of Hartshorne. Schaefer's article, published in 1953 in *Annals of Association of American Geographers*, was not only reciprocation to Hartshorne's chorological viewpoint; it was also a call for scientific approach to geography based upon the search for geographical laws (the ultimate form of a scientific generalization). Schaefer died before his article even appeared in print, and so he was never able to elaborate his argument, nor defend himself from Hartshorne's subsequent attack in 1959. But the article became a rallying point for the younger generation who were intent on reinventing the discipline as a science, or *spatial science* as it was later called. Schaefer's paper was the first to challenge Hartshorne's presentation and also interpretation of the works of Hettner and others who had tremendous support for chorological viewpoint in geography. This paper was published 14 years after Hartshorne's *Nature of Geography*. His intent was to criticize the 'exceptionalist' claims made for regional geography, and to present the case for geography adopting the philosophy and methods of the positivist school of science. Once again, as stated by Johnston, geography's identity and methodology were called into question.[3] The exchanges between Richard Hartshorne and F. K. Schaefer provided fuel for continued questioning in the nature of geography. To understand each point of view, an explanation of ideologies of both must be given. Hartshorne, according to Johnston, described geography as: "...a science that interprets the realities of areal differentiation of world as they are found, not only in terms of the differences in certain things from place to place, but also in terms of the total combination of phenomena in each place, different from those at every other place." Hartshorne tried to prove that the nature of geography is regional and its evolution can be historically traced. This was the paradigm accepted by geographers until contested by Schaefer, up to the mid-20th century. Schaefer's view of geography, as indicated by Johnston, was that: "...geography has to be conceived as the science concerned with the formulation of laws governing the spatial distribution of features on the surface of the earth and these spatial arrangements of phenomena, not the phenomena themselves, should be the subject of geographers' search for law-like statements." These debates regarding geography's nature, in the post Second World War period, helped evolve its specific areas of focus. Now, not only was geography a science concerned with regional interest but contained laws that enabled definition of its methodology. However, with this evolution came continued debate regarding the significance between two main geographic concerns: idiographic-chorological and nomothetic-systematic perspectives. Dissatisfaction developed with chorological or regional approaches. The new generation of geographers was increasingly drawn to the systematic studies. Main reason was that the chorological viewpoint was promoting the concept of uniqueness of areas, neglecting theory and generalizations. The chorological paradigm had become sterile in geography and was not yielding laws and theories. This pent up feeling of disappointment with the state of affairs in contemporary geography of 1950s had found voice in the forceful paper by Schaefer. Basically, Schaefer favoured the view that without the formulation of laws a science cannot be recognized. Schaefer criticized the exceptionalist claim made for geography as a chorological science, and put forward a strong case for the discipline to adopt the philosophy and methodology of scientific positivism[4]. Geography is concerned with the spatial arrangement of phenomena and not the phenomena themselves. Geographical laws are the 'pattern laws', as contrasted to the 'process laws' of other social science disciplines. Schaefer's paper posed a challenge to the concept of geography as projected and popularized by Hartshorne's *"Nature of Geography"* (1939). Hartshorne's second book *"Perspectives on the Nature of Geography"* (1959) was written as a reaction to Schaefer's allegations. This book was a supplement to his previous book. Hartshorne, in his second book, presented a wide-ranging review on the nature of subject, including discussions on and clarifications about many points of criticism regarding the concept of geography as areal differentiation. Although Hartshorne stuck to essentially the chorological viewpoint in his second book as well, he redefined, clarified many of the previously stated concepts. The essential difference between the positions taken by Hartshorne and Schaefer was that Hartshorne's was a positive view of geography, i.e. geography is what geographers have made it. Schaefer's view, on the other hand, was a normative one, i.e. what geography should be irrespective of what it had been. The fact, however, was that since Hartshorne's view was fast losing ground, geographers in increasing numbers had, by the end of 1950s, turned round to Schaefer's view of geography as a spatial (locational) science. For this they used methods of other systematic sciences and were increasingly concerned with quantification and development of theory, so that by the time Hartshorne's *"Perspectives on the Nature of Geography"* had appeared in 1959, the Quantitative Revolution had reached its zenith. The net outcome of Schaefer-Hartshorne Debate was that geography, by the end of the 1950s had come increasingly to be viewed as a science requiring the use of the scientific method so that like other sciences, it could also develop laws and theories relevant to its field of study. This brought about a distinctive shift in emphasis from 'regional' to 'systematic' studies. This meant that geography thereafter began increasingly to be viewed in a nomothetic perspective, i.e. it required developing the habit of seeking the general in the particular. This also involved a shift from 'areal' to 'locational' studies; from 'absolute' to 'relative' locations; and from 'areal differentiation' to 'spatial interaction'.

QUANTITATIVE REVOLUTION: BEGINNINGS AND END

As a consequence of a long methodological debate upon its scope and content an apparently new perspective has been opened up under the impact of so-called Quantitative Revolution during the mid-20th century. The quantitative revolution occurred during the 1950s and 1960s and marked a rapid change in the method behind geographical research, from regional geography into a spatial science.[5] In the early

1950s, there was a growing sense that the existing paradigm for geographical research was not adequate in explaining how physical, economic, social, and political processes are spatially organized, ecologically related, or how outcomes generated by them are evidence for a given time and place. A more abstract, theoretical approach to geographical research has emerged, evolving the analytical method of inquiry. The analytical method of inquiry led to the development of generalizations that are logically valid about the spatial aspects of a small set of closely defined events embodied in a wide range of natural and cultural settings. Generalizations may take the form of tested hypotheses, models, or theories, and the research is judged on its scientific fit and its validity. Adoption of the analytical approach had helped geography become a more law-giving science, and the conception of the discipline as an idiographic field of study has become less acceptable. The changes during the 1950s through 1970s were not the introduction of mathematics into geography, but mathematics as a tool for explicit purposes and for statistical methodology and formal mathematical modelling. The Quantitative Revolution led to an increased use of computerized statistical techniques, in particular multivariate analysis, in geographical research. The newly adopted methods reflected an array of mathematical techniques that improved precision. Some of the techniques that epitomized the quantitative revolution include: descriptive statistics; inferential statistics; basic mathematical equations and models, such as gravity model of social physics; stochastic models using concepts of probability, such as spatial diffusion processes; and deterministic models, e.g. Von Thünen's and Weber's location models. The common factor, linking the above techniques, was a preference for numbers over words, plus a belief that numerical work had a superior scientific pedigree. Proponents of quantitative geography tended to present it as bringing science to geography. Burton, for instance, described the Quantitative Revolution as "a radical transformation of the spirit and purpose of geography".[6] By this he meant a newfound enthusiasm for the use of numerical techniques of some kind, directed towards elucidating the details of Earth surface patterns. The quantitative revolution was geography's attempt to redefine itself as a science. The revolutionaries declared that the purpose of geography was to test general laws about the spatial arrangement of phenomena. They adopted the philosophy of positivism from natural sciences and turned to mathematics—especially statistics—as a way of proving hypotheses. The mathematical and statistical methods were introduced to attain a desired level of objectivity for the search of laws and theories. All these efforts were devoted to the fundamental conception of geography as a spatial science. Although, this movement is over now, it did lead to mathematization of much of our discipline, with an attendant emphasis on the construction and testing of theoretical models. The Quantitative Revolution did lead to a shift from an ideographic to nomothetic geography.

For long in history, geography has been characterized by much physical, cultural and economic descriptive analysis, developing what we now call regional geography. The pinnacle of regional geography occurred during the mid-20th century following Richard Hartshorne's *The Nature of Geography* (1939). Debate raged predominantly where regional geography was the major philosophical school. All of these events presented a great threat to geography's position as an academic subject and thus geographers began seeking new methods to counter critique. Under the banner of the scientific method, the Quantitative Revolution began. Four main criticisms were raised against Regional geography. Firstly, it was over descriptive following a purely inventory format for the physical and social features at a specific site. Secondly, it was almost purely moulded for education purposes. Regions aren't really geographical entities; they are just convenient for descriptive purposes. Thirdly, it failed to actually describe why the patterns were the way they were, thus providing no real geographical understanding. Fourthly and most importantly, regional geography was considered unscientific. The Quantitative Revolution began not only as a response to regional geography paradigm, but also as a result of the 1950s crisis. During the late 1940s and early 1950s a series of events nearly led to the downfall of geography as an academic subject. This is known as the 1950s crisis in geography. The crisis occurred for several reasons as: the closing of many geography departments and courses in universities, e.g. the abolition of the geography programme at Harvard University in 1948; the continued division between Human and Physical Geography; geography being seen as overly descriptive and unscientific and the claim that there was no explanation of why processes or phenomena occurred within the regional framework; geography was seen as exclusively educational and there were few if any applications of contemporary geography; and continuing question of what geography is—science, art, humanity or social science. Then, during the Second World War technology became very influential in our society. This culminated in the Quantitative Revolution, during which science and mathematics enjoyed unrivalled importance. Nomothetic aspects of geography heightened in importance and quantitative methods and techniques were applied to virtually all investigation and experimentation. The revolution in geography has come largely as a result of the impact of work by non-geographers, a process shared by many other disciplines, where an established order has been overthrown by a rapid conversion to a mathematical approach. It has expanded to transform first the physical and then the biological sciences. It is now strongly represented in most of the social sciences including geography. Physicists and mathematicians began the movement that led to the revolution in geography. However, in geography, the Quantitative Revolution is contemporaneous with the appearance of neo-determinism. Although its antecedents can be traced far back, the quantitative revolution in geography has begun actually in the late 1940s. It reached its culmination in 1960s and is over now.

In geography, the quantitative techniques were first used in physical geography, and they entered later in human studies. A.N. Strahler, the prominent physical geographer, saw an immense scope of application of quantitative techniques to geomorphology in the late 19th century itself when 'Davisian Geomorphology' was popular. Strahler criticized the traditional approaches to physical geography and emphasized on the use of quantitative techniques in physical studies, especially including rate of erosion, soil creep, and various other hill and valley forms. According to Strahler, instead of advancing the scientific study of geomorphological process, the contemporary geographers (late 19th and early 20th century) made splendid contribution to descriptive and regional geomorphology. The possible reason for this was considered to be the confusion or fear regarding the use of statistical techniques in geography, i.e. related to the loss of descriptive quality and a fear of gaining false impression of objectivity. In spite of this, in 1950s and 1960s several geomorphologists started using quantitative methods and the practice spreaded gradually. They were, for instance, Strahler, Chorley, Dury, Mackay, Wolman, etc. As far as the field of climatology is concerned, there has been a little argument about the widespread application of quantitative techniques to it. This branch of geography embraces the most apparently manageable and quantifiable continuum that geographers have been concerned to study. The scholars like Koeppen, Thornthwaite, Mather, Hare, Bryson and others have been applying quantitative techniques to climatic studies with great effect. By far the greatest struggle for the acceptance of quantitative methods has been in human geography. It is because of the prevalent notion that the human behaviour is unpredictable. It is here that the revolution runs up against the notion of freewill and the unpredictability of human behaviour.

Among all the modern schools of thought, it was the British geography that showed the earliest signs of change during the quantitative revolution. Initially, it was through the publication of S. Gregory's book *"Statistical Methods and the Geographers"* in 1963. This was followed in 1965 by *"Frontiers in Geographical Teaching"* by P. Hagget and R.J. Chorley and was the first of the whole series of books they have written or edited about the search for laws in geography. Also in 1965 Hagget's book *"Locational Analysis in Human Geography"* attempted to provide a theoretical framework for the study of regularities in human behaviour. In 1967 another book edited by Hagget and Chorley: *"Models in Geography"* attempted to review the entire theoretical framework which geographers had used to study locational patterns.

However, within two to three decades of the beginning of Quantitative Revolution, the overwhelming focus on statistical modelling itself eventually became responsible for the undoing of this revolution. Many geographers became increasingly concerned that these techniques simply put a highly sophisticated technical gloss on the approach to study that was barren of theory. Other critics argued that it removed the 'human dimension' from a discipline that always prided itself on studying the human and natural world alike. As the 1970s dawned, the Quantitative Revolution came under direct challenge. The mood changed with a significant number of geographers beginning to question the wisdom and even the morality of this quantitative turn. Meanwhile, the opposition to the Quantitative Revolution came in several ways[7]: (i) There were those who thought that the whole idea was a bad one and that quantification would mislead geography in a wrong and fruitless direction; if such critics are still among us they have not made themselves heard for some time; (ii) There were traditional scholars like Dudley Stamp[8] who argued that geographers had spent too long perfecting their tools (maps, cartograms and other diagrammatic representations) and should get on improving these tools further rather going for something new like quantification; (iii) A third kind of opposition holds that statistical techniques are suitable for some kind of geography; but not all geography, because there are certain things that cannot be measured; a variant of this argument is that the variables with which geography is concerned are too numerous and complex for statistical analysis; (iv) Another class of objection is that although quantitative techniques are suitable and their application to geographic problems is desirable, they are nevertheless being incorrectly applied; ends are confused with means; quantitative analysis has failed on occasion to distinguish the significant from trivial; the alleged discoveries of the quantifiers are not very novel; and so on; and (v) A final kind of criticism to note is that quantification is all right, but the quantifiers are not; they are perky, suffer from over-enthusiasm, vaulting ambition, or just plain arrogance. That these criticisms have a grain of truth cannot be denied, but to the valid, correct use of quantitative methods they are irrelevant.

THE IMPACT

As far as the consequences of Quantitative Revolution are concerned, the effects were two fold, viz. spread of positivist thinking and the emergence of counter-positivist responses. The counter-positivist response from human geography was created in a form of behavioural, radical and humanistic geography. Dissatisfaction with idiographic geography lies at the root of the quantitative revolution. The development of theoretical, model-building geography has been the major consequence of the quantitative revolution. The emergence of geography as an abstract, theoretical science appears to have been the most overriding development in geographic research during the 1960s. Attention, since then, has focused on the spatial organization of economic, social, physical, political and urban processes, and on the outcomes generated by these processes at given times and in particular places. Special attention has been placed on the spatial aspects of social processes. Problems related to environmental perception and control have also been researched. Accompanying these developments has been a widespread acceptance of the need for more geographically oriented quantitative methods for use in model building. The Quantitative Revolution has greatly influenced the practice of Physical and Human geographers alike, as well as vastly improving the scientific credibility of the discipline. Continued progress has been made in the observation and handling of data related to geographic problems. The Quantitative Revolution pioneered statistical and mathematical methodologies to improve geographers' understanding of complex processes. New technology began to be used, such as remote sensing and satellite imagery from beyond the Earth's atmosphere to improve the precision and analysis of mapping. The fusion of quantitative revolution and new technology has given birth to a new field of study, called 'geomatics'[9]. The greater use of computers in geography and the creation and application of GIS and remote sensing, in fact, is the result of the developments in geomatics. These new developments have allowed geographers for the first time to assess complex models on a full-scale model and over space and time. The development of geomatics led to geography being reunited as the complexities of the human and natural environments could be assessed on new computable models. Further, advances also led to a greater role of spatial statistics and modelling within geography. Eventually, the quantitative revolution had its greatest impacts on the fields of physical, economic and urban geography. The true value of mathematical and statistical methods lies in the role that they play in the testing and verifying of ideas established, however tentatively, within a theoretical framework. According to Burton (1963) 'quantification is inextricably intertwined with theory'. The quantitative methods are especially helpful using models in geography. A model is the formal presentation of a theory, often in mathematical form, and is constructed for the specific purpose of enabling the relationships embodied in the theory to be tested. The quantitative techniques provide precise tools to test theories and analyse data. They could, however, be used effectively for making precise predictions. Prior to quantitative revolution, the geographers attempted to use their experiences to formulate geographical problems and then to solve them by the direct examination of empirical evidence. Now an increasing number of geographers translate their problems into mathematical terms and then manipulate the equation until they obtain a mathematical solution. They then re-translate this solution into geographical language or statement. The capacity to convert problems into mathematical terms has enabled geographers to tackle for more difficult problems, to formulate them more precisely, and to obtain, with the use of statistics, much more soundly based conclusions than was the case in pre-revolution phase.

To conclude, the Quantitative Revolution has been called one of the greatest periods of intellectual achievement in the whole history of the discipline. The quantitative techniques provide most appropriate methods for the development of the theory. The quantitative era lasted as long as its methods proved to be aiding in the development of theory. It follows that any branch of geography claiming to be scientific has need for the development of theory, and any branch of geography that has need for theory has need for quantitative techniques. The development and testing of theory is the only way to obtain new and verifiable knowledge and new and verifiable understanding. However, the efforts towards quantification do not discourage 'qualitative' view of study. First task of geographer is always to find out whether different geographical elements, phenomena, place, etc. represent any relationships, differences, or any other type of connections. Later comes the use of quantitative methods, i.e. in analysis, to measure the degree of relationships or inter-connections. The geographers are not going to explore irrelevant relationships.

NOTES

1. Livingstone, 1992.
2. Dikshit, R.D. (1997), 118-121.
3. Johnston, 1979, 50-8.
4. Scientific or Logical Positivism is a philosophy that combines *positivism*—which states that the only authentic knowledge is scientific knowledge—with a version of *a priorism*—the notion that some propositional knowledge can be had without, or "prior to", experience. Logical positivism denied the soundness of *metaphysics* and traditional philosophy, and affirmed that statements about metaphysics, religion and ethics are devoid of cognitive meaning; only statements about mathematics, logic and natural

sciences have a definite meaning. Logical positivism originated in *the Vienna Circle* in the 1920s, where Rudolf Carnap, Otto Neurath, and others divided meaningful statements into those which are analytic (true *a priori*), and those which are synthetic (verified by sensory experience, *a posteriori*). Logical positivism refuted synthetic *a priori* knowledge: an evident criticism to Kantian philosophy.

5. *'Quantitative revolution'*, Wikipedia, the *Free Encyclopaedia.*
6. Burton, 1963.
7. Davies, 1972.
8. Sir Laurence Dudley Stamp (1898-1966), was professor of Geography at Rangoon and London, and one of the internationally best known British geographers of the 20th century. He specialised in the study of geology and geography and taught at the Universities of Rangoon (1923-26) and London (1926-45). From 1936 to 1944 he directed the compilation and publication of the report of the Land Utilisation Survey of Britain. He worked on many official enquiries into the use of land and planning.
9. Geomatics is the modern scientific term referring to the integrated approach to measurement, analysis, and management of the discription and location of Earth-based data. These data come from many sources, including earth-orbitting satellites, air and sea-borne sensors and ground based instruments. Geomatics has applications in all disciplines which depend on spatial data, including forestry, environmental studies, planning, engineering, navigation, geology and geophysics. It is fundamental to all the areas of study which use spatially related data, such as Surveying, Remote Sensing and Photogrammetry, Cartography, GIS, Property or Cadastral Studies and Global Positioning.

REFERENCES

Burton, I. (1963), "The Quantitative Revolution and Theoretical Geography", *The Canadian Geographer,* Vol. 7, pp. 151-162.

Davies, W.K.D. (1972) *Conceptual Revolution in Geography,* Rowman and Littlefield, New Jersey.

Dikshit, R.D. (1997), *Geographical Thought: A Contextual History of Ideas,* Prentice Hall of India.

Harvey, David (1969), *Explanation in Geography,* Edward Arnold, London.

Johnston, R.J. (2004), *Geography and Geographers: Anglo-American Human Geography Since 1945,* 6th Edition, Edward Arnold, London.

Livingstone, David N. (1992), *The Geographical Tradition: Episodes in the History of a Contested Enterprise,* Blackwell Publishers.

Wikipedia, the *Free Encyclopaedia.*

CHAPTER-9

Laws, Theories and Models

Few would deny the fact that the last few decades have been one of the greatest periods of intellectual change in the trends of geographic development. Much of these changes, the questioning of the past approaches, looking at old problems with new eyes, have been of a methodological nature involving, in virtually every instance, the substitution of quantitative approaches to problems formerly treated in descriptive ways. Throughout its history geography has been characterized by an unceasing methodological debate upon its scope and content. Today an apparently new perspective has been opened under the impact of so-called quantitative revolution. Statistical methods have been introduced to attain a desired level of objectivity, and search for laws and theories and proceed apace. All are devoted to the fundamental conception of geography as a science. These laws and theories cover the general behaviour of the empirical events or objects with which the science is concerned, and thereby enable the geographers to connect together the knowledge of the separately known events and to make reliable predictions of events as yet known. Of all the problems of contemporary concern in the thinking of geographers, the most disturbing has been the question whether geography 'like the science' could develop the knowledge of principles, laws and general truths, or its function has been merely to describe the innumerable unique things. Geography has not always been considered as a scientific discipline. But, various scholars, like Humboldt, Ritter, Davis and Blache have all evoked the necessity of considering its scientific nature. This demanded from geographers to search for similarities, and especially functional similarities among unique things. With this understanding, therefore, a rapprochement with modern science has taken place at an accelerating pace within the last few decades. It has been realized that social and human phenomena can also be regarded as generalizations, along with the physical. The work of Hartshorne (1939, 1959) may be considered as the last in the chain of traditional writers in geography. The Concepts of Geography elucidated by Hartshorne and accepted by many practicing geographers began to come under attack from the early 1950s onwards. Many began to feel that geographers should attempt to discover laws to explain the regular distribution of phenomena on the earth's surface, as much effort had went into the description of unique areas. Basically a growing number of geographers felt that while regional description could be progressively refined and made more sophisticated it was not carrying geography any closer to understanding why certain areas had a particular character. One of the most extreme statement of the view that geographers should seek for laws which govern the location of phenomena on the earth's surface, came from F.K. Schaefer, an American scholar, who attacked the views of Hartshorne in a paper entitled "Exceptionalism in Geography: A Methodological Examination", published in 1953. Schaefer attempted to answer the question of why geography seemed to be failing to make progress. Geographers in his view had concentrated on describing particular places or regions when they should have been building up a cumulative body of theory to explain why features on the earth's surface, whether natural or man-made, were located where they were. Another factor that encouraged the spread of Schaefer's ideas was the spread of quantification. A growing number of geographers became aware that mathematics and statistics could be applied to geographical problems. These provide tools to test theories and analyze data. They could, however, only be used effectively if they were supported by carefully constructed theories, which made precise predictions, and which could then be tested. The process of intellectual change led geographers to concentrate less and less on describing the differences between particular areas or places and more and more on the study of uniformities and the production of theories about the spacing of phenomena on the earth's surface. This change led the current geographical thinking to a stage where considerable emphasis is being placed on discovery and formulation of laws and theories. This is a logical outcome of continued search for scientific explanation of the spatial relations of phenomena. Instead of concentrating on the question of how and where phenomena are located and arranged, the emphasis is now directed toward the general question why the facts are as alleged. Since geography counts itself among the sciences, then such an emphasis on 'nomothetic approach" is in right direction. The recent presentation of Hagget and Bunge (1965 and 1966) are also indicative of this trend clearly, and many more examples could be cited. Generally speaking, there are two alternative routes or methods to be followed in establishing a scientific explanation.[1] The first is by 'induction'-proceeding from numerous particular instances to universal statements; and the second that of 'deduction'-proceeding from some *a priori* universal premise to statements about particular sets of events. The laws, theories, models and systems analysis provide important tools of explanation in geography.

The quest for explanation is a quest for a theory. The development of theory is at the heart of all explanations. Undoubtedly, the observations and descriptions cannot be theory free. To state a fact entirely divorced from theoretical interpretations is not justified. Theories represent generalizations, used for explanations. They can make precise predictions. The quantitative techniques can be used effectively if they are supported by carefully constructed theories. Since, methodological debates for a long time hampered the process of theory-building in geography, even till today the discipline is short in theories and long on facts. Schaefer's arguments in 1950s brought great motivation among geographers and helped them to break away from the traditional geography of distribution and classification of objects and to move a step further to explain and predict the distributions. Awareness towards theory-building was already there before, but a number of works, particularly in the 20th century have played a great role in encouraging the use and formulation of laws and theories in geography[2]. These studies prove that there is 'some hidden order within chaos' and the geographers' task is to search for this order. This order may be induced empirically, i.e. based on personal observations; proceeding for numerous particular instances to universal statements; moving from particular to general; or deduced theoretically, i.e. proceeding from some *a priori* universal premise to statements about particular sets of events or phenomena, i.e. proceeding from general to particular. In the latter case, the explanations are based on already established theory (*a priori*), whereas in the former the explanations arrive at some theory, i.e. formulation of a new or original theory (*a posteriori*). However, the theories in either of ways are used for explanation and are the highest order generalizations. The great advantage of a law or of generalization is that it allows us to predict from what is known about a part of a class to the whole class, even though the whole class has not been examined. In other words, a law allows an extension of knowledge beyond what has actually been observed.

LAWS

Law is a system of rules; a rule of being or of conduct. Theory is the framework within which one seeks explanation and by constitution, laws should form the part of theory. A law is seen as a universal statement of unrestricted range, i.e. unrestricted in its application over time and space.[3] The numerous analyses by logicians and philosophers suggest that two criteria are of major significance in identifying laws. The first is the universality of the statement. The second is the relationship of a statement to the surrounding statements and in particular the way in which one statement fits into a whole collection of statements which themselves form a scientific theory. A scientific law may be interpreted most rigidly as a generalization that is empirically universally true, and one that is also an integral part of a theoretical system in which we have supreme confidence. A substantial part of scientific explanation is concerned with establishing how laws are related to a surrounding structure of theory. In other words, it is impossible to determine whether a statement is or is not a law simply by reference to the truth or falsity of the generalization it contains. Consider, for instance, the statement that 'towns of similar size and function are found at similar distances apart'. Although, we may have reservations about regarding this as a law (the notions of similarity and function are too inexplicit for example) we certainly regard it as being more law—like than a statement that 'all towns contain collections of buildings.' The significant difference between statements is not basically a function of either empirical status or universality. It is simply that the first statement has some place within the structure of central place theory that, however crudely it has been formulated, has a putative deductive structure in the formulations of Christaller and Losch, whereas the second statement has no theoretical structure of explanations surrounding it and is in any case trivially true. Accordingly, a major criterion in determining whether a statement is or is not a law is the relationship of that statement to the system of statements that constitutes a theory.

At least five different types of laws may be formulated by geographers, such as: (i) Cross-Section; (ii) Equilibrium; (iii) Historical; (iv) Developmental; and (v) Statistical.[4] A cross-section law states the functional connection between the values that several variables have at the same time. If we regard a system as consisting of an infinite number of "states"—each state representing the condition of the system of a given time period—then a cross-section law is one relating to the specific state of a system. The cross-section law is found in geography in the form "where A there B". Crude attempts at forming such laws can be found in the works of the environmentalists. An example of an attempt to formulate a cross-section law can be paraphrased from the work of Christaller: *"Wherever there occurs an urban place of size A, then there will be urban functions B associated with it"*. Whereas, a cross-section law states a functional connection that exists under some given circumstances, an equilibrium law states that some change will occur if the connection the equilibrium formula states is not not obtained. These laws are also static and, in a sense, they are imperfect—i.e., they state what will occur if certain conditions are fulfilled, but do not say what will occur if the conditions are not fulfilled. An example of the scheme of an equilibrium law is: *If X and Y, then if there is to be equilibrium, X = Y*. This type of law is exemplified by the law of the lever in physics and by the demand-supply relationships in economics. It appears in geography in research on interaction between places, and it has also been used in defining trade areas, hinterlands, and so on. For the most part the 'equilibrium' situation has been used to define boundaries where the attractive force of one centre equals that of another centre. Commonly such lines of indifference, or market-supply area boundaries, are defined empirically. Both laws discussed so far are static laws. The general format of a dynamic law is somewhat different from the schemes already shown. Reduced to its simplest elements, the dynamic law states that, if a system exemplifies at a certain time a certain character, certain other character a (and conversely). That is, at a time 't_0' a system has the character 'a', then it will have at some later time 't_4' the character 'b'. Conversely, if a system at time 't_4' has the character of 'b' then it must have had 'a' at a specified earlier time 't_0'. The classic example of this was the early attempt to formulate a five-stage law of demographic transition (the so called demographic 'cycle'). The Historical law allows the construction of a schema for laws that provide the future not from the present alone but from the present in conjunction with some information about the past. The historical law is the name of a certain type of generalization which is based not so much on the fact that instances on which it is structured lie in the past, but rather that the law conforms to a certain schema with respect to time. For example, given states of a system, A, B and C, then the schema for the historical law is: *If 'B' now and 'A' earlier, then 'C' later*. Examples of studies which may be interpreted to fit this framework are numerous in geography. The most obvious examples occur in studies of population growth, and in time-oriented studies between physical features (e.g. slope of land), occupancy types (crop types), and cultural and technological attributes of populations. Reconstructions of the changing locations of economic and social activities through time, and some migration studies, have provided a mass of empirical evidence, which may contain a number of historical laws. Often it is difficult to distinguish between historical and developmental laws, since both involve temporal change. The essential difference is found in the schema the laws follow. As opposed to the 'if B now and A earlier then C later' form of the historical law which may use past and present information to predict the future, a developmental law takes the form: 'If B now then A earlier and successively C and D later'. This is the four stage developmental law anchored at the second stage. Freely translated, this states that if a system of a certain kind has at a certain time the character B, then it will under normal conditions at some earlier time have had the character A and at some later time will successively have the characters C and D. Particular examples of attempts to formulate generalizations of this type can be found where longitudinal development is stressed, as Rostow's stages of economic growth, the law of demographic transition, and the diffusion process. A statistical law, on the other hand, states that if each member of a class of objects has the character A, then a certain proportion P ($O < P < 1$) of the objects have the character S. If such a class has N members, then $N_1 = (P.N)$. For example, given $N = 100$ and $P = 0.5$, then in a certain number of K classes (in which each member exemplifies A), each of the K classes should have $N_1 = 50$. Yet we do not expect successive samples from the population to yield N_1 is exactly equal to 50. In general we consider the law confirmed if the N_1s cluster or scatter around 50 in the a certain fashion. The antecedents of such laws invariably contain a probability law to the effect that the frequency of a certain character in successive samples from a large population converges towards a certain number in a random manner. Attempts to generalize from sample studies or small scale studies to large populations or universe or phenomena are a common part of the statistical procedure. This constitutes an attempt to extend the result of the sample study to a more general level of knowledge. Thus it is possible to build on the results of a local study (such as Hagerstand's studies of international migration in Sweden) to produce a more comprehensive statement.

THEORY

A theory is defined as "a system of ideas explaining something"; or "a system of ideas based on general principles independent of the facts or phenomena to be explained"; or "a scientific statement or a group of scientific statements". In order to understand the meaning of theory, the difference between a 'simple' and a 'scientific' statement needs to be made clear. Consider, these two statements, for instance: (1) Delhi lies across the river Yamuna; and (2) One finds the big cities generally located across the rivers in the world. Of these two, the former is a 'simple statement' and the latter a 'scientific statement', because the 'scientific statements' are based on generalizations, derived from a number of simple statements (facts). After searching out some relationship/order, we state it or express it in the form of scientific statements. The lowest order statements are 'generalizations' followed by 'laws' and 'theories' at higher order of explanation. Thus, the theories are the highest order scientific statements or the universal statements. They state some rule of action, behaviour, process or development. If the form of explanation is empirically inductive, it generates original theories. But in the case of theoretically deduced explanations the process is reversed. In this situation, the theory already exists; only its testing or verification is required. Before the theory is tested and verified in real world situation, it is stated in the form of 'hypothesis'. The difference between this term and the scientific law is simply a matter of confirmation. After confirmation it becomes a generalized statement or generalization.

The scientific theory has a formal structure, which basically includes its Calculus and Text. The terms, statements and rules constitute the calculus of a theory. Various words that form the specific vocabulary of a theory are its 'terms'. These terms are the building block of a theory. There are two types of 'terms', viz. 'Axioms' and 'Derived Terms'. The 'axioms' are the primitive terms that are basic, original and not derived, e.g. 'point' or 'line' in geometry; and 'river', 'plain', 'settlement', 'market', 'desert', 'road', etc. in geography. The derived terms, on the other hand, need further definition, as they may have several connotations. They are formed from the primitive terms. The terms like 'distance', 'network', 'region', 'space', 'long', 'short', 'high', 'low', 'up', 'down', etc. fall in this category. They are required to be defined and explained within some given context. Their meaning would change or vary in different contexts or references. The original and derived terms combine together to make 'statements', the scientific sentences. Again, there are two types of statements, viz. 'Axiomatic' and 'Derivative'. The axiomatic statements are primitive statements. For example: *'Delhi lies across the river Yamuna'* or *'Thar Desert lies on the western margin of Indian Subcontinent'*. The derivative statements are derived from axiomatic statements, and the explanations, at times, are sought from an existing theory. For example: *'The important cities of the world lie across major rivers'* or *'The western margins of continents are deserts'*. In addition to the primitive terms and axiomatic statements scientific theories also possess certain Rules or Laws that govern the formulation of the derivative sentences. Five types of laws have already been discussed in the previous section. The axioms, statements and rules (laws) make up the 'Calculus' of a theory. But a theory is useful in empirical science only if it is given some interpretation with reference to empirical phenomena. Thus in Euclidean geometry, for instance, primitive terms such as 'point' and 'line' may be interpreted by 'dots' and 'pencil lines'. By elaborating a formal structure we ensure the logical truth of the propositions contained in the theory. These propositions are linked to empirical phenomena by a set of interpretative sentences—called a 'Text'. The text of a theory tells about its scope, i.e. where and how the theory should be applied and also its limitations in explanation. Thus, the text of a theory performs two important functions, viz. (1) It provides a translation from completely abstract theoretical language to the language of empirical observation. Without such a translation there is no possibility of empirical support for the theory. Or it identifies an abstract symbol with a particular class of real world phenomenon. For example, in a correlation-regression model 'X' and 'Y' represent the real world phenomenon of

Independent (e.g. 'rainfall') and Dependent (e.g. 'floods') variables respectively; and (2) Another important function of the text of a theory is to identify the domain of objects and events to which theory can be applied. This domain may simply be defined by a set of spatial and temporal co-ordinates. Domain, in its simplest form is the field of application of theory. It is the section, aspects, reality that the theory adequately covers, including its limitations in application.

The theory itself is simply an abstract set of relata; the text states how and under what circumstances such an abstract system may be applied to actual events. The extent of domain of the theory varies according to the number of terms within it that have to be given a specific translation in terms of a specific subject matter. A theory without a text and a well-defined domain is useless for prediction. To a greater or lesser degree theories are provided with appropriate texts. No text can be absolutely perfect, but undoubtedly the provision of a text for theoretical structures accounts for their greater predictive success. Thus, a theory with a complete calculus and text is a Formal Theory. But most theories, particularly in social sciences like geography, are incomplete. It has already been suggested that it is comparatively rare for theories in either the natural or social sciences to be stated in a completely formal manner. In some cases this may simply be because sufficient information is not available for such a formal statement to be made. This raises, therefore, the problem of how theories are in fact stated, how far such theories can be partially formalized, and what criteria we need to employ in distinguishing speculative fantasies from scientific theory. In fact, in a continuum of theoretical formulations, at the one end of which lies the pure formal theory and the at the other end lies the purely verbal speculative statement. Thus, there is a whole range of theories in-between. Here, an attempt is made for a brief classification of theoretical structures according to their degree of formulation, i.e. based on the degree of precision, and the extent to which the theories are structurally complete or incomplete.[5]

There are four main types and these range from completely formal Type-I theories, through the Type-II theories and Type-III theories which involve pre-supposition and quasi-deduction respectively, to the more nebulous Type-IV theories, which scarcely conform in any respect to the standards of scientific theory. Type-I are Deductively Complete Theories: Such theories possess completely formal structures. Their axioms are fully specified. All steps in the deductive elaboration are fully stated. For example, a textbook in Euclidian Geometry exhibits this kind of structure. The probability theories and the theorems in geometry fall in this category. They are perfect, but do not have empirical content, as it is not actually required. In other words, they have perfect calculus, but the text is missing. Type-II are Theories with Systematic Presupposition: They involve the reference to another set of theories. Such theories may or may not be deductively complete. These have two sub-types: *Elliptical Formulations* and *Common Sense Presupposition.* The Elliptical Formulations are deduced completely, but proof is not given. Proof of deduction lies in ellipses, i.e in dark; whereas in Common Sense Presuppositions the deductive part of theory is missing, because it is technically difficult to give a deduct and prove. Otherwise, the theories appear complete. Such theories appear so sound that the need to look for original source is rarely felt. Type-III is Quasi-Deductive Theories: They may be regarded as incomplete theories. Because the primitive terms of the theory, the deductive elaborations of it, do not conform to the standards of formal theory. The terms or theorems have not been derived properly and completely; and have been substituted in various forms. Accordingly, three sub-types may be distinguished: (1) *Inductive systematization*: i.e. the theory substituted by assumptions; (2) *Incomplete Deductive Elaboration*: where certain steps have been left out, because they appear too complicated for explicit deductive procedures to be employed; (3) *Theories with Relative Primitives*. In this case the theories do not make use of original primitive terms or axioms; instead use only derived (parallel) terms or theorems. Lastly, Type-IV theories, on the other hand, are Non-Formal Theories: They may be regarded as statements made with theoretical intensions, but for which no theoretical language has been developed. They are verbal explanations, ranging in sophistication from carefully thought-out system of linked statements to the kind of 'explanation sketch'. Historians frequently use such theories. They are also called as Verbal Explanations, where systematic explanation is given but without any calculus or text; or Pseudo-theories/ Speculative statements, where no systematic relative statements are made. The discipline having Type-I theories is considered as the most advanced. In geography, the theory building has reached, at the maximum, at Type-II level. However, the most common in geography are Type-III. In practice, geographical theory varies a great deal in its degree of formulation. Given the nature of geographic concepts the development of formal theory in geography appears to be a very restricted possibility. For the most part we must at best rest content with varying degrees of partial formulation. In the most cases the systematic pre-supposition or the quasi-deduction involved are not of the 'harmless' variety in which a full proof or full theory is available but not stated.

An explicitly developed scientific theory requires a number of axiomatic statements from which the theorems can be derived. To achieve this empirical status these axiomatic statements (containing primitive terms) require translation to either observable classes of events or to theoretical concepts from which the behaviour of observable classes of events can be derived. The concepts that correspond to the axioms of the theory are called its basic postulates. Geographers have in the past developed 'concepts' and 'principles' to facilitate explanations and to justify the way in which they organize a particular set of facts. Such concepts may ultimately act as the basic postulates for theory. In most cases explicit theory has not been developed from them. In some instances particularly in location theory, geometric analysis, and in analysis in physical geography—the basic postulates have been more explicitly developed and function more in accordance with standard scientific theory. Such concepts and principles (or the postulates) those have been used or may be used by geographers in theory building can be classified into two, viz.: Derivative Concepts and Indigenous Concepts.

The use of derivative concepts involves the "consumption" of some theoretical structure from another discipline. A number of such concepts have entered in geography from other disciplines, as economics, psychology, sociology, geometry, physics, chemistry, biology, etc. (1) 'Economic Concepts' have frequently been used as the foundation for geographic theory. Economics has been the most successful of the social sciences in developing formal theory (even if the empirical status of that theory is open to doubt). Many of the postulates and theorems of economics have been absorbed into geographic theory. In particular, the location theory, which has been "especially concerned with the development of the theoretical-deductive method in geography", can be related to economic postulates. Out of many such cases, Central Place Theory by Christaller has been described as the "one relatively well-developed branch of theoretical economic geography, using economic geographical laws. The fundamental spatial concept—'the range of a good' is basically derived from economics. Similarly, Losch (1954) treated the location of settlement as part of the general location problem and, grounding his analysis firmly on 'Chamberliman Economic Theory', gave a more powerful theoretical foundation for the settlement theories of Christaller. Dacey (1965) has provided a geometric version of a probabilistic central-place system. The articulation of a theoretical structure in this may again be traced back to the basic economic postulation. All the above examples are just few out of many to demonstrate how geographical theory may be derived from the basic postulates of economics. (2) Psychological and sociological postulates have also been introduced in the construction of geographical theory. Human geographers (e.g. Bruhnes, Sauer, Wolpert) have long recognized that geographical patterns are the end-product 'of a large number of individual decisions made at different times for often very different reasons' and that it was necessary to employ some psychological notions in explaining those patterns. Traditional notions regarding the importance of individual and group behaviour in the creation of geographic patterns can be sharpened by reference to the psychological literature. Psychological postulates, particularly behavioural ones, have been employed directly by geographers with profit. The use of sociological postulates in human geography is equally as widespread as that of psychological. For instance, the concepts of fertility, mortality, migration, etc. all have a deep grounding in sociology. (3) The relationship between geography and geometry is of special relevance. As a branch of mathematics, Geometry provides an abstract language for discussing sets of relationships. Geography maps many of its problems into this abstract language. The various forms of geometry appear to be a peculiarly appropriate language for theorizing about spatial relationships, about morphometry, and about spatial pattern. From this language we may derive the 'morphological laws' that help to explain geographical distributions. (4) Of all the derivative postulates, the physical postulates are of tremendous significance to research in geography, and especially physical geography. These postulated have been mainly derived from the sciences like physics, chemistry and biology. Much of the studies on desert erosion, coastal erosion, glacial erosion, make direct use of the basic postulates and known relationships of physics. Similarly, works in meteorology are related to the postulates of physics, while work on soil formation, weathering processes, and so on, refers to the concepts of chemistry and biology. In fact, any work on process in physical geography can be related, directly or indirectly, to the postulates of the various physical sciences. The Davisian system in geomorphology needs a special mention here.

There are plenty of 'concepts' and 'principles' developed by geographers that could function as postulates for theory. But few have been developed in indigenous manner. In fact, we have not sufficient experience of theory-construction in geography to discuss indigenous postulates with any uncertainty. But, on the basis of the limited experience we possess in this direction, together with some *a priori* notions regarding the nature of geographical enquiry, some clues may be provided as to the nature of such indigenous postulates. One of such postulates, and the one that served as the central concept of geography for a long time is 'the concept of region'. This has been repeatedly used to explain the areal differentiation of earth's surface and the human spatial organization. The region is not the only concept of this type. Some other concepts well may form a set of indigenous postulates for the development of geographic theory. These concepts are often related to what are often called 'spatial processes'—or rather sets of spatial relationships. These concepts have essentially to do with 'location', 'distance', 'pattern', and 'morphology'.

MODELS

The practical problem that follows theory-building is related to the presentation of information. One significant and popular way is the employment of model building or analogue-theory in geography. The quest for models is a recurrent theme in research and it has become very fashionable in geographic research.[6] But there is no consensus of opinion among philosophers of science as to what is meant by a 'model', or as to its function in scientific research. The general confusion surrounding the term is reflected in geographic research where very different opinions can be found. The meaning and function of the term *model* thus require some careful methodological investigation. However, in general, model building is concerned with simplification, reduction, concretization, action, extension, globalization, theory formulation, theory testing, explanation, etc. The models link generalizations with theories.

The term 'model' is conventionally used in a number of different ways. In its simplest form a 'model' is the representation of reality in an idealized form. The process of model building is actually a process of idealization. The traditional reaction of man to the apparent complexity of the world around him has been to make for him a simplified and intellectual picture of the real world. The mind decomposes the real world into a series of simplified systems. The system is viewed from a certain scale; details that are too microscopic or too global are on no interest to us. A model is thus a simplified structuring of reality that presents supposedly significant features of relationships in a generalized form. Models are highly subjective approximations in that they do not include all associated observations or measurements, but as such they are valuable in obscuring incidental detail and in allowing fundamental aspects of reality to appear. This selectivity means that models have varying degrees of probability and a limited range of condition over which they apply. The most successful models possess a high probability of application and a wide range of condition in which they seem appropriate. Indeed, the value of a model is often directly related to its level of abstraction. However, all models are constantly in need of improvement as new information or vistas of reality appear, and the more successfully the model was originally structured the more likely it seems that such improvements must involve the construction of a different model.[7] Scientific models are utilized to accommodate and relate the knowledge we have about different aspects of reality. They are used to reveal reality and more than this to serve as instruments for explaining the past and present, and for predicting and controlling the future. According to another viewpoint, 'a model is a skeletal representation of a theory'. This implies that a theory may imply more than one models but a model cannot have many theories to be represented. Thus, we can say that models are of lower order than the theories. Otherwise, there is no distinction between theory and model. The only distinction is that theory is an abstract and model is a concrete. Thus, a model may be regarded as a formalized expression of a theory. A 'model' is a simplified structuring of reality that presents supposedly significant features or relationships in a generalized form. Models are highly subjective approximations as they do not include all associated observations and measurement, but as such they are valuable in obscuring incidental detail and in allowing fundamental aspects of reality to appear. This selectivity means that models have varying degrees of probability and limited range of conditions over which they apply. The most successful models possess a high probability of application and a wide range of conditions in which they seem appropriate. However, all models are constantly in need of improvement as new information appears. Models are different from reality, as they are approximations of the reality. Therefore, they are called analogous. The term 'true' or 'false' cannot be applied in the explanation of models. Instead, the ones like 'appropriate', 'stimulating' or 'significant' should replace them.

The term 'model' is conveniently employed in a number of different ways. It is used as a noun implying a representation, as an adjective implying a degree of perfection, or as a verb implying to demonstrate or to show what something is like. In fact, models possess all these properties. The most fundamental feature of models is that their construction has involved a highly selective attitude to information, wherein not only noise but less important signals have been eliminated to enable one to see something of the heart of things. Models can be viewed as selective approximations, which, by the elimination of incidental detail, allow some fundamental, relevant or interesting aspects of the real world to appear in some, generalized form. Thus, models can be thought of as selective pictures and a direct description of the logical characteristics of our knowledge of the external world shows that each of these pictures gives undue prominence to some features of our knowledge and obscures and distorts the other features that rival pictures emphasize. Each of them directs such a bright light on our part of the scene that it obscures other parts in a dark shadow. Only by being unfaithful in some respect can a model represent its original. Another important model characteristic is that models are structured, in the sense that the selected aspects of the 'web of reality' are exploited in terms of their connections. It is interesting that what is often termed a model by logicians is called a 'structure' by econometrician. The model feature leads immediately to the suggestive nature of models, in that a successful model contains suggestions for its own extension and generalizations. This implies, firstly, that the whole model structure has greater implications than a study of its individual parts might lead one to suppose, and, secondly, that predictions can be made about the real world from the model. Models have thus been termed 'speculative instruments', and a promising model as 'one with implication rich enough to suggest novel hypothesis and speculations in the primary field of investigation'. A good model is regarded as experimentally fertile, suggesting further questions, taking us beyond the phenomena from which we began, and tempting us to formulate hypotheses. The 'intuitive grasp' of the capacities and implications of a model is thus the key to the exploitation of a suggestive character. Selectivity implies that models are different from reality in that they are approximations of it. A model must be simple enough for manipulation and understanding by its users, representing enough in the total range of the implications it may have, yet complex enough to represent accurately the system under study. In another sense, too, models represent compromises in that each has a circumscribed range of conditions within which it has relevance. Because models are different from the real world they are analogies. The use of hardware models is an obvious example of the general aim of the model builder to reformulate some features of the real world into a more familiar, simplified, accessible, observable, easily formulated or controllable form, from which conclusions can be deduced, which, in turn, can be reapplied to the real world. Application is a prerequisite for models in the empirical sciences. Although some mathematical model builders disclaim responsibility for the degree to which their idealization may represent the real world, claiming that their responsibility is discharged completely and with honour if they avoid internal error; most geographical model builders would judge the nature of a model almost entirely in terms of its reapplicability to the real world.

Once devised, the models are of tremendous utility for their users as Models simplify the otherwise complex relationships of the phenomena in the real world; Models represent the reality not only in simple but also in systematic/orderly manner. In geography, models are taken as generalizations, as they encourage the role of nomothetic approach in making observations; Models help in making prediction of trends; and Models act like a bridge between the observational and theoretical levels. Thus, the models may be used to connect theory and experience, experience with imagination, theories with other theories, imaginative creations with formal theory and so on. Based on the various properties and uses of a model, its nine functions may be identified, viz. Acquisitive/Organizational, Psychological, Logical, Normative, Systematic, Constructional, Selective, Interpretative and Cognitive.

(1) *Acquisitive/Organizational*: The model provides a framework wherein the information may be defined, collected, ordered and manipulated. A model acquires the information that could be defined in its framework or provides framework for defining certain kind of information.
(2) *Psychological*: A model acts as a psychological device that facilitates complex interactions to be more easily visualized, i.e. a kind of picturing device.[8] This function enables some group of phenomena to be visualized and comprehended more easily that could otherwise not be because of its magnitude and

complexity. A model helps to understand the reality in a simpler manner than otherwise it would have been.

(3) *Logical*: A model explains the situation rationally, accounting for how a particular phenomenon comes about, or how a particular relationship among component parts works about.

(4) *Normative*: The model represents reality in an idealized form, i.e. with the help of certain norms, conditions, assumptions, etc. The normative function of a model allows broad comparisons to be made, by comparing some less known phenomena with more familiar ones.

(5) *Systematic*: A model functions like a system. The systematic function of model stresses that the 'web of reality' should be viewed in terms of inter-locking systems. This leads to the constructional function of the system.

(6) *Constructional*: It means that a model forms stepping-stone to the building of theories and laws. As a constructional device it helps in searching for geographic theory or the extension of the existing theory.

(7) *Selective*: Model is a selçctive approximation, allowing some fundamental, relevant or interesting aspects of the real world to appear in some generalized form.

(8) *Interpretative*: An important function of the model is to provide an interpretation of the theory in the sense that every sentence occurring in the theory is a meaningful statement.

(9) *Cognitive*: Finally, there is the cognitive function of model, promoting the communication of scientific ideas.

Unfortunately, there is no common and firm classification for models. All are suggested typologies. This is mainly because of a number of meanings and functions of models, as understood by differently by different scholars. The term 'model' has been used in such a wide variety of contexts that it is difficult to define even the broad types of usage without ambiguity. However, some general types include: *A priori* and *A posteriori* Models; Descriptive and Normative Models; and Hardware and Software Models. Two ways of the construction of a formal theory have been described. The first route begins with empirical observation from which a number of regularities of behaviour may be extracted. To explain these regularities a theory is proposed which may contain theoretical abstract concepts an eventually the theory may be given axiomatic treatment and may be verified. This theory may then be represented by some structured model, which can be used to facilitate deductions and simplify calculations. In this case the model developed in order to represent the theory is *'a posteriori'*. On the other hand, the important form of the second route to theory construction lies through giving interpretation to a completely abstract calculus. In this case the model used is *'a priori'*. This distinction is based on the type of procedure used in employing models in scientific explanations. In case the model is developed in advance in order to represent a theory/explanation, it becomes *a priori*. In such a situation the function of the model is simply to represent something that is already known, and the only arising question is that of the appropriateness of a model for a given purpose and this can be fully defined only if the appropriate theory is referred to. But, when the process is reversed, i.e. observations or theory precede a model, the model is the end result, an after product of explanations, it is *a posteriori*. However, *a priori* models are more common in geography. The 'Descriptive' models are behavioural, i.e. they suggest how things exist in reality, whereas the 'Normative' models explain how they ought to be. The former are concerned with some stylistic description of realiy and the latter with what might be expected to occur under certain stated conditions.[9] Descriptive models can be 'static', concentrating on equilibrium structural features, or 'dynamic', concentrating on processes and functions through time. Where, the time element is particularly stressed 'historical' models result. Descriptive models may be concerned with the organization of empirical information, and be termed 'data', 'classificatory' (taxonomic), or 'experimental design' models. The Normative models, on the other hand, often involve the use of a more familiar situation as a model for a less familiar one, either in a time (historical) or a spatial sense, and have a strongly predictive connotation. The 'Hardware' models are based on the use of some hard/concrete material, e.g. physical, planning or defence project models. But, the non-physical, conceptual, symbolic or statistical models are categorized as 'Software' models. Besides, this general categorization of models, some specific typologies have also been presented by the scholars like Ackoff, Hagget and Chorley in 1960s. Of these the attempts of Chorley are most elaborate. In fact, Chorley devised two classifications. He regarded all models as being analogous of some kind, and suggested his first classification in 1964. In the later presentation (1967) Chorley revised and extended this classification system. This new classification incorporated all those types of models he discussed earlier and also included the ones those have been devised by his predecessors and fellow scholars. Hence, this classification is relatively the most extensive and complete. It consists of three major categories of models with a number of sub-types[10]:

1. Natural Analogue System Model
    (a) Historical Analogue
    (b) Spatial Analogue
2. Physical System Model
    (a) Hardware Model
        (i) Scale (Iconic)
        (ii) Analogue
    (b) Mathematical Model
        (i) Deterministic
        (ii) Stochastic
    (c) Experimental Design model
3. General System Model
    (a) Synthetic
    (b) Partial
    (c) Black Box

The first group of models involves searching for analogous situations or events at different times or in different places, and drawing some conclusions. The second group of models corresponds to the more conventional notion of a model in the sciences. The third is a newer concept that treats the structure of a landscape as an assemblege of interacting parts, and attempts to represent the processes as such.

The Natural Analogue System Models are simplified models to be used as a basis for further analysis and prediction by their translation into some similar natural circumstances. The explanations are sought from the natural world phenomena. This group of models involves searching for analogous situations or events at different times or in different places, and drawing conclusions. Accordingly, it has two major sub-types, viz. historical analogue and spatial analogue models. The Historical Analogue Models represent analogous events at different times. An example of such a procedure is Rostow's schematic representation of the economic growth process that is derived from historical analysis and searching for analogies between different countries at different times. The model of Demographic Transition Theory serves another example. The Spatial Analogue models represent analogous situations at different places. A number of examples are available: The shift of highways being compared with the shift of rivers (by Bunge); or The growth and shrinkage of ice-crystals representing growth and shrinkage of market areas (by Christaller); or The study of Natural Cycle of Erosion of a river based on Human Life Cycle (by Davis).

The group of Physical System Models corresponds to the more conventional notion of a model in sciences. The relevant properties of real world are represented by the same properties in the model as well. They simply mean picturing the real world phenomena. Three sub-types have been identified, viz. Hardware, Mathematical and Experimental Design. As the term suggests, the Hardware Models use some concrete material, and may be scale models or simply analogue models. The Scale or Analogue, both the sub-types, means some kind of figurative representation. The Scale Models use the same material as in the real world phenomena, but with only change in scale. In Ackoff's classification such models have been designated as 'iconic'. Iconic or Scale models are generally three-dimensional models—made of same material but on different scale. For example, the globe is an iconic model of the earth. The Analogue Models have real-world properties represented by different properties.[11] Besides, the change in scale, the analogue models also involve a change in the materials used in building the model, e.g. an electric circuit being assumed as an analogue for traffic system. Ackoff (1962) calls them as 'Simulation Models'. The Mathematical Models represent reality by some symbolic system, such as a system of mathematical equations or statistics. They have also been called as Symbolic Models. The mathematical models can further be classed according to the degree of probability associated with their prediction into 'Deterministic' and 'Stochastic'. The model when used with certainty of the effect becomes deterministic. In this case, the outcome or results of the exercise are pre-conceived or are more of less sure to come true. The laws of probability strictly govern a Stochastic model, and there is doubt about the exact effect of a given cause. The Experimental Design Models, on the other hand, involve some practical procedures, as in laboratory or in field. The models used in defence or planning are the experimental design models.

The third category of General System Models represents a newer concept of the times treating the structure of geographical landscape as an assemblage of interacting parts and attempting to represent the process as such. Three sub-types, viz. Synthetic, Partial and Black Box, are discussed here. Of these, the Synthetic System Models (i.e. in synthesis) try to bring the reality and its representation in perfect harmony or synthesis, such models stimulate reality in a structured way, i.e. a perfect correspondence is expected between reality and the model used to represent it. Synthetic systems are artificially built to stimulate reality in a structured way and, as Chorley points out, such models may be similar to experimental design models. The partial systems are concerned with workable relationships and attempt to derive results without complete knowledge of the internal workings of the system, i.e. there may be some chance of correspondence between reality and its representation through the selected model. The 'black box' approach, on the other hand, attempts to derive results from a situation in which we have no knowledge of the internal workings of the system.

There are still more varied approaches to model-classification—demonstrating the extraordinary breath, flexibility and potential of the model concept. Moreover, certain types of disciplines are concerned with certain types of models (sometimes almost exclusively). In fact, in practice it is a failure to develop a complete mutually exclusive classification of models. To conclude, the concept of models poses considerable methodological difficulty. There is a multiplicity of model types performing a multiplicity of functions associated with a multiplicity of definitions. Each particular model exhibits a different logical capacity for performing the function required of it. In fact, the type, nature, use, quality and significance of models—all depends on the types of theories to which they serve.

NOTES

1. Harvey, 1969, 32-36.
2. Rana, L. (2008), p. 317.
3. Braithwaite, 1960, 12.
4. Davies, 1972.
5. Harvey, 1969, 96-100.
6. Harvey, 1969, 141.
7. Chorley and Hagget, 1967, 22.
8. Chorley and Haggett, 1967, 24.
9. Chorley, 1964.
10. Harvey, 1969, 155.
11. Ackoff, 1962.

REFERENCES

Ackoff, R.L. (1962), *Scientific Method: Optimising Applied Research Decisions*, New York.

Braithwaite, R.B. (1960), *Scientific Explanations*, Harper Torchbooks, New York.

Bunge, W. (1966), *Theoretical Geography*, Lund Studies in Geography. C No. 1 (Second Edition).

Chorley, R.J. (1964), "Geography and Analogue Theory", *Annals of Association of American Geographers*, Vol. 54, 127-37.

Chorley, R.J. and P. Hagget (1967), *Integrated Models in Geography*, Part IV, Mathuen and Co. Ltd., London.

Dacey, M.F. (1965), "The Geometry of Central Place Theory", *Geogrphical Annlr Series*, B, 47, pp. 111-124.

Davies, W.K.D. (1972), *Conceptual Revolution in Geography*, Rowman and Littlefield, New Jersey.

Haggett, P. (1965), *Locational Analysis in Geography*, London.

Hartshorne, R. (1939), *The Nature of Geography: A Critical Survey of Current Thought in the Light of the Past*, Association of American Geographers, Lancaster.

Hartshorne, R. (1959), *Perspective on the Nature of Geography*, Rand McNally and Co., Chicago.

Harvey, D. (1969), *Explanation in Geography*, Arnold Publishers, New Delhi.

Rana, L. (2008), *Geographical Thought: A Systematic Record of Evolution*, Concept Publishing Company, New Delhi.

CHAPTER-10

Systems' Analysis

There is nothing very new about the concept of the system as such. Newton wrote on the solar system, economists have written on economic systems, biologists on living systems, plant and human ecologists have used system's concept and geographers have certainly made considerable use of the notion of a system ever since the discipline originated. The thread connecting all these uses is the notion of a complex interrelated whole. All-round us, we find every phenomenon, every event and every feature assigned to a system, e.g. economy is a system, politics is a system, nature is a system, and even an individual human being is also a (biological) system in himself. Each part of the system or each individual over the earth is significant not only in terms of functions it is performing independently, but also in terms of its relationships with others; and unless and until these individuals are studied together cannot form a system. Hence, it is within the framework of the systems that we are studying each and every component of the world or the Earth. The system's approach can be suggested as a method of comprehending the world as a whole. The modern emphasis on system as an explicit item for analysis may be seen as apart of a general change in emphasis from the study of very simple situations in which the interactions are few, to situations in which there are interactions between very large numbers of variables. The interest in these complex systems has grown very rapidly in the 20th century. Given the multivariate nature of most geographic problems, it is hardly surprising that systems analysis provides an appealing framework for discussing these problems. In geography, the awareness towards the use of system's approach has developed because of the realization that: Firstly, the earth's surface (world) is made up of different types of areas, regions, or places; and these, besides having an individual significance, also are part of a 'whole' as conceived by Ritter (in his *Erdkunde*) and Humboldt (in his *Cosmos*); and Secondly, these parts or sub-parts are not only inter-related with each other, they also form independent sub-systems of their own.

DEFINITION

"A system is a functioning whole with various sub-systems interlinked with each other". The 'system', contrary to chaos, is the name of an order. In other words, it is 'the way, sequence in which various components are organized into a whole, into a totality." There is a whole range of systems from microscopic to micro, meso and macro. Systems analysis treats both the phenomena and the concepts used to organize phenomena. There exists some kind of organization, interaction, inter-dependence and integration of parts within the system. It assumes the existence of some structure that constitutes an organized whole. It is the way in which the parts are connected and interact with each other that makes it a system.[1] Various characteristics of a System may be summarized as: (i) A system has an order of or sequence of functions; (ii) Although each part of a system plays an individual role in the system's operation, no part is entirely independent of others; (iii) A change in the operation of one part will have significant repercussions throughout the system; (iv) Systems are generally open-ended, accordingly, a system has some inputs and some outputs (Figure 10.1); (v) The system is not a juxtaposition of various elements; it is rather a functioning whole; (vi) There is always some stimulus (driving force) behind the functioning of a system; (vii) Systems are generally at balance or at equilibrium; Malfunctioning of one part disturbs the balance of whole system; and (viii) Within macro systems there are micro systems (the sub-systems). A system is not merely the assemblage of various components; rather it is the functioning of those components together and independently as well. The 'whole' is greater than the parts. Any little change leads to various corresponding changes in the whole system. For instance, continuous flow of smoke and gases from the factories and mills have greatly increased the amount of carbon dioxide in the atmosphere; and this increase has disturbed the ecological balance of CO_2 already present in the atmosphere. As a result there is decrease in the total amount of rainfall, increase in the temperature, etc., which in turn affects the crop production.

Essential Features of a System

The essential features of a system define its basic functional characteristics in terms of its environment, behaviour, state of existence, information and organisation. All are interrelated[2].

The Environment of a System

The environment of a system is the whole of which the system is only a part. For example, the economy constitutes the environment of a firm; or a farm system has its environment in the biosphere. The changes in this environment bring about direct changes in the values of the elements contained in the system under examination. Environment changes from system to system, even of the same time, because it is not the time more considered here, rather it is the manner or way in which the elements are combined and functioning together. This flexible approach to the concept of environment in systems analysis is particularly useful to geography. It is useful at this juncture to clarify the usual meaning of the terms 'open' and 'closed' systems. The open system interacts with the environment. It means then an open system is not isolated from its environment, but exchanges materials or energies with it. A closed system, on the other hand, operates without any kind of exchange. The main difference between the open and closed systems ban be illustrated by the example of a domestic heating system of the hot-air variety. Such a system consists, in essential, of a number of ducts (tubes), gratings (arrangement of bars), fans and filters, all linked to a furnace by the flow of air. As separated unconnected elements (the ducts, furnaces and so on) have different meaning than as a connected whole. Since energy enters the system in the form of gas for the furnace and leaves it as warmed air, the energy transference occurs across the limits of the system. A closed system/has no energy transference; it can change only as a result of innate (inborn) elements.

The Behaviour of a System

When we speak of the 'behaviour' of some system we are simply referring to flows, stimuli, and responses, inputs and outputs, and the likes. We can examine both the internal behaviour of some system or its transactions with the environment. A study of the internal behaviour accounts for a study of functional 'laws' that connect behaviour in various parts of the system. Most analyses of behaviour tend to concentrate on the latter aspect. For instance, a system has one or more of its elements related to some aspect of the environment, and the environment undergoes a change. Then, at least one element in the whole system is affected and the effects are transmitted throughout the system until all connected elements in the system are affected. This constitutes a simple stimulus-response, or input-output system without feedback to environment (Fig. 10.1).

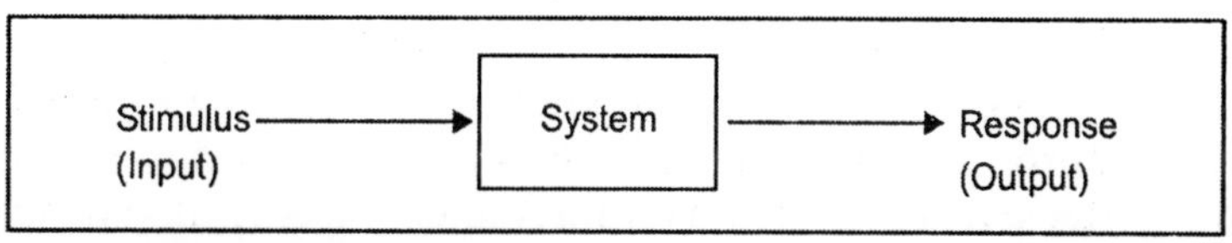

FIGURE 10.1 Behaviour of a System

In other words, the behaviour of a system is described by its flows that connect the inputs (stimulus) with the outputs (responses). The simplest example of this is provided by input-output analysis of economics, in which a vector of final demands (e.g. derived from imports, home consumption, or however) is related to a vector of final outputs in various sectors in the economy.

The State of a System

In general the state of a system may be thought of as the values which the variables take on within the system at any particular point of time. Now it is possible for the variables to take on a large number of values, so that the term 'state' is often used in a more restricted sense to refer any well defined condition or property that can be recognized if it occurs again. It is, therefore, useful to differentiate between the *transient* or *transitional* states and the various types of *equilibrium* states, which have distinctive properties. 'Equilibrium' refers to a system that maintains some kind of balance instead of being in a 'transient' or ever-changing state. 'Homeostasis' implies that the balance is at a fixed point or level. A 'steady state' is an equilibrium that does not depend on a fixed point or level. 'Morphogenesis, is the process that leads to changes in a system's form, structure, or state, so that it comes to exist at a new and more complex level of equilibrium. Normally, we can recognize two kinds or categories of equilibrium, viz. *stable* and *dynamic*. The stable equilibrium includes both homeostasis and steady states, as defined above. In a homeostatic social system there is always activity, but it does not alter (change) the balance between the systems' components. A social system that was in a steady state would equally stable, but it would also change in an orderly way. Dynamic equilibrium refers to the process by which a slight disturbance engenders constant change throughout the system. If feedback is present this kind of change can go on indefinitely, at least in theory.

Organization and Information in System

The twin concepts of *organization* and *information* are exceedingly important in systems analysis. The concept of 'organization' can best be examined by way of an example. Consider a system, containing 'n' elements, that behaves in such a way that if we know the value of one element in the system we can predict the values of all the others. Such a system is highly organized. Consider a similar system in which even though we know the values of '1-n' elements, we still cannot predict the value of the 'nth' element. Such a system is disorganized. 'Information' may be regarded as 'the measure of the amount of organization' in the system. The term 'entrophy' and 'negentrophy' are closely related to the organization and information in the system. Entrophy (a measure of unavailable energy) is often referred to as a measure of disorder or disorganization. Basically, it is an expression for the degree to which energy has been unable to perform the work. As such, it is closely linked to the second law of thermodynamics (the science of heat as a mechanical agent). It states "systems can only proceed to a state of increased disorder". Negative entrophy or Negentrophy, on the other hand, is a measure of order. For an organism, keeping alive is a matter of being able to extract negentrophy from the environment. Maximum entrophy, the state towards which all organisms move, is death. These two concepts are best illustrated by an inanimate example. If we put sugar in coffee, there will be an increase in entrophy, since the sugar crystals, which would otherwise have held their shapes indefinitely, will dissolve. If we heat the coffee, entrophy will increase further, since the increase in heat motion will dissolve the sugar faster. Any closed system tends to increase its entrophy in this way, and will finally approach the inert (inactive) state of maximum entrophy. An open system, on the other hand, can maintain a fairly low level of entrophy, by interacting with its environment. As a result, it will tend to develop a more complex structure. It is useful to think of entrophy and negentrophy when studying socio-economic systems because it makes us ask how organized they are. Whereas, entrophy brings disorder, information is thought to bring order or negentrophy. In a social system, entrophy means not knowing what to expect; if entrophy is maximized, then, as we say, anything can happen. At the other extreme, in a perfectly negentropic system there would be maximum circulation of information, and we would be absolutely certain about the behaviour of the system.

Structure of a System

Structure of a system refers to the arrangement of component parts (elements) in it. This structure is composed essentially of the 'elements' and the 'links' between elements.

Element: It is the basic unit of a system. The definition of an element depends on the scale at which we conceive of the system. The international monetary system, e.g. may be conceptualized as containing countries as elements; an economy may be thought of as being made up of firms and organizations; organizations themselves may be thought of as systems made up of departments; a department may be viewed as a system made up of individual people; each person may be regarded as a biological system, and so on. In substantive terms, therefore, we face the problem that systems may be embedded in systems, and that what we choose to regard as an element at one level of analysis may itself constitute a system at a lower level of analysis. The only way for the problem to be solved is to simply group elements into a 'hierarchy' of 'classes' with each higher-order class forming an element in a higher order system (Fig. 10.2).

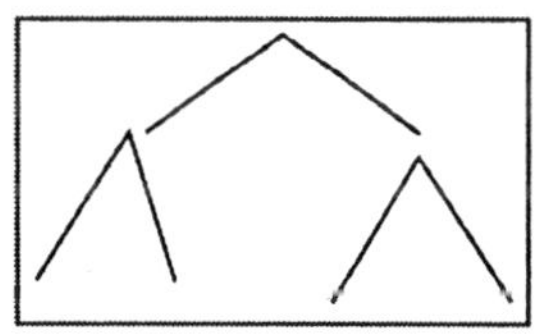

FIGURE 10.2: Hierarchy of Elements

Links: The relationships or links within the elements provide the other component in the structure of a system. Generally, the five forms of relationships can be defined [Figs. 10.3 (a, b, c, d and e)].

(i) ***Cause and Effect Relationship***: Also known as 'Series' relationship, it is the simplest and is the characteristic of elements connected by an irreversible link. For example, rainfall affects the rate of soil erosion but soil erosion apparently has no effect on rainfall.

(ii) ***Parallel Relationship***: It is similar to multiple-effect structures in that both a_i and a_j are affected by some other element 'a_k'. In other words, for both the sets of elements there is common cause or stimulus to function. Taking the same example as in the previous case, the rainfall is the cause of soil erosion on the hill slopes and also of the increased runoff in the streams.

(iii) ***Feedback Relationship***: It is a kind of link that has newly been introduced into analytic structures. It describes a situation in which one element influences itself. For instance, if crop-production of a year is not up to the desired level of production, then there will be a kind of feedback to the stimulus, i.e. to improve

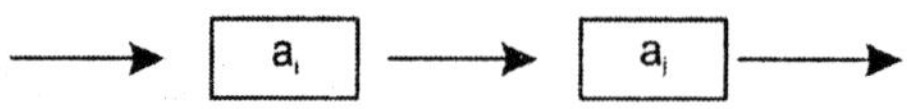

FIGURE 10.3(a): Cause and Effect Relationship

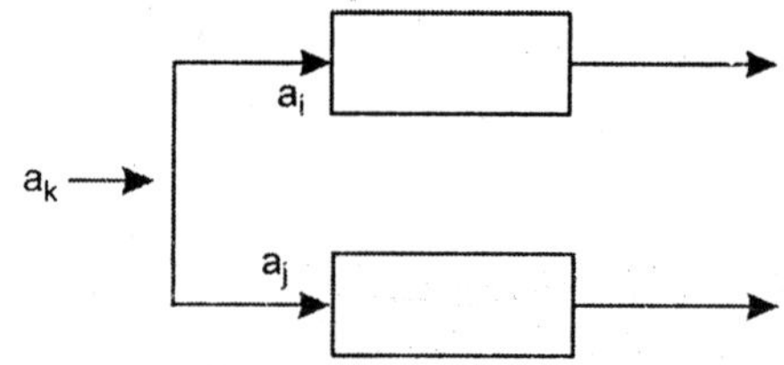

FIGURE 10.3(b): Parallel Relationship

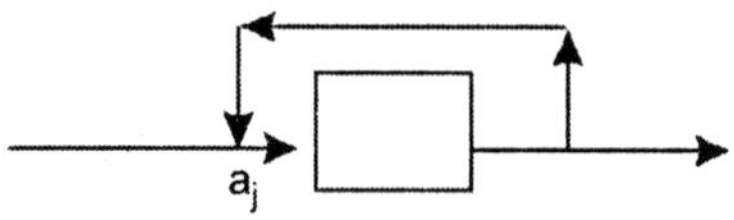

FIGURE 10.3(c): Feedback Relationship

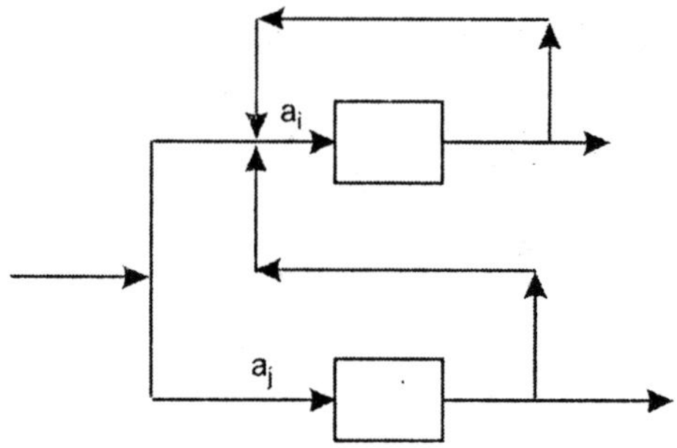

FIGURE 10.3(d): Simple Compound Relationship

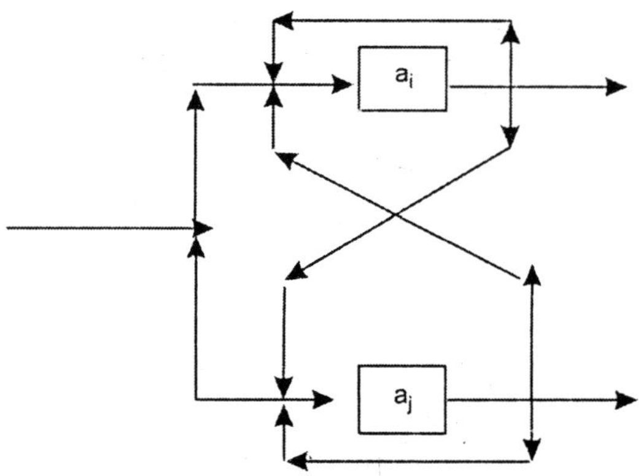

FIGURE 10.3(e): Complex Compound Relationship

the ways and means of production. Then the whole system of the crop-production will either modify itself or will be discarded. Feedback in a system is essentially a way whereby the output is used to control its working so that it may achieve its desired goal. It is a self-steering mechanism.

(iv) ***Simple Compound Relationship***: Where a set of components is affected by two ways, i.e. by feedback and also because of the influences from other set of components, working simultaneously. The present Indian Society can be an example cited here, because today it is not only changing and modifying its norms and values by discarding certain old ones but also adopting certain characteristics of the western society. It is all for the betterment of human beings.

(v) ***Complex Compound Relationship***: Where there are influences and changes from all sides and within each and every component, modifying and influencing each other. In this system, there are influences and changes from all sides and within each and every component—modifying and influencing each other. Our environment, consisting of physical and cultural environment, may be cited as an example here. Such systems are very difficult to be interpreted.

All the above five types of links form a kind of 'wiring system' connecting the elements in various ways.

TYPES OF SYSTEM

There are various ways in which we could classify systems. We would differentiate between open and closed systems; between man-made and natural systems; and so on. Rather than to attempt an exhaustive classification of systems, the attention has been concentrated upon those types of systems which have something new to tell us regarding the analysis of complex interactions. Those systems, such as 'simple action systems', which express traditional modes of analysis, such as cause and effect, in a systems framework are, therefore, be ignored. Most that is new in systems analysis has to do with systems which are homeostatic, self-regulatory, adaptive, and particularly with systems that incorporate some form of feedback.

(i) *Homeostatic System*: It is the system that maintains a constant operating environment in the face of the random external fluctuations. Such systems resist any alteration in the environmental conditions and exhibit a gradual return to equilibrium or steady-state behaviour after such an alteration. The displacement of a spring, e.g. will be followed by a series of oscillations until eventually the spring returns to a stationary state. Human body itself is a homeostatic system, i.e. it maintains its equilibrium at about 98.2 degrees Fahrenheit. Generally, the homeostatic systems refer to open-systems' analysis, and are associated with the important concept of steady-state—a concept which has a great significance in the study of fluvial and other geomorphologic processes.

(ii) *Adaptive System*: It is similar to homeostatic system in many respects, but possesses some special characteristics. An adaptive system is one for which there exists for each possible input a set of one or more preferred states, or preferred outputs. The adaptive character of the system means that if the system is not initially in a preferred state, the system will so act as to alter its state until one of the preferred ones is achieved. The study of such systems provides a mode of approach to systems that are usually throughout of as 'goal-seeking'. Such systems clearly rely upon feedback mechanisms of some kind in order to achieve the preferred state. Example here can be taken of the 'wheat-production'. Suppose a nation has set the goal for the production of wheat during certain plan year. If the required figures are not achieved in that year, in the next year more sophisticated and advanced techniques will be used and something more will be done in order to increase the production of the wheat up to the desired level, i.e. there is a kind of feedback to the system.

(iii) *Dynamic System*: It may be regarded as a separate class of systems. Both homeostatic and adaptive systems show a change of state over time as they move towards steady or preferred state. In a truly dynamic system, however, feedback operates to keep the state of the system changing through a sequence of unrepeated states usually termed as the 'trajectory' or 'line of behaviour' of the system. Feedback may, e.g. cause new preferred states to be identified. Economic growth models, such as the circulation and cumulative causation models may be regarded as dynamic systems.

(iv) *Controlled System*: In this kind of system, the operator has some level of control over the inputs. Such controlled systems are, of course, of great interest in systems engineering and cybernetics (the study of communication and control mechanisms in machines and living beings). Systems control theory provides a good deal of insight into the behaviour of systems, and is not irrelevant to the application of geography to substantive problems. Particularly in the field of planning, government of both national and local levels, controls some of the inputs into the economic systems and manipulates (handles, manages) these in order to try and achieve some desired level of output. Monetary or budgetary policy are thus used to stimulate house demand, while at the local level the investment in roads, utilities, public housing and so on which is controlled by local government, provides an important means for varying the inputs in order to achieve certain goals (outputs).

In most situations we have control over certain inputs while others are impossible or too expensive to manipulate. In seeking to maximum agricultural input, e.g. we may be able to control water input by irrigation, but we must do so in a situation where other aspects of the biosphere remain uncontrolled. Partial controlling systems are thus of great interest.

SYSTEMS ANALYSIS IN GEOGRAPHY

During the last century the scientific knowledge has depended heavily upon the analytic method. Basically, this method consists of isolating all the variables that influence an object or event, determining the precise effect of each variable by means of controlled experiments and producing the general relationships between objects and variables. Consequently, the mathematics has developed as a language capable not only of giving objectivity and precision to the various relationships, but also of being used as a deductive tool in its own right[3].

Historical Aspect

Geography, throughout most of the last century has tended to lag behind in scientific progress. Strenuous efforts are being made to catch up with the main body of science by the wholesale adoption of analytic methodology. The application of formal systems concepts, thus, has been a relatively recent feature of research in geography. But, the notion of a system is not, however, in any way new to geographic thought. It has a long history as the concept of system in science as a whole. The history of systems thinking in geography is very much bound up with the functional approach, with the organismic analogy, with the concept of regions as complex inter-related whole, with the ecological approach in geography. It is possible to identify elements of systems thinking in the work of geographers such as Ritter, Vidal de la Blache, Bruhnes, Sauer and so on. But as in the rest of science, the concept of a system has tended to remain in the periphery of geographic thought rather than at its very core.

Changing Attitudes

Stoddart and Chorley (1965 and 1962), in particular, have shown how certain ideas that are currently important in terms of modern scientific methodology, made an early appearance in the literature, but lay dormant for years until a more modern revival occurred. During the last few decades the focus has changed to make the concept of the systems of much greater significance. The search for generalizations based on the whole rather than on individual parts is, therefore, a complementary method of modern science known as systems analysis. The broad framework of concepts lying behind it is called General Systems Theory[4]—the one of the methods employed by science. Since all systems, whether physical or human or combination of both, consist of a set of objects and the relationships binding these objects together into some organization, it is not surprising that the approach is especially useful in dealing with functional aggregates. Indeed, now the main focus of the scientific enquiry has moved away from the study of objects or substances to the study of relationships and organizations. And, as all organizations are recognized as being particularly complex, systems analysis proves to be a particularly appropriate framework of study in geography. By carefully defining the structure, functioning and general process change in any situation, these three major perspectives may be integrated into one system of study. Moreover, the approach may also be said to be particularly relevant to those studies attempting to bridge the gap between the natural and social sciences, the portion long claimed, but rarely effectively occupied by geographers.

Hence, the objective of the systems approach is not to provide a unified substantive theory—rather it is to provide a synthesis of the functional and structural relationships, to emphasize the comparative study of systems and to point out the way to future research into the existing gaps in our understanding. In this way the cross-fertilization of scientific disciplines will be preserved and the individual pieces of knowledge thrown up by purely analytical research will be inter-related, particularly those dealing with the functioning and organization of the whole system. At the very outset it must be repeated that the systems approach is not a replacement for the analytic method, but it is an additional line of modern scientific enquiry designated to break down the barriers between inter-disciplinary enquiries. It thus represents one of the major current research frontiers in geography.

A Brief Review of the Works of the Geographers, applying Systems Approach in their Study

The works of various scholars, as Chorley (1962), Ackerman (1963), Berry (1964) and

Stoddart (1965) have been pioneering in the application of system's approach to geographical problems.

Chorley, in his paper "Geomorphology and General Systems Theory" represents a great step forward in the development of modern scientific method in geography (Chorley, 1962). It was the first major paper devoted exclusively to a systems theme and to the framework of General Systems Theory. Systems analysis draws upon these ideas exclusively and illuminates many of the fundamental issues. By interpreting the Davisian cycle of erosion in systems terms, Chorley provides a valuable example of these points and incidentally demonstrates that knowledge of these frameworks of thought is as important to the interpretation of the theoretical-deductive system of enquiry as to any empirical-deductive one. A major part of Chorley's paper is devoted to the description of the differences between the two major systematic frameworks—'open' and 'closed' systems—and the application of these ideas to geomorphology. In a 'closed system' definite and rigid boundaries occur through which no material or energy can pass. The development of such systems leads to the eradication (entire destruction) of the existing differences or gradients in the system, with the initial conditions determining the final equilibrium form. 'Open systems', on the other hand, are systems which need an energy supply for their maintenance and preservation; and can maintain their organization within the geometry of their form by adjustments within the geometry of the system. Moreover, different initial conditions in such systems may lead to similar end results.

Ackerman stressed, in one of his papers, "Where is the Research Frontier" in 1963, the need for geographers to develop Systems analysis. Systems, he considered as one of the most pervasive (vast) and characteristic phenomena in nature, and they are specially been created for geography as such. He has described two characteristics of the system. The one is 'the flow of information within the system'. 'Information' here refers to any mechanism that holds together the interdependent, interacting parts of a system. As far as geography is concerned, the connectivity within its system is the most important characteristic. He has suggested eight processes which generate this connectivity—four in physical and four in cultural. Among the cultural, they are: demographic movement, organizational evolution, the resource converting techniques, and the Space adjusting techniques; and among the physical they are: dynamics of soil mantle, movement of water, climatic processes, and biotic processes. The second important characteristic, he mentioned, is 'the existence of sub-systems within the system'. It is the functional sub-systems that are generally the significant ones.

Berry in his paper "Cities as Systems within System of Cities" not only demonstrates the utility of the systems approach to the study of urban areas but also provides a vivid illustration of the direction in which modern scientific method is moving (Berry, 1964). Berry attempts to provide a basis for the study of "Cities as Systems within System of Cities" by the use of the twin concept of organization and information in spatial form. The basic contention of the paper is that studies dealing with the geography of urban areas have fallen into two main groups, one group providing inductive generalizations, the other providing logical constructs. Berry observed that if a sound urban-theory is ever going to develop these two lines of enquiry must be integrated. In this process of integration the inductive generalization will be translated into theory and logical constructs will be quantified and tested in the real world.

Lastly, Stoddart's article "Geography and the Ecological Approach" is an additional contribution to the concept of systems in geography (Stoddart, 1965). Stoddart justifies the ecological approach not by empiricism (by the use of series of case studies to show the need for a synthesis between organisms and their habitat) but on methodological grounds. He shows that the search for synthesis is not new, but geographers in particular have rarely achieved much success in this direction, because results were either framed in a qualitative way or on a regional basis, while the true nature of geographic relationships involved was obscured by naïve (natural, pure) deterministic statements. Stoddart has suggested ecology as the systems approach to geography. It means that the geography should be studied in terms of man and environment relationships and these relationships can be considered within the framework of 'ecosystem'. An ecosystem is a functioning, interacting system composed of one or more organisms and their effective environment, both physical and biological. The description of an ecosystem may include its spatial relations; inventories of its physical features, its habitats and ecological niches, its organisms, and its basic reserves of matter and energy; its patterns of circulation of matter and energy; the nature of its income (or input) of matter and energy; and the behaviour or trend of its entropy level.

CRITICISM

Systems analysis is extremely useful to economic geographers because it encompasses a number of key concepts, such as interactions and spatial relationships, in terms of processes within socio-economic systems. Nor is this the only advantage, it can also encompass value-directed behaviour such as decision-making and goal-seeking. Nevertheless, it is vulnerable to certain criticism, which, must therefore be discussed here. It seems that the employment of systems concepts and systems analysis has not yet achieved powerful operational status in geography. In part this must be attributed to the complexity of systems analysis itself, which, if it is to be fully employed, involves mathematical techniques beyond the reach of most geographers. The solution to this difficulty is, of course, for the geographers to learn more mathematics, but this is perhaps easier said than done. But there are other difficulties. Operationalising systems analysis involves many evaluative judgements regarding the closure of the system and so on. The greater our experience of some problem and the more information we possess, the easier it is to make such evaluative judgements with some degree of confidence. Our general lack of experience with systems analysis, together with the relative weak development of theory, does not allow us to make such evaluations with any degree of confidence, except in those cases where we can easily make assumptions regarding the structure and behaviour of some system. We are, in short, very much in the stage of *a priori* model use in our attempts to apply systems concepts to geography.

One of the criticisms is that the systems approach tends to crowd out all the others. Other criticism comes from those who believe systems analysts claim too much, both scientifically and ideologically. On the other level, the concepts of entropy and homeostasis have been called in question on the grounds that systems are not things but abstractions from people's observed behaviour—abstractions that exist only in the minds of the people who are concerned with it.

Another frequent objection is that the systems approach somehow carries with it the assumption that all social systems either are or should be in a state of equilibrium. Some analysts believe that this would be true of closed systems, and that nearly all societies are open systems. The objection has some force. Obviously, many societies are from harmoniously integrated, social institutions, are changing radically (completely), traditional types of political organization, marriage, economic production and consumption are all breaking down. It is only too easy to assume that the components of a system must be functionally related, i.e. related in such a way that each component somehow contributes to the stability of the system. Such an approach has been called the functional paradigm, its exponents are generally known as functionalists.

Fortunately, it is not necessary to be functional in order to use the systems approach. Any description of a system necessarily catches it at one point of time, and what seems functional today may be a source of conflict tomorrow. Society is not static or stable in a functionalistic sense; it is continually changing, and discloses few simple cause-effect relationships. It may well be structured in some way, but not necessarily in any particular way.

Systems' analysis has, undoubtedly, yielded insights into the structural characteristics and the behaviour of complex interacting phenomena, and systems concept, therefore, provides an appropriate conceptual framework for handling substantive geographical problems. In conclusion it is worthwhile to stress that there is no single path to scientific understanding. All scientists search for order in the real world, whatever their disciplinary perspective is in organizing the search. Systems approach is found to be most fruitful source of ideas and concepts. In such circumstances, this approach cannot be dispensed with lightly.

NOTES

1. Harvey, 1969, 447-455.
2. Harvey, *op. cit.*, 455-459.
3. Harvey, *op. cit.*, 467-480.
4. General Systems Theory (GST) is an interdisciplinary practice that describes systems with interacting components, applicable to biology, cybernetics, and other fields. Karl Ludwig von Bertalanffy (1901-1972) was an Austrian-born biologist known as one of the founders of GST. Bertalanffy proposed that the laws of thermodynamics applied to closed systems, but not necessarily to "open systems," such as living things. His mathematical model of an organism's growth over time, published in 1934, is still in use today. GST attempted to provide alternatives to conventional models of organization. GST defined new foundations and developments as a generalized theory of systems with applications to numerous areas of study, emphasizing holism over reductionism, organism over

mechanism. Bertalanffy's contribution to systems theory is best known for his theory of open systems. The system theorist argued that traditional closed system models based on classical science and the second law of thermodynamics were untenable. We may well suspect that many characteristics of living systems which are paradoxical in view of the laws of physics are a consequence of this fact. Although potential applications exist in other areas, the theorist developed only the implications for biology and cybernetics. In the social sciences, Bertalanffy did believe that general systems concepts were applicable, e.g. theories that had been introduced into the field of sociology from a modern systems approach that included "the concept of general system, of feedback, information, communication, etc." The theorist critiqued classical "atomistic" conceptions of social systems and ideation "such as 'social physics' as was often attempted in a reductionist spirit." Bertalanffy also recognized difficulties with the application of a new general theory to social science due to the complexity of the intersections between natural sciences and human social systems. However, the theory still encouraged for new developments from sociology, to anthropology, economics, political science, and psychology among other areas. Today, Bertalanffy's GST remains a bridge for interdisciplinary study of systems in the social sciences.

REFERENCES

Ackerman, A.E. (1963), "Where is a Research Frontier", *Annals of Association of American Geographers,* Vol. 53, pp. 429-440.

Berry, B.J.L. (1964), "Cities as Systems within Systems of Cities", *Pap. Regional Science Association,* 13, 147-63.

Bertanlanffy, L. von. (1968), *General System Theory: Foundation, Developments, Applications,* George Braziller, New York.

Chorley, R.J. (1962), *Geomorphology and General Systems Theory,* U.S. Govt. Print. Off., Washington.

Harvey, D. (1969), *Explanation in Geography,* Arnold Publishers, New Delhi.

Stoddart, D.R. (1965), "Geography and the Ecological Approach", *Geography,* 50, 252-251.

CHAPTER-11

Behavioural Geography

Behavioural geography sought to counter the perceived tendency of quantitative geography to deal with humanity as a statistical phenomenon. It flourished briefly during 1970s and sought to provide a greater understanding of how people perceived places and made locational decisions and sought to challenge mathematical models of society, in particular the use of econometric techniques. But the lack of a sound theoretical base left behavioural geography open to critique.[1] Geography in the first half of the 20th century was primarily an empirical science concentrating on the regional accounts. Man was to follow the tyrannical dictums of Nature and Act accordingly. He was considered an organism without any 'values' of his own in interplay with Nature. Geography as a discipline severely suffered from theoretical vacuum.[2] The 1950s was primarily a period of theory building based on hypotshetico-deductive method (pure reasoning) of positivism, with quantitative techniques as a tool. By 1960s, the researchers began stressing that factors such as access of information, images of environment, the stages of learning about stimulus situations, attitudes towards place, degree of risk aversion, place utility and revealed preferences were important in undertaking space activity. Geographers started treating man as a social being essentially endowed with certain values which affect his subjective vision of reality and consequential behaviour. Geographers, like other behavioural scientists, arrived at the conclusions that the behaviour of an individual is a result of the perceived reality rather than the objective reality. The man's activities were being examined as responses to attendant environmental conditions and his social-cultural fabric. The approach was so widely and rapidly propagated that it earned the phraseology of behavioural revolution. This subfield having emerged within geography has been called 'behavioural geography'. Behavioural geography is an approach to human geography that examines human behaviour using a disaggregate approach. Behavioural geographers focus on the cognitive processes underlying spatial reasoning, decision-making and behaviour. Behavioural geography is that branch of human science, which deals with the study of cognitive processes with its response to its environment, through behaviourism. Behavioural geography draws from early behaviourist works such as Tolman's concepts of "cognitive maps". More behaviourally oriented geographers are materialists and look at the role of basic learning processes and how they influence the landscape patterns or even group identity. The cognitive processes include environmental perception and cognition, wayfinding, the construction of cognitive maps, place attachment, the development of attitudes about space and place, decisions and behaviour based on knowledge of one's environs, and numerous other topics. The approach adopted in behavioural geography is closely related to that of psychology, but draws on research findings from a multitude of other disciplines including economics, sociology, anthropology, transportation planning and many others.

People-environment interaction has been the focus of geographical inquiry for a long time. Some scholars have adopted a normative perspective and described the behaviour that should result under specified environmental conditions; while others ignored people and have concentrated on the geometry of landscape. This is most clearly demonstrated in the Spatial Analysis School of Thought that came to dominate human geography in the 1960s and which seeks to dissect the environment into nodes, networks, and surfaces. The behavioural approaches, on the other hand, argue that none of these approaches is satisfactory and that understanding of people-environment interaction is best achieved by looking at both the psychological processes through which individual "actors" come to know the environment and the way in which these processes influence the nature of resultant behaviour. In other words, the central theme of the behavioural approach is that the focus of human geographical inquiry should be on people as human beings. Although this perspective requires some attention to be paid to the evolutionary adaptation of humans to their environments and to intellectual development, it necessitates above all a close look at how individuals derive information from the environment, how they use public and private information channels, how they build up cognitive images, how they evaluate environmental information in making decisions, and how overt behaviour is manifest in the form of activity systems that are highly constrained in both time and space. Great attention has been paid by behavioural researchers in human geography to environmental cognition and, in particular, to environmental imagery.

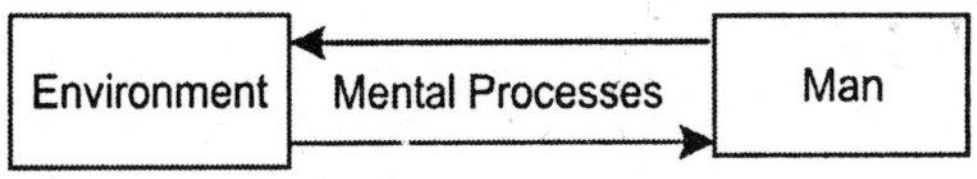

FIGURE 11.1 Man and Environment Link

Behavioural geography, as a branch of geography, has emerged in the late 20th century. It is the geographical expression of behaviouralim. Behaviouralism is a general movement that spreaded into the social sciences in recent years from the behavioural sciences. In geography, it may be considered as an approach, which looks at man-environment relationships with a new and sharp outlook. The aim of behaviouralism is to replace the simplistic and mechanistic conception that previously characterized man-environment theory, with new version that explicitly recognizes the complexities of human behaviour, i.e. in a given set of environment, how human beings behave, the degree and the impact on space, i.e. how they organize it or use it. The 'behaviour' may be defined as "a response to some stimulus". The behaviouralist views the human subjects of his studies as thinking beings, whose actions are mediated (intervened) by cognitive (mental) processes. In between the environmental influences and the man's actions there are mental processes of various kinds, which govern his behaviour, i.e. how to react to a given set of environmental elements. Because, within the same set of elements, the response of various human beings may not necessarily be the same or similar.

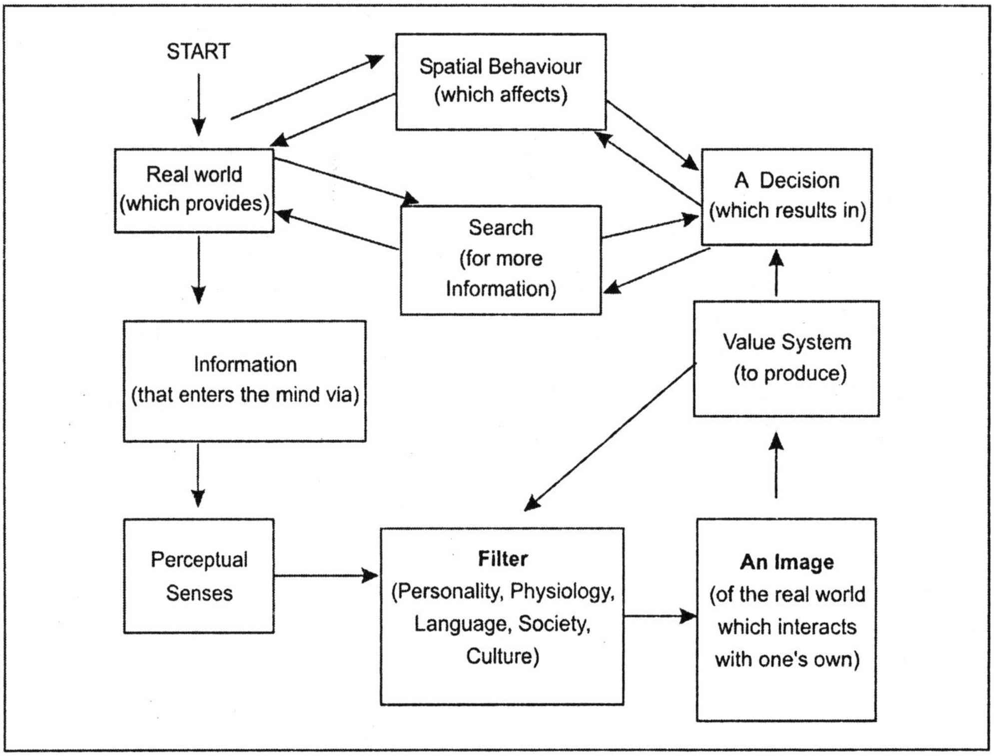

Source: Huckle, J.(ed.) (1983) pp. 43-45.

FIGURE 11.2 The Relationship among Perception, Cognitive Images, Values and Behaviour

The behaviouralist is interested in the way in which people come to terms with their physical and social milieu. He is also interested in the factors that influence the interrelationships of thought and action of man. Moreover, in doing so, he in no way assumes that all human decisions are based upon sequence of reasoned thought with systematic evaluation of all available options, i.e. all human actions may not necessarily be systematic and sequential all over. Because, human behaviour at times is unpredictable. Some responses are mean, selfish, slavish, whereas others are more planned and systematic, i.e. made with utmost care and forethought. The full spectrum, from sub-conscious impulse to self-conscious deliberation, is of concern to the behaviouralist. Seen in this light, behavioural geography may be regarded as the most common name for that part of geography that adopts the behaviouralist approach to man-environment relationships, where explanations of spatial patterns of behaviour is sought primarily in cognitive process, which underpinned that behaviour. This may be elaborated as a model for the study of the relationships between perception, cognitive images, values and behaviour as shown in the diagram in Fig. 11.2.

BEHAVIOURAL MOVEMENT

The developmental perspective on people-environment interaction generally held sway in the second half of 20th century in behavioural approaches in human geography. Though several geographers had, from time to time, drawn attention to the role and importance of perception in the study of human behaviour, and the process of creating landscapes (e.g. Sauer, Wright, White and Kirk), the work that proved rudimentary in the starting of the movement towards behaviouralism as an 'alternative' to the normative approach to theory (then so much in circulation) was a paper by Julian Wolepart entitled "The Decision Process in a Spatial Context" published in the *Annals of Association of American Geographers* in 1964. Julian Wolpert was a professor of Geography, Public Affairs and Urban Planning at the Woodrow Wilson School of Public and International Affairs, Princeton University, U.S.A. The paper was based on his Ph.D. Thesis submitted at the University of Wisconsin, America. The paper was built around the results of a field study of agricultural land use in a part of Sweden. The author compared the actual labour productivity of farmers with what could have been achieved under optimizing decision-making by the farmers as rational economic persons, and found that there was up to 40 per cent less than the expected level of productivity. The lower than expected level of productivity, was explained as a result of the farmers' lower aspiration level which was itself a reflection of their economic and social environment, the context which motivated them to act differently from the theoretical economic man. Wolepert's various papers (1964, 1965, 1967 and 1970) on the satisficing model of decision-making behaviour had contributed to heralding the arrival of a new behavioural movement in geography. By the second half of 1960s, a new branch of geography, termed as 'Behavioural Geography' had begun to be identified. So great was the excitement at the behavioural thinking in geographical work that many were led to believe that geography had experienced yet another revolution—the behavioural revolution.

Though the behavioural movement of the post-quantitative revolution phase had been triggered by the writings of Wolpert, the attention of the profession towards the role of perception in the creation of spatial patterns and forms studied by geographers had been drawn in a focused manner by at least two geographers much before Wolpert, and they were J.K. Wright and W.K. Kirk, the American and British by origin respectively. J.K. Wright, in his presidential address to the Association of American Geographers at the 1946 annual conference, had emphasized that although man had explored every area of the earth, there still lay a territory that remained almost completely unexplored—*terrae incognitae*—the territory to be found in the minds of men, the private world specific to particular individuals, but which could have features shared by others belonging to the same social, cultural or occupational group. Wright proposed a new branch of geography, which he named '*geosophy*', to be devoted to the examination of geographical ideas "both true and false, of all manner of people, i.e. a field of study concerned with subjective conceptions". Wright underlined that acquisition of geographical knowledge is conditioned by a complex interplay of cultural and physical factors, and "nearly every activity in which man engages...is to some extent affected by the geographical knowledge at his disposal." These words could, indeed have heralded a much earlier start of behavioural geography, but apparently owing to the lack of references to them in the works of later writers they failed to have any significant impact.

A new comprehensive attempt to introduce behavioural postulates in geography was made by the British geographer W.K. Kirk (1951, 1963). He had presented one of the first behavioural models to serve as an alternative to the environmental concepts of man-environment relationship. Kirk suggested that people act according to the way they perceive their environment, and that the environment as perceived might markedly differ from objective reality (i.e. the environment as exists). He maintained that people's perception of the environment is greatly influenced by their social and cultural values. Thus, "Gestalt Psychology"[3] inspired the views of Kirk, so that he distinguished between the *Objective* or phenomenal environment and the *Behavioural* environment. Kirk maintained that since geographers are concerned with decision-making and the consequences of decision-making, they must pay attention to the role of behavioural environment in their research.

Through the 1970s it had been generally asserted that geography was experiencing a second revolution—the behavioural revolution—which was expected to bring about a major transformation in the discipline but experience shows that any expectation of a 'revolution' was at best premature. Nevertheless, behavioural geography did represent a gradual emergence and consolidation of a set of approaches that aimed to increase the scope of geographical explanation by seeking a fuller understanding of the processes that underpin our real world. This certainly represents "a significant milestone in geographical thought" (Gold, 1980)[4].

Achievements of Behavioural Movement

Following Gold (1980), the major achievements of behavioural movement in geography may be listed as under:

1. It asserted the role of the individual in geography, which at that time had allowed strong signs of wishing to reduce human activities to point patterns and spatial preferences and indifference curves.

2. It led to thorough reappraisal of the methodological approaches to man-environment relations. It countered the environmental and neo-environmental doctrines by recognizing the true complements of human behaviour.
3. It acted as a forum for new philosophies, approaches, and methods, and it revived interest in the study of some older themes, such as cultural landscape, and idiographic analysis of places.
4. It opened up new channels of dialogues and debate with other disciplines, and thereby, contributed to a trend toward cross-fertilization of ideas.
5. Lastly, researchers in behavioural geography had contributed much to the debate in social relevance of geographical research, by helping to explode the myth of value-freedom, by promoting interest in problems of social concern, and by supporting the geographers' involvement in public policy issues.

PARADIGMS IN BEHAVIOURAL GEOGRAPHY

The impact of environment on man had echoed from the ancient to modern period and human behaviour was seen as a response to the environmental stimuli. Various paradigms emerged one after the other in order to study this relationship. The commonly held paradigm about human behaviour during the first half of 20th century was a deterministic one. 'Environmental Determinism' as such developed by Ratzel (1882, 1891). Semple (1911) treated the man as a product of the earth's surface. This school continued until mid-20th century (Huntington 1945; Taylor 1940) but it had already reached its zenith by 1920s. A counter doctrine of 'Posssibilism' was put forward by Blache, Bruhnes and Febvre (1926) who saw man as an active agent and the environment as a permissive forum for his activities. Blache proclaimed that there are "no necessities, but everywhere possibilities and man as a master, is the judge of their use." A pragmatic thesis was initiated by Taylor (1951) who propounded 'Neo-Determinism'. This thesis acknowledged the key role of environment but also provided limited choice to man. He anticipated that failure to understand the nature given limitations must face disaster. A neo-environmentalism: 'economic determinism' was propagated in the spatial scientific approach in 1950s and early 1960s. This approach has stemmed from the works of Christaller (1933); Losch (1940) and Isard (1956, 1960) who treated man as an economic man and extended the range of economic analysis by incorporating the space as an element. This economically oriented thesis treated the behaviour of a man as total subject of his economic environment. The shackles of economic man and tyranny of space were broken in 1960s. The rapid development of science and technology reduced the world to a village and a month to a minute. It was a state of the 'recession of the physical environment' which presented man as a satisficier rather than an optimizer. Wolpart (1964) concluded that the satisficier concept was more accurate of behavioural pattern of the farmers than could be inferred from the notion of economic Man. The information availability, skill to employ these information, personality variables and system of values were found to be relevant in decision-making process (Pred, 1967 and 1969). Pocock (1978) developed a conceptual model of man-environment interaction which shows how a perceiver interacts with an environment to create a mental image of that environment. The information status of the environment influences what the individual comes to know about the real world while the individual's physiological, psychological fabric determines how that knowledge contributes to the development of an image. Although individual formulation varies from author to author, this model is a characteristic of a large volume of research. Man has been depicted as a thinking being whose behaviour with the environment is mediated by mental processes and 'cognitive representation' of the external environment. The model propounded by Pocock can alternatively by a simple conventional paradigm about man-environment relationship (Fig. 11.3).

The complexity of human behaviour depends upon the different roles of man, i.e. imbued with different mental environments. Simon[6] has presented four important models of man, each eliciting a different behaviour in the same environmental stimulus:

1. *Psychoanalytic Man*: This model treats man as a non-rational adult whose behaviour is largely determined by the resolution of psychological conflicts experienced earlier in life. This model had its seminal impact in behavioural thoughts in the early 20th century.
2. *Computer Man*: A man acts as a logical theorist, objectively screens and processes information without inhibition and with reference to long and short term memory stores. This model is opposed to the Psychoanalytical model.
3. *Economic Man*: An economic man as a consumer is dominated by the wish to maximize utility and as an entrepreneur to maximize the profit. His behaviour is totally influenced by objective economic factors of which he is aware and therefore, predictable in given conditions. An economic man is always considered as an optimizer.
4. *Bounded Rational Man*: This model is also called as a 'satisficing man' which is usually committed to satisfying a course of action which is good enough in terms of his aspiration level in a given situation but is not necessarily the best from a purely economic point of view. The concept of as a satisficier rests on an essentially practical and realistic view of the likely alternatives. If the outcome of his choice does not meet his aspirations, the next time when he is faced with a similar situation he will presumably aim higher. Certainly he will not satisfy for less than the previous solution. This model man operates intermediately between his respective psychological, behavioural and logical roots.

The behavioural geography concentrates on how a man views the environment, extracts information and develops his cognitive representation and reflects the same in the form of a response. Behavioural geography assumes that the human mind acts as a 'black box' in a decision-making process. A black box is considered as a place where information processing system takes place. The black box is governed by value system, learning, thinking, deciding and the whole range of motor responses. It is the mind which forms cognitive structures. Indeed, the black box can be characterized as a hidden world.[7] This, of course, is a notion familiar to geographers who have long recognized that the *terrae incognitae* (unexplored/unknown space) in the mind may influence behaviour to a much greater extent than actual environmental stimuli.[8] Essentially what is being argued in this view is that it is not so much what we see, but how we feel about what we see, that is crucial to understanding behaviour.[9] In other words, 'man has the peculiar aptitude of being able to live by notions of reality which may be more real than reality itself.[10] What the black box concept emphasizes is that spatial behaviour involves not only stimuli and responses but also a complex process of relationship and feedback based on cognitive and physiological processes.

The environmental information is fundamental to survival, to everyday living and to the maintenance of a sense of well-being. In fact, it is central to the basic human needs. Understandings of the ramifications and functions of environmental system is fundamental to human behaviour, therefore, the first and foremost postulate of behaviouralism is that there are different levels of environments within which man operates. The emphasis has been laid primarily on the difference between the 'objective environment' and the 'perceived environment'. The latter exists within the former but it is the perceived environment which is actually a meaningful determinant in mediating behaviour of an individual or a group of a people. The terms, 'objective reality' and 'perceived reality' are also frequently used as substitutes as for 'objective environment' and the 'perceived environment' respectively. Sonnenfeld (1972) has proposed a nested hierarchy of four environments (Fig. 11.3).

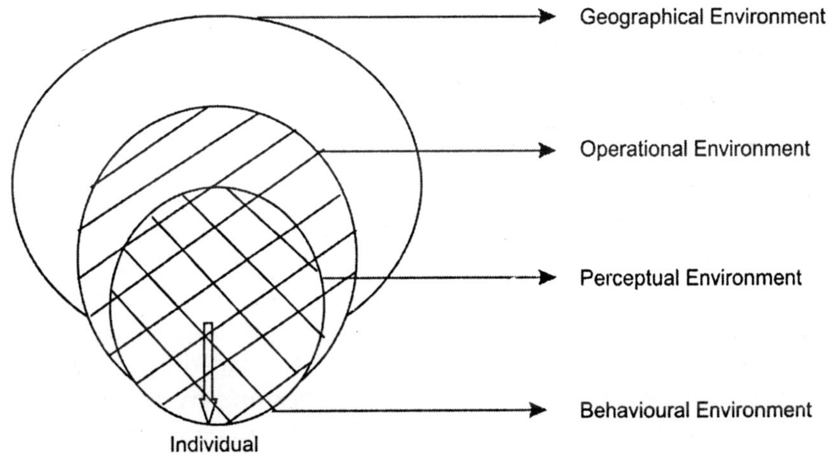

FIGURE 11.3 Levels of Environments within which Man operates

1. *Geographical Environment* encompasses the whole external environment which includes the totality of all physical facts that exist at a certain time.
2. *Operational Environment* is the functional portion of geographical environment. It impinges on man as an individual or a large group and acts as a filter to coming information and influences behaviour in some way or the other.
3. *Perceptual Environment* is that part of the operational environment, of which man is conscious either through physical sensitivity or learning and experience. A man has his own framework of information selection in terms of selective

exposure, selective perception and selective retention. Spatially this environment is expressed by spatial perception and, generally in terms of mental map of Reality (Gould, 1966, 3).

4. *Behavioural Environment* is that part of environment which 'elicits a behavioural response towards which behaviour is directed, such as results in a conscious utilization or transformation of environment'.[11] This represents the totality of facts which determine the behaviour of an individual at a certain moment. The core of the behavioural environment is the activity space which refers to all those peoples, organizations or places with which man is in direct and regular contact. This activity space is the most dominant component in the behavioural environment.

BASIC CONCEPTS OF BEHAVIOURAL GEOGRAPHY

The 'Environmental Perception' and 'Mental Maps' are recognized as the basic concepts of Behavioural Geography.

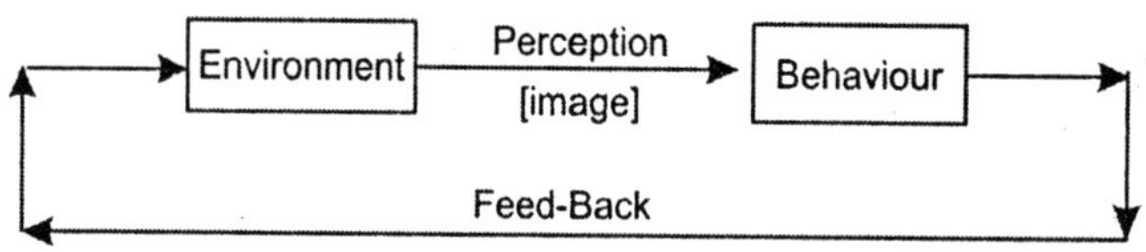

Figure 11.4 Environment and Behaviour Link

Environmental Perception

The concept relates to the way in which an individual perceives the environment (or attains awareness and understanding of the environment); it is the process of evaluation and storing information about the environment. It is the perception of the environment that concerns the human geographers because decision makers base their judgement on the environment they perceive. It is the total perception an individual has of his surroundings, which creates mental maps. For instance, if someone has really to make a choice to select his place of dwelling, he would be assailed by images of faraway places, of different climates and different landscapes and by his personal feelings towards cultures other than his own. He would also become very sensitive to the affection of old friends, and the familiarity of his present surroundings. He would be aware of the pull between 'here' and 'there' as places in which to live. Differences between the attributes of 'here' and 'there' have always been of great interest to geographers, because it is precisely the differences between places that generate movement of goods, people and information. Such movements produce complex spatial patterns. The traffic jam in cities, crowds on summer time beaches, vast flows of letters and electronic messages and the great networks of pipelines that criss-cross nation after nation. To analyse such complex patterns produced by man on the surface of the earth, geographers are increasingly looking at the questions of 'relative location-questions which consider places not in any absolute, old fashioned, latitude and longitude sense, but in terms of their costs and times to all other places with which they might exchange goods, money, people and messages. In geographers' language these patterns are known as 'spatial interaction'. Geographers are especially concerned with the effect of distance on movement—that is the effect of the location of an individual relative to his proposed destination. Modern geographical studies of environmental perception look at the ways in which people form images of other places and how these images influence many decisions—including to the one to move. Decision-makers operating in an environment, base their decisions on the environment as they perceive it, not as it is. The action, resulting from decision, on the other hand, is played out in a real environment.

'Environmental Perception' is the perception either of the whole environment, or of specific selected elements within it, or even of 'space' in the abstract: essentially environmental perception is a property of the mind rather than a construct of that mind. Thus the mind may have perception of drought, the administrative system or the advantages of seashore climate; but these are only elements in the perceived environment, selected by the individual in response to particular stimuli. The simple and obvious facts of Environmental Perception have long been recognized, and been given some place in the methodology of cultural and especially historical geography. In geographical methodology as a whole, however, realization that this is significant in our work came rather later. The room of Environmental Perception in geographical studies has been made due to the anthropocentric character of the discipline.

The Range and Limits of Personal Knowledge of the World: Whether we stay or move about, our environment is subject to sudden and often drastic change. In consequence, we must be able to see things not only as they are, but also as they might become. Our private milieus are, therefore, flexible, plastic, and somewhat amorphous. We are psychologically equipped for a wide range of environments, including some of those that we create. As individuals, we learn most rapidly about the world not by paying close attention to a single variable, but by superficially scanning a great variety of things. Everyday perception tends to be selective, creative, temporary, inexact, generalized and stereotyped, just because imprecise, partly erroneous impressions about the world in general often convey more than exact details about a small segment of it. The observant are not necessarily most accurate: effective observation is never unwaveringly attentive. Changing perceptions are necessary to preserve mental alertness and normal powers of thought. Essential perception of the world, in short, embraces every way of looking at it: conscious and unconscious, blurred and distinct, objective and subjective, inadvertent (unintentional) and deliberate (well considered), literal and schematic. Perception itself is never unalloyed: sensing, thinking, feeling, and believing are simultaneous, interdependent processes.

Types of Environment perceived: Different kinds of environments around an individual may be delimited (Fig. 11.3). The environment immediately next to an individual is the 'Local' or 'Behavioural' Environment. Usually these environments are the places of dwelling, e.g. the urban areas. Interest, here, has focused on the way which people perceive certain landmarks, routes, boundaries and neighbourhoods. It has been noted that some people in a city always seemed to have a good sense of orientation, while others are usually subject to confusion. Some people have informal, imaginary maps in their heads centred upon locations of their homes. In local environment an individual performs most of his daily functions. The 'Perceptual Environment' is immediately next to the Local Environment, e.g. the recreational centres. The environment in which man operates, the 'Operational Environment', is created, e.g. by the people working in neighbouring towns or cities, but dwelling in rural areas. Then, ultimately, is the 'Geographical Environment', i.e. the one whose only a fraction or a part has been so far perceived by the individual. The most important of all these environments remains the personal (behavioural) environment or the personal space, which no one likes to be violated by any other individual.

Uniqueness of Private Environment: Despite their congruence with each other and with the world as it is, private milieus do diverge markedly among people in different cultures, for individuals within a social group, and for the same person as a child and as adult, at various times and places. The life of each individual constitutes an original and irreversible perceptive experience. Each private worldview is unique, to begin with, because each person inhabits a different milieu. The fact that no two human beings can occupy the same point at the same time and that the world is never precisely the same on successive occasions—means that the physical world is idiosyncratic for each individual. Experience is not only unique; more significantly it is also self-centred. Each private worldview is also unique because everyone chooses from and reacts to the milieu in a different way. We elect to see certain aspects for the world to avoid others. Moreover, because everything that we know about an object affects the way in which it appears to the eye, no object is apt to seem quite the same to any two percipients. Thus, in some respects each man's appraisal of an identical situation is peculiarly his own. The private worldviews diverge from one another even within the limits set by logical necessity, human physiology, and group standards. In any society, individuals of similar cultural background, who speak the same language, still perceive and understand the world differently. So what one thinks depends on what is familiar to him and his proclivities (inclinations).

The way a landscape looks depends on all the attendant circumstances, for each sense is affected by the others. Velvet looks soft, ice sounds solid, and red feels warm because experience has conformed these impressions. The individual carries with him his characteristic sensory abilities, intelligence, interests, and temperamental qualities, and his responses are coloured and to some extent determined by these inherent individual qualities. People at home in the same environments, e.g. habitually select different modes of orientation.

Another reason why private worldviews are irreducibly unique is that all information is inspired, edited, and distorted by feeling. Coins look larger to the children of the poor; the feast smells more fragrant to the hungry; and the mountains loom higher to the lost. Had our perceptions no connection with our pleasures we would soon close our eyes on this world? We seldom differentiate among people, places, or things until we have a personal interest in them. One American town is much alike another to an individual, unless one has a good motive for telling them apart. The most exhaustive study of photographs and ethnological evidence does not enable us to distinguish among individuals of another race with the ease, speed, and certainty generated by strong feeling. All Chinese may look the same to some, but not to the man—however foreign—with a Chinese wife.

Mental Maps

A person's perception of the world is known as a Mental Map. A mental map is an individual's own internal map of their known world. A Mental Map can be defined as an internal representation of a place based on memory. There are two types of mental maps. The First type of Mental Maps is that formed from our personal experiences. For example, a trip to 'Disneyworld' in Florida might conjure up images of Mickey Mouse, fun and sun provided that the trip was a pleasant experience. This example deals more with our perception of the attributes of a place. Provided the tourist paid attention to the direction of travel and the coastal features, they probably would also have a good perception as to the relative location of Florida stored in their memory. So it is easy to see that our minds our very efficient at storing locational as well as attribute information. A mere mention of a place will not necessarily lead to the formation of a mental map by a person. Ordinarily, the person needs some frame of reference in their minds from either past experience or knowledge of the place. Of course, Mental Maps are capable of stirring up emotions about a place whether good or bad. Unfortunately sometimes we have perceptions of places based on what we have heard or perhaps what our fears about unfamiliar places dictates to us. The Second type of Mental Maps is the one formed from references to maps we have seen. This can sometimes be dangerous if maps are inaccurate. A concept a person has of the shape of the United States is an example of a mental map based on the printed map. If it were not for maps we would have no idea of the shape of our country. Generally, it can be said that the maps we store in our minds are full of distortions.

Mental maps not only provide a better visual conception of our surroundings but also shape our attitudes about people and places. The aspect of behavioural analysis that has been most enthusiastically adopted by geographers is the concept of mental maps. The concept has been referred to as the 'geographies of human mind'. The seminal influence in this regard came from a paper by Peter Gould (1966). Through his work, he argued that since locational decisions are guided by how the decision-maker perceives the quality of his environment, it is necessary for the geographer to map out the mental image of the environment on the basis of which a decision was taken. The theme was fully developed in the book entitled: "*Mental Maps*" by Gould and White (1974).

Mental Map studies were part of the primary paradigms of man-environment relations in which man is depicted as thinking individual whose transactions with the environment are mediated by mental processes. Various concepts were used as surrogates for the cognitive representations underlying the work on mental maps, but the one most commonly employed was 'the concept of image'. Over a period of time, individuals develop the mental impressions (i.e. images) of the world through their everyday contact with the environment. These images later guide the individual or group behaviour. Images can be investigated at various scales but the dominant trend of enquiry has focus upon the urban environment. Discussion of the links between images and behaviour has centred upon areas of research interest, most prominent being the issues like: Response to natural hazards; spatial behaviour in urban areas; and the role of neighbourhood preferences in Intra-urban migration.

SALIENT FEATURES OF BEHAVIOURAL GEOGRAPHY

The approach to man-environment relationship that many geographers adopted in the late 1970s as part of the movement toward behaviouralism in geography has five distinguishing features. They are:

(1) First, it is argued that the environmental perception upon which people act may well differ markedly from the true nature of the real world. Space, here, can be said to have dual character.
   (i) The objective environment, (i.e. real, observable, measurable, physical, etc.). It is the reality as exists in nature and may be studied by direct means (observation).
   (ii) The behavioural environment (i.e. the world of mind). It can be studied only by indirect means. It is the reality as perceived by the decision-maker. Obviously, the behavioural geographer focuses attention on the environment as perceived.

(2) Second, in order to understand the human behaviour, the meaning of environment should be broadened to include both the physical environment and the social environment. Earlier, only the concept of physical environment has been emphasized more in man-environment relationships. The social environment includes the elements like Human taste, Aspirations, Attitudes, Culture, Available resources, and Socio-economic background. Over the years, geographical analysis has primarily considered the ways in which physical environmental influences man, but it is clearly only one aspect of the matter. The concept of environment, therefore, needs to be widened to cover social environment as well. Because it is the social environment within which an individual lives, acts, thinks or behaves (Fig. 11.5). Thus, an individual shapes and responds both the physical and social environment.

(3) Although the behavioural geographers do recognize the importance of social environment, within this social environment their focus of study is even more precise and sharp, i.e. on the individual, rather approaching the problem at the level of social group. The 'operational environment' is the environment human beings live in, affecting them in some way or the other, whether or not they are conscious of it. The 'perceptual environment' is the portion of the operational environment of which human beings are aware. This awareness may result directly from physical sensitivity to environmental stimuli or indirectly from learning or experience. Although one can maintain that all

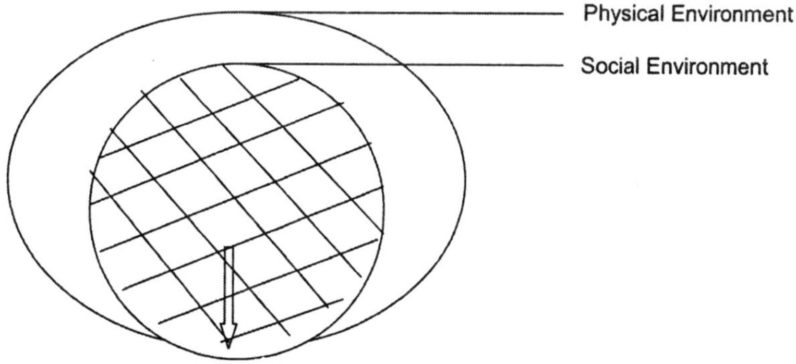

FIGURE 11.5 The Milieu of An Individual

individuals have their own perceptual environment, it is more important to note that persons belonging to the same culture or sub-culture most likely share basic ideas about and attitudes toward the environment. In other words, the perceptual environment is the external world as culturally conceived. The 'behavioural environment', on the other hand, is the environment for which every individual is very conscious of. Thus, within the perceptual environment one can focus more narrowly on the portion that draws out a behavioural response or towards which behaviour is directed. This behavioural environment receives much attention in current research in social science disciplines (including geography). It is in this behavioural environment that the man acts and interacts most closely. It is in this behavioural environment that physical features acquire values and potentialities that attract or repel human actions.

(4) One distinguishing feature of behavioural geography is its multi-disciplinary outlook. An understanding of behavioural geography particularly depends on the disciplines like psychology, sociology, anthropology, ethnology (racial geography), planning, etc. The multi-disciplinary outlook is not a new problem for geography, a discipline that has always derived much of its methodological inspiration from other subjects, but it is probably more pressing now than ever before, as the range of disciplines concerned is so much wider.

(5) The central theme of behavioural approach is that the focus of geographic enquiry should be on people as human beings. The behaviouralist views the human subjects of his studies as thinking beings, whose actions are mediated b cognitive processes. The field thus concentrates on how a man views the environment, extracts information and develops his cognitive representation and reflects the same in the form of a response. The complexity of human behaviour depends upon the different roles of man, i.e. imbued with different mental environments. Behavioural geography assumes that human mind acts as a 'black box' in a decision-making process. What the black box concept emphasizes is that spatial process of relationship and feedback based on cognitive and physiological processes. There are different levels of

environments within which man operates. But, the emphasis has been laid primarily on the 'perceived environment'.

SCOPE OF BEHAVIOURAL GEOGRAPHY

As the behavioural perspective in geography became generally accepted as a useful analytical device, the scope of behavioural geography widened. Five dominant areas of behavioural research in human geography have been identified. These are:

1. Studies of decision-making and choice of behaviour (e.g. locational choice, route selection and patronage patterns);
2. Analysis of information flows (in relation to innovation—diffusion);
3. Models of learning (derived from theories in psychology);
4. Perception—research (focusing on hazard perception, image formation and mental maps); and
5. Studies in voting behaviour (geography of elections).

To conclude, behavioural geography may be considered as an approach to study man-environment relationships; and not as a viewpoint opposite to Determinism, Possibilism or Ecological Approach. It has rather a sharper outlook. But, the lack of a sound theoretical base left behavioural geography open to criticism.

NOTES

1. Wikipedia, *The Free Encyclopaedia.*
2. Gurjar, R.D. (2005).
3. Gestalt psychology (also Gestalt theory of the Berlin School) is a theory of mind and brain that proposes that the operational principle of the brain is holistic, parallel, and analog, with self-organizing tendencies. The classic Gestalt example is a soap bubble, whose spherical shape (its Gestalt) is not defined by a rigid template, or a mathematical formula, but rather it emerges spontaneously by the parallel action of surface tension acting at all points in the surface simultaneously. This is in contrast to the "atomistic" principle of operation of the digital computer, where every computation is broken down into a sequence of simple steps, each of which is computed independently of the problem as a whole. The Gestalt effect refers to the form-forming capability of our senses, particularly with respect to the visual recognition of figures and whole forms instead of just a collection of simple lines and curves.
4. Gold, J.R. (1980).
5. Gould, P.R. (1966).
6. Simon (1957, 287).
7. Walmsley *et al.*, 1982, 3.
8. Wright, 1947.
9. Lowenthal, 1961.

10. Watson, 1975.
11. English and Mayfield, 1972, 623.

REFERENCES

English, P.W. and R.C. Mayfield (1972), *Man, Space and Environment*, Oxford University Press, London.

Gold, J.R. (1980), *An Introduction to Behavioural Geography*, Oxford University Press.

Gould, P.R. (1966), "On Mental Maps", *Michigan Inter-University Community of Mathematical Geographers*, Discussion Paper 9, reprinted in Downs, R.M. and D. Stea (Eds.) (1973), pp. 182-220.

Gould, P.R. and R. White (1974), *Mental Maps*, Hammondsworth: Penguin Books.

Gurjar, R.D. (2005), "Cognitive System and Paradigms in Behavioural Geography", in *Trends in Geographical Research*, H.S. Sharma, Department of Geography, Rajasthan, Jaipur, pp. 1-19.

Huckle, J. (ed.) (1983), *Geographical Education: Reflection and Action*, Oxford University Press, Oxford, pp. 43-45.

Lowenthal, D. (1961), "Geography, Experience and Imaginations: Towards a Geographical Epistemology", *Annals of the Association of American Geographers*, Vol. 51, pp. 241-260.

Simon, H.A. (1957), *Models of Man*, Wiley, New York.

Sonnenfeld, J. (1972), "Geography, Perception and the behavioural Environment", in *Man, Space and Environment*, Oxford University Press, London, pp. 244-250.

Walmsley, D.J. and Lewis, G.J. (1982), *Human Geography: Behavioural Approaches*, Longman, London.

Watson, J.W. (1975), "Perception and Place", *Geographical Journal*, Volume 141, pp. 271-274.

Wikipedia, *The Free Encyclopaedia*.

Wright, John K. (1947). Terrae Incognitae: The Place of Imagination in Geography", *Annals of the Association of American Geographers*, 37: 1-15.

CHAPTER-12

Marxist Geography

Karl Heinrich Marx (1818-1883) was a German sociologist whose ideas played a significant role in the development of social science and the socialist political movement. He published various books during his lifetime, the most notable being *The Communist Manifesto* (1848) and *Capital* (1867-1894). *The Communist Manifesto* is Marx's most widely read work and this was jointly written with Friedrich Engels, his friend and fellow German revolutionary socialist.[1] Karl Marx was born in Trier, in the German Rhineland. A precocious school child, Marx studied law in Bonn and Berlin. Marx's theories about society, economics and politics—collectively known as Marxism—hold that all societies progress through the dialectic of class struggle: a conflict between an ownership class which controls production and a lower class which produces the labour for goods. Heavily critical of the current socio-economic form of society, capitalism, he called it the "dictatorship of the bourgeoisie", believing it to be run by the wealthy classes purely for their own benefit, and predicted that, like previous socio-economic systems, it would inevitably produce internal tensions which would lead to its self-destruction and replacement by a new system, socialism. He argued that under socialism society would be governed by the working class in what he called the "dictatorship of the proletariat", the "workers state" or "workers' democracy". He believed that socialism would, in its turn, eventually be replaced by a system of stateless, classless society—communism. Along with believing in the inevitability of socialism and communism, Marx actively fought for the former's implementation, arguing that both social theorists and underprivileged people should carry out organised revolutionary action to topple capitalism and bring about socio-economic change. Revolutionary socialist governments espousing Marxist concepts took power in a variety of countries in the 20th century, leading to the formation of such socialist states as the Soviet Union in 1922 and the People's Republic of China in 1949. Many labour unions and worker's parties worldwide were also influenced by Marxist ideas. Various theoretical variants, such as Leninism, Stalinism, Trotskyism and Maoism, were developed. Marx is typically cited, with Émile Durkheim and Max Weber, as one of the three principal architects of modern social science. Marx has been described as one of the most influential figures in human history. In addition to his overtly philosophical early work, his later writings have many points of contact with contemporary philosophical debates, especially in the philosophy of history and the social sciences, and in moral and political philosophy. Historical materialism—Marx's theory of history—is centred on the idea that forms of society rise and fall as they further and then impede the development of human productive power. Marx sees the historical process as proceeding through a necessary series of modes of production, characterized by class struggle, culminating in communism. The analysis of history and economics come together in Marx's prediction of the inevitable economic breakdown of capitalism, to be replaced by communism.

The intellectual climate within which the young Marx worked was dominated by the influence of Hegel[2], and the reaction to Hegel by a group known as the Young Hegelians. The Economic and Philosophical Manuscripts of Marx cover a wide range of topics, including much interesting material on private property and communism, and on money, as well as developing Marx's critique of Hegel. However, the manuscripts are best known for their account of alienated labour. Here, Marx famously depicts the worker under capitalism as suffering from four types of alienated labour: (1) from the product, which as soon as it is created is taken away from its producer; (2) in productive activity (work) which is experienced as a torment; (3) from species-being, for humans produce blindly and not in accordance with their truly human powers; and finally (4), from other human beings, where the relation of exchange replaces the satisfaction of mutual need. Essentially he attempts to apply a Hegelian deduction of categories to economics, trying to demonstrate that all the categories of bourgeois economics—wages, rent, exchange, profit, etc.—are ultimately derived from an analysis of the concept of alienation. Consequently, each category of alienated labour is supposed to be deductible from the previous one. Marx combines the insights of both traditions to propose a view in which human beings do indeed create—or at least transform—the world they find themselves in, but this transformation happens not in thought but through actual material activity; not through the imposition of sublime concepts but through the sweat of their brow, with picks and shovels. This historical version of materialism, which transcends and thus rejects all existing philosophical thought, is the foundation of Marx's later theory of history. As per Marx's views on Economics, a commodity is defined as a useful external object, produced for exchange on a market. Thus two necessary conditions for commodity production are the existence of a market, in which exchange can take place, and a social division of labour, in which different people produce different products, without which there would be no motivation for exchange. In Marx's analysis labour power is the only commodity which can produce more value than it is worth, and for this reason it is known as variable capital. Other commodities simply pass their value on to the finished commodities, but do not create any extra value. They are known as constant capital. Profit, then, is the result of the labour performed by the worker beyond that necessary to create the value of his or her wages. This is the surplus value theory of profit. It appears to follow from this analysis that as industry becomes more mechanised, using more constant capital and less variable capital, the rate of profit ought to fall. For as a proportion less capital will be advanced on labour, and only labour can create value. Fundamentally, Marx assumed that human history involves transforming human nature, which encompasses both human beings and material objects. Humans recognise that they possess both actual and potential selves. For both Marx and Hegel, self-development begins with an experience of internal alienation stemming from this recognition, followed by a realisation that the actual self, as a subjective agent, renders its potential counterpart an object to be apprehended. Marx further argues that, by moulding naturein desired ways, the subject takes the object as its own, and thus permits the individual to be actualised as fully human. For Marx, then, human nature—Gattungswesen, or species-being—exists as a function of human labour. Fundamental to Marx's idea of meaningful labour is the proposition that, in order for a subject to come to terms with its alienated object, it must first exert influence upon literal, material objects in the subject's world.

Marx's thoughts on labour were related to the primacy he gave to the economic relation in determining the society's past, present and future (economic determinism). Accumulation of capital shapes the social system. Social change, for Marx, was about conflict between opposing interests, driven, in the background, by economic forces. This became the inspiration for the body of works known as the conflict theory. In his evolutionary model of history, he argued that human history began with free, productive and creative work that was over time coerced and dehumanised, a trend most apparent under capitalism. Marx noted that this was not an intentional process; rather, no individual or even no state can go against the forces of economy. The organisation of society depends on means of production. Literally those things, like land, natural resources, and technology, necessary for the production of material goods and the relations of production, in other words, the social relationships people enter into as they acquire and use the means of production. Together these compose the mode of production, and Marx distinguished historical eras in terms of distinct modes of production.

Marxism

Marxism is an economic and socio-political worldview and method of socio-economic inquiry that centres upon a materialist interpretation of history, a dialectical view of social change, and an analysis and critique of the development of capitalism. Marxism was pioneered in the early to mid-19th century by two German philosophers, Karl Marx and Friedrich Engels. Marx addressed the matters of alienation and exploitation of the working class, the capitalist mode of production, and historical materialism. He is famous for analysing history in terms of class struggle, summarised his Communist Manifesto (1848): "The history of all hitherto existing society is the history of class struggles". His ideas were influential in his time, and it was greatly expanded by the successful Bolshevik October Revolution of 1917 in Imperial Russia. Friedrich Engels was Marx's co-developer of communist theory. Marx and Engels met in 1844, discovering that they shared like views of philosophy and socialism. After Karl Marx's death, Friedrich Engels became the editor and translator of Marx's writings. Engels made intellectually significant contributions to feminist theory and Marxist feminism. Marxism encompasses Marxian economic theory, a sociological theory and a revolutionary view of social change that has influenced political movements around the world. The Marxian analysis begins with an analysis of material conditions, taking at its starting point the necessary economic activities required by human society to provide for its material needs. The form of economic organization, or mode of production, is understood to be the basis from which the majority of other social phenomena—including social relations, political and legal systems, morality and ideology—arise (or at the least by which they are greatly influenced). These social relations form the superstructure, for which the economic system forms the base. As the forces of production, most notably technology, improve, existing forms of social organization become inefficient and stifle further progress. These inefficiencies manifest themselves as social contradictions in society in the form of class struggle. Under the capitalist mode of production, this struggle materializes between the minority (the bourgeoisie) who own the means of production, and the vast majority of the population (the proletariat) who produce goods and services. Taking the idea that social change occurs because of the struggle between different classes within society who are under contradiction against each other, the Marxist analysis leads to the conclusion that capitalism oppresses the proletariat, which leads to a proletarian revolution. Capitalism according to Marxist theory can no longer sustain the living standards of the population due to its need to compensate for falling rates of profit by driving down wages, cutting social benefits and pursuing military aggression. The socialist system would succeed capitalism as humanity's mode of production through workers' revolution. In a socialist society private property in the means of production would be superseded by co-operative ownership. A socialist economy would not base production on the creation of private profits, but would instead base production and economic activity on the criteria of satisfying human needs—that is, production would be carried out directly for use. Eventually, socialism would give way to a communist stage of history: a classless, stateless system based on common ownership and free-access, superabundance and maximum freedom for individuals to develop their own capacities and talents. As a political movement, Marxism advocates the creation of such a society. A Marxist understanding of history and of society has been adopted by academics studying in a wide range of disciplines. The term 'Classical Marxism' is used to denote the theory propounded by Karl Marx and Friedrich Engels and it is based on the concepts of historical materialism, socialism and communism. Some variants of Marxism may be discussed as Trotskyism, Western Marxism, Structural Marxism, Autonomist Marxism, Marxist Humanism and Marxist Feminism.

Historical Materialism

The historical materialist theory of history, also synonymous to "the economic interpretation of history" looks for the causes of societal development and change in the collective ways humans use to make the means for living. The social features of a society (social classes, political structures, ideologies) derive from economic activity; "base and superstructure" is the metaphoric common term describing this historic condition. The base and superstructure metaphor explains that the totality of social relations regarding "the social production of their existence", i.e. civil society forms a society's economic base, from which rises a superstructure of political and legal institutions, i.e. political society. A conflict between the development of material productive forces and the relations of production provokes social revolutions, thus, the resultant changes to the economic base will lead to the transformation of the superstructure. This relationship is reflexive, the base determines the superstructure, in the first instance, and remains the foundation of a form of social organization which then can act again upon both parts of the base and superstructure, whose relationship is dialectical, not literal. Marx considered that these socio-economic conflicts have historically manifested themselves as distinct stages (one transitional) of development: (1) Primitive Communism: as in co-operative tribal societies; (2) Slave Society: a development of tribal progression to city-state; aristocracy is born; (3) Feudalism: aristocrats are the ruling class; merchants evolve into capitalists; (4) Capitalism: capitalists are the ruling class, who create and employ the proletariat; (5) Socialism: workers gain class consciousness, and via proletarian revolution depose the capitalist dictatorship of the bourgeoisie, replacing it in turn with dictatorship of the proletariat through which the socialization of the means of production can be realized; and (6) Communism: a classless and stateless society.

Socialism

Marxists believe that the transition from capitalism to socialism is an inevitable part of the development of human society. Marxists believe that a socialist society is far better for the majority of the populace than its capitalist counterpart. The term "socialism" could be used to describe two fundamentally different ideologies—democratic socialism and Marxist-Leninist socialism. The Marxist-Leninists sought to work towards the workers' utopia in Marxist ideology by first creating a socialist state, which historically had almost always been a single-party dictatorship. On the other hand, democratic socialists attempt to work towards an ideal state by social reform and are often little different from social democrats, with the democratic socialists having a more leftist stance. The Marxist-Leninist form of government has been in decline since the dissolution of the Soviet Union and its satellite states. Very few countries have governments which describe themselves as socialist. As of 2011, Laos, Vietnam, Nepal, Cuba, and the People's Republic of China had governments in power which describe themselves as socialist in the Marxist sense. On the contrary, electoral parties which describe themselves as socialist or democratic socialist are on the rise, joined together by international organizations such as the Socialist International and the Fourth International. Parties described as socialist are currently dominant in the democracies of the developing world and serve as the ruling party or the main opposition party in most European democracies. Eco-socialism, and Green politics with a strong leftist tinge, are on the rise in European democracies.

Communism

Communism is an ideology that seeks to establish a future classless, stateless social organization, based upon common ownership of the means of production and the absence of private property. It can be classified as a branch of the broader socialist movement. A number of states declared an allegiance to the principles of Marxism and have been ruled by self-described Communist Parties, either as a single-party state or a single list, which includes formally several parties, as was the case in the German Democratic Republic. Due to the dominance of the Communist Party in their governments, these states are often called "communist states" by Western political scientists. However, they have described themselves as "socialist", reserving the term "communism" for a future classless society, in which the state would no longer be necessary (on this understanding of communism, "communist state" would be an oxymoron)—for instance, the USSR was the Union of Soviet Socialist Republics. Communist governments have historically been characterized by state ownership of productive resources in a planned economy and sweeping campaigns of economic restructuring such as nationalization of industry and land reform (often focusing on collective farming or state farms.) While they promote collective ownership of the means of production, Communist governments have been characterized by a strong state apparatus in which decisions are made by the ruling Communist Party.

Trotskyism

Trotskyism is the theory of Marxism as advocated by Leon Trotsky[3]. Trotsky advocated proletarian revolution as set out in his theory of "permanent revolution", and he argued that in countries where the bourgeois democratic revolution had not triumphed already (in other words, in places that had not yet implemented a capitalist democracy, such as Russia before 1917), it was necessary that the proletariat make it permanent by carrying out the tasks of the social revolution (the "socialist" or "communist" revolution) at the same time, in an uninterrupted process. Trotsky believed that a new socialist state would not be able to hold out against the pressures of a hostile capitalist world unless socialist revolutions quickly took hold in other countries as well, especially in the industrial powers with a developed proletariat. On the political spectrum of Marxism, Trotskyists are considered to be on the left. They fervently support democracy, oppose political deals with the imperialist powers, and advocate a spreading of the revolution until it becomes global.

Western Marxism

Western Marxism is a term used to describe a wide variety of Marxist theoreticians based in Western and Central Europe (and more recently North America), in contrast with philosophy in the Soviet Union, the Socialist Federal Republic of Yugoslavia or the People's Republic of China.

Structural Marxism

Structural Marxism is an approach to Marxism based on structuralism, primarily associated with the work of the French theorist Louis Althusser and his students. In Althusser's theory, the structural order of the capitalist mode of production is distinct both from the actual, real agents involved in its relations and from the ideological forms in which those relations are understood. Structural Marxism was influential in France during the late 1960s and 1970s, and also came to influence philosophers, political theorists and sociologists outside of France during the 1970s.

Autonomist Marxism

Autonomism is a term applied to a variety of social movements around the world, which emphasizes the ability to organize in autonomous and horizontal networks, as opposed to hierarchical structures such as unions or parties. Autonomist Marxists, including Harry Cleaver, broaden the definition of the working-class to include salaried and unpaid labour, such as skilled professions and housework; it focuses on the working class in advanced capitalist states as the primary force of change in the construct of capital. Modern autonomist theorists such as Antonio Negri and Michael Hardt argue that network power constructs are the most effective methods of organization against the neoliberal regime of accumulation, and predict a massive shift in the dynamics of capital into a 21st century Empire.

Marxist Feminism

Marxist feminism is a sub-type of feminist theory which focuses on the dismantling of capitalism as a way to liberate women. Marxist feminism states that private property, which gives rise to economic inequality, dependence, political confusion and ultimately unhealthy social relations between men and women, is the root of women's oppression. According to Marxist theory, in capitalist societies the individual is shaped by class relations; that is, people's capacities, needs and interests are seen to be determined by the mode of production that characterises the society they inhabit. Marxist feminists see gender inequality as determined ultimately by the capitalist mode of production. Gender oppression is class oppression and women's subordination is seen as a form of class oppression which is maintained (like racism) because it serves the interests of capital and the ruling class. Marxist feminists have extended traditional Marxist analysis by looking at domestic labour as well as wage work in order to support their position.

Thus, Marxism is the philosophy, social theory and political practice based on the works of Karl Marx, a 19th century German socialist philosopher.[4] Marxism is centrally concerned with the ways in which the production of space, place and landscape is implicated in the reproduction of specific 'social formations'. It developed against capitalism[5]. The main concern of Marxism is uneven development, social justice and regional inequalities as a result of division and exploitation of labour resources and environment, through surplus value. In Marxist philosophy, economic classes and private property are the main cause of historical change. These two factors determine the man and environment relationship also. Marx writes: It is not consciousness (ideas) that determines life, but life that determines ideas. It is not the ideas that change the world, but the development of actual reality (space and place) which changes the ideas. Marx believed that society develops in stages in accordance with the development in the factors of production (materials and instruments of labour). In other words, consciousness develops with productivity, increase in needs, and with increase in the number of people. It emerges as a dynamic process because in transforming the bonds of our experience we transform ourselves. It develops specially with the division of labour, particularly the division between the 'material' and 'mental' labour. Hence, the key to the understanding of the structure of social life and consciousness is the mode of production of the material basis of that life and consciousness. A given society or a cultural landscape is organized on the basis of several modes of production. According to Marx, the man and environment relationship or the man and space relationship changes with the change in the mode of production. For example, from the stage of nomads, mankind turned to the stage of settled living, hunting, fruit gathering and sheep rearing gave way to domestication of plants and animals. Then came the urban culture and a rich diversity of vocations. For the first time economic classes within human societies were recognisable distinct, and the relationship between them formed the basis of their belief systems, social hierarchy, codes of behaviour, rules of punishment for crime and misconduct and institutionalised arrangements for the same, worship, recreation, family obligations and ties, the apparatus of ruling authority, etc. These constituted what is called the cultural landscape and the cultural pattern of society. The division of people into economic classes had the effect of initiating class struggle. Polarized class interests could only set class against class. Thus, there was a class of employees, another of employed. In familiar communist terminology, this would be called *exploiters* versus *exploited*. Thus, according to the Marxism, the key interaction between man and nature is labour. Man is confronted with a natural world that cannot be transcended (surpassed), and must be appropriated (possessed) in order to survive. His mode of appropriation is labour. Marxism refuses to accept that the scientific laws of society are eternal (which always existed and will always exist). This view, here, contrasts sharply with the claim of positivists who assert that scientific laws are universal and eternal in space and time. The essential difference between Positivism and Marxism is that Positivism simply seeks to understand the world whereas Marxism seeks to change it.

Marxist Geography

Marxist Geography is a strand of critical geography that uses the theories and philosophy of Marxism to examine the spatial relations of human geography. In Marxist geography, the relations that geography has traditionally analyzed—natural environment and spatial relations—are reviewed as outcomes of the mode of material production. To understand geographical relations, on this view, the social structure must also be examined. Marxist geography attempts to change the basic structure of society. Marxist geography is radical in nature and its primary criticism of the positivist spatial science centred on the latter's methodologies, which failed to demonstrate or account for the mechanisms of capitalism and exploitation that underlie human spatial arrangements. As such, early Marxist geographers were explicitly political in advocating for social change and activism; they sought, through application of geographical analysis of social problems, to alleviate poverty and exploitation in capitalist societies. Marxist geography makes exegetical claims regarding how the deep-seated structures of capitalism act as a determinant and a constraint to human agency. Most of these ideas were developed in the early 1970s by dissatisfied quantitative geographers; David Harvey is generally regarded as the primary trail-blazer of the Marxist movement in human geography. In order to accomplish such philosophical aims, these geographers rely heavily upon Marxist social and economic theory, drawing on Marxian economics and the methods of historical materialism to tease out the manner in which the means of production control human spatial distribution in capitalist structures. Marx is also invoked to examine how spatial relationships are affected by class. The emphasis is on structure and structural mechanisms. This emphasis has yielded results but also criticism.

Marxist geography attempts to explain the world and also to change it. Marxism sees human beings gradually transforming themselves from stage to stage until they reach social perfection, and this transformation is seen as an aim towards which societies should be moving. This change is brought about by 'dialectical' processes—conflict between opposing forces—bringing forth a new synthesis which again is contradicted, and so on. The forces shaping society are seen as entities, which include capital, labour, capitalism and other modes of production, the state, class, society, and the market. Marxist geography highlights the dialectical relationships between social processes and the natural environment and spatial relationships. It is concerned with the modes of production which underlie the superstructure of society. It sees spatial and environmental problems, such as the destruction of habitats or uneven development, originating deep within the social formations of capitalism. It is aimed at changing the fundamental operations of social processes by changing the workings of production. Marxist geography is the study of the inherent contradictions of capitalism as they appear in the landscape and as they relate to each other. Examples may be given of the changing structures and contradictions of capitalism. There is an inherent contradiction in a capitalist state which seeks to generate better conditions funded by taxation. Higher demand for raw materials generates higher costs. More output leads to more pollution and, in an environment where the authorities require strict controls on pollution, costs again rise. The result of these contradictions is the movement by multinational firms from established industrial regions in search of new environments to develop (despoil) and of new, politically virgin, labour to hire (exploit).

Marxist geography has been widely criticized: for the ethnocentric and patriarchal nature of its assumptions, for the passive role apportioned to individuals, who have been turned into non-decision-makers, for its preoccupation with class and industry and its neglect of environmental issues, and by postmodernists, who reject overarching meta-theory.

NOTES

1. Friedrich Engels (1820-1895) was a German-English industrialist, social scientist, author, political theorist, philosopher, and father of Marxist theory, alongside Karl Marx.
2. Georg Wilhelm Friedrich Hegel (1770-1831) was a German philosopher, one of the creators of German Idealism. His historicist and idealist account of reality as a whole revolutionized European philosophy and was an important precursor to Continental philosophy and Marxism.
3. Leon Trotsky (1879-1940), born Lev Davidovich Bronshtein, was a Russian Marxist revolutionary and theorist.
4. Husain, 1995, 380-86.
5. Capitalism generally refers to an economic system in which the means of production are mostly privately owned and operated for profit, and in which distribution, production and pricing of goods and services are determined in a largely free market. It is usually considered to involve the right of individuals and groups of individuals acting as "legal persons" or corporations to trade capital goods, labour and money. Capitalism has emerged as the Western world's dominant economic system since the decline of feudalism. Since the Industrial Revolution, capitalism gradually spread from Europe, particularly from Britain, across political and cultural frontiers. In the 19th and 20th centuries, capitalism provided the main means of industrialization throughout much of the world.

REFERENCES

Avineri, Shlomo (1968), *The Social and Political Thought of Karl Marx*. Cambridge University Press.

Craig J. Calhoun (2002), *Classical Sociological Theory*. Wiley-Blackwell.

Dikshit, R.D. (1997), *Geographical Thought: A Contextual History of Ideas*, Prentice Hall of India, pp. 165-168.

Harvey, M.E. and B.P. Holly (1989), *Themes in Geographical Thought*, Rawat Publications, Jaipur and New Delhi.

Husain, M. (1995), *Evolution of Geographical Thought*, Third Revised Edition, Rawat Publications, Jaipur.

Jon Elster, (1986), *An Introduction to Karl Marx*. Cambridge, England.

"Karl Marx—Stanford Encyclopaedia of Philosophy".

McLellan, David (1973), *Karl Marx: His life and Thought*. New York: Harper and Row.

McLellan, David (2007), *Marxism After Marx*. Basingstoke: Palgrave MacMillan.

Oxford Dictionary of Geography.

Peet, R. (1998), "Radical Geography, Marxism and Marxist Geography" in *Modern Geographical Thought*, Blackwell, pp. 67-111.

Social Science: Marxist influences—Britannica Online Encyclopedia.

Soja, E.W. (1997), "Marxist Geography and Critical Social Theory" in *Postmodern Geographies*, Rawat Publications, pp. 43-75.

Stanford *Encyclopedia of Philosophy.*

Tom Rockmore (2002), *Marx after Marxism: The Philosophy of Karl Marx*. John Wiley and Sons.

Wheen, Francis (2002), *Karl Marx: A Life*. New York: Norton.

Wikipedia, *The Free Encyclopedia.*

CHAPTER-13

Feminist Geography

Gender is a particularly important facet of identity that refers to the social differences between men and women rather than biological differences related to sex. We use the term 'gender' to refer to socially created distinction between faminity and masculinity, while the term 'sex' is used to refer to biological differences between men and women. However, when we talk about the geography of gender, we generally refer to feminist geography. Feminism, as such, is a diverse collection of social theories, political movements, and moral philosophies, largely motivated by or concerned with the experiences of women. Feminism aims at defining, establishing, and defending equal political, economic, and social rights for women. In addition, feminism seeks to establish equal opportunities for women in education and employment. A feminist is a "person whose beliefs and behaviour are based on feminism." The feminists are especially concerned with social, political and economic inequality between men and women; whereas gendered identities of 'men' and 'women' are socially constructed. Feminists differ over the sources of inequality, how to attain equality, and the extent to which gender and sexual identities should be questioned and critiqued. In simple terms, feminism is the belief in social, political and economic equality of the sexes and the movement organised around this belief. One illustration of the feminst approach to geography is to study gender differences in terms of personal access, mobility and safety, especially in respect to the design and use of urban space and open places such as public parks and footpaths. The experience of walking alone through a city centre late at night or Public Park even in daytime is likely to be different for males and females and also for people of different ages, cultures and so on.

Historical Perspective

Depending on historical moment, culture and country, feminists around the world have had different causes and goals. The history of the modern western feminist movements is divided into three "waves".[1] Each wave dealt with different aspects of the feminist issues. The first wave comprised women's suffrage movements of the 19th and early 20th centuries, promoting women's right to vote. The second wave was associated with the ideas and actions of the women's liberation movement beginning in the 1960s. This wave campaigned for legal and social equality for women. The third wave is a continuation of, and a reaction to, the perceived failures of second-wave feminism, beginning in the 1990s. First-wave feminism was a period of activity during the 19th and early 20th century in the United Kingdom and United States. It focused on the promotion of equal contract, marriage, parenting and property rights for women. Roots of the movement in US and UK include the Women's Suffrage movement of the early 1900s and the Women's Liberation (or "Second Wave Feminist") movement of the 1960s and 1970s. The Equal Rights Amendment, which proponents claimed would address the inadequacies of the Fourteenth Amendment concerning women and citizenship, was proposed in the US in 1923. In the United States, first-wave feminism is considered to have ended with the passage of the Nineteenth Amendment to the United States Constitution (1919), granting women the right to vote in all states. The term first wave was coined retroactively to categorize these western movements after the term second-wave feminism began to be used to describe a newer feminist movement that focused as much on fighting social and cultural inequalities as political inequalities. The feminist movement in the West evolved in the 1980s with the rise of so-called Post-Feminism (also called "Third-Wave" feminism), which stresses that women have many rights that go unrecognized, often by women themselves, in everyday life, and in the American legal structure. Second-wave feminists see women's cultural and political inequalities as inextricably linked and encourage women to understand aspects of their personal lives as deeply politicized and as reflecting power structures. Second and third-wave feminism in China has been characterized by a re-examination of women's roles during the communist revolution and other reform movements, and new discussions about whether women's equality has actually been fully achieved. In Latin America, revolutions brought changes in women's status in countries such as Nicaragua, where feminist ideology during the Sandinista Revolution aided women's quality of life but fell short of achieving a social and ideological change. In the early 1990s in USA, third-wave feminism began as a response to perceived failures of the second wave and to the backlash against initiatives and movements created by the second wave. Third-wave feminism seeks to challenge or avoid what it deems the second wave's essentialist definitions of femininity, which, they argue, over-emphasize the experiences of upper middle-class white women. Third-wave feminists often focus on "micro-politics" and challenge the second wave's paradigm as to what is, or is not, good for women, and tend to use a post-structuralist interpretation of gender and sexuality. Feminist leaders rooted in the second wave, such as Gloria Anzaldua, Chela Sandoval, Cherrie Moraga, AudreLorde, Maxine Hong Kingston, and many other black feminists, sought to negotiate a space within feminist thought for consideration of race-related subjectivities. Since, the 1980s standpoint feminists have argued that the feminist movement should address global issues and culturally specific issues in order to understand how gender inequality interacts with racism, homophobia, classism and colonization in a "matrix of domination." Third-wave feminism also contains internal debates between difference feminists, who believe that there are important differences between the sexes, and those who believe that there are no inherent differences between the sexes and contend that gender roles are due to social conditioning. The term post-feminism is used to describe a range of viewpoints reacting to feminism since the 1980s.

Feminist Theory

Feminist theory, which emerged from these feminist movements, aims to understand the nature of gender inequality by examining women's social roles and lived experience. Feminist activists campaign for women's rights—such as in contract law, property, and voting. Feminist campaigns have changed societies, particularly in the West. Feminist theory is the extension of feminism into theoretical or philosophical fields. It encompasses work in a variety of disciplines, including anthropology, sociology, economics, history and philosophy. Feminist theory aims to understand gender inequality and focuses on gender politics, power relations, and sexuality. While providing a critique of these social and political relations, much of feminist theory also focuses on the promotion of women's rights and interests. Themes explored in feminist theory include discrimination, stereotyping, objectification (especially sexual objectification), oppression and patriarchy[2]. Many overlapping feminist movements and ideologies have developed over the years.

Some branches of feminism closely track the political leanings of the larger society, such as liberalism and conservatism, or focus on the environment. Liberal feminism seeks individualistic equality of men and women through political and legal reform without altering the structure of society. Radical feminism considers the male-controlled capitalist hierarchy as the defining feature of women's oppression and the total uprooting and reconstruction of society as necessary. Conservative feminism is conservative relative to the society in which it resides. Libertarian feminism conceives of people as self-owners and, therefore, as entitled to freedom from coercive interference. Ecofeminists see men's control of land as responsible for the oppression of women and destruction of the natural environment. The materialist feminism grew out of Western Marxist Thought and have inspired a number of different (but overlapping) movements, all of which are involved in a critique of capitalism and are focused on ideology's relationship to women. Marxist feminism argues that capitalism is the root cause of women's oppression, and that discrimination against women in domestic life and employment is an effect of capitalist ideologies. Socialist feminism distinguishes itself from Marxist feminism by arguing that women's liberation can only be achieved by working to end both the economic and cultural sources of women's oppression.

During much of its history, feminist movements and theoretical developments were led predominantly by middle-class white women from Western Europe and North America. However, women of other races have proposed alternative feminisms. This trend accelerated in the 1960s with the civil rights movement in the United States and the collapse of European colonialism in Africa, the Caribbean, parts of Latin America, and Southeast Asia. Since, that time, women in developing nations and former colonies and who are of colour or various ethnicities or living in poverty have proposed additional feminisms. Postcolonial feminists argue that colonial oppression and Western feminism marginalized postcolonial women but did not turn them passive or voiceless. Third-world feminism is closely related to post-colonial feminism.

In the late 20th century various feminists began to argue that gender roles are socially constructed, and that it is impossible to generalize women's experiences across cultures and histories. Post-structural feminism draws on the philosophies of post-structuralism and deconstruction in order to argue that the concept of gender is created socially and culturally through discourse. Post-modern feminists also emphasize the social construction of gender and the discursive nature of reality.

Feminist theology is a movement that reconsiders the traditions, practices, scriptures, and theologies of religions from a feminist perspective. Some of the goals of feminist theology include increasing the role of women among the clergy and religious authorities, reinterpreting male-dominated imagery and language about God, determining women's place in relation to career and motherhood, and studying images of women in the religion's sacred texts. Christian feminism is a branch of feminist theology which seeks to interpret and understand Christianity in light of the equality of women and men, and that this interpretation is necessary for a complete understanding of Christianity. While there is no standard set of beliefs among Christian feminists, most agree that God does not discriminate on the basis of sex, and are involved in issues such as the ordination of women, male dominance and the balance of parenting in Christian marriage, claims of moral deficiency and inferiority of women compared to men, and the overall treatment of women in the church. Islamic feminists advocate women's rights, gender equality, and social justice grounded within an Islamic framework. Advocates seek to highlight the deeply rooted teachings of equality in the *Quran* and encourage a questioning of the patriarchal interpretation of Islamic teaching through the *Quran*, *Hadith* (sayings of Muhammad), and *Sharia* (law) towards the creation of a more equal and just society. Although rooted in Islam, the movement's pioneers have also utilized secular and Western feminist discourses and recognize the role of Islamic feminism as part of an integrated global feminist movement. Jewish feminism is a movement that seeks to improve the religious, legal, and social status of women within Judaism and to open up new opportunities for religious experience and leadership for Jewish women.

Generally, the distinction is sought between 'radical feminism' and 'socialist feminism'. Linking the root cause of women oppression in patriarchal gender relations falls under radical feminism, whereas the socialist feminism attempts to link gender relations to the wider framework of social relations existing in society as a whole. Radical feminism is a branch of feminism that views women's oppression (or patriarchy) as the basic system of power upon which human relationships in society are arranged. It seeks to challenge this arrangement by rejecting standard gender roles and male oppression. Most forms of feminism characterize patriarchy as an unjust social system that is oppressive to women. As the feminist and political theorist Carole

Pateman writes: "The patriarchal construction of the difference between masculinity and femininity is the political difference between freedom and subjection."[3] In feminist theory the concept of patriarchy often includes all the social mechanisms that reproduce and exert male dominance over women. The term 'Militant feminism' is a pejorative term which is often associated, usually by detractors, with radical feminism. Often, radical feminism is seen by people other than adherents as a form of identity politics. The term 'radical' in radical feminism is used as 'an adjective meaning of or pertaining to the root or going to the root'. Radical feminists locate the root cause of women's oppression in patriarchal gender relations. The root cause of the oppression of women in society is traced to the system of 'patriarchy', which functions in favour of male domination over females. The radical feminism considers the subordination of women by men as the primary inequality in human society. It is argued that the division of labour between men and women had proceeded (and given birth to) the division of labour between class and race. The radical feminists insist that elimination of sexual oppression (subordination) should form the cornerstone of attempts to eliminate other types of oppressions in society. From such a perspective, men and women appear to form somewhat antagonist classes irrespective of divisions on either side on the basis of age, race and income status. The adherents to radical feminism emphasize that male domination of the man-woman relationship has resulted in restricting women largely to the private sphere where they are made to carry out many personal services for men—services that remain largely unreciprocated—apart from having to carry the burden of child rearing. Patriarchy, therefore, works to ensure that the vast majority of women remain legally and economically dependent on their men. This process of subordination through patriarchy begins early in life so that little girls are allowed far less freedom of movement than boys of the same age, and the process becomes all the more visible at the adolescent stage. Radical feminists appear united in viewing men as the enemy in their struggle for women's liberation; they insist on examining women's and men's behaviour in terms of interrelationships between the two genders, setting aside other types of cleavages in society. The Socialist feminism, on the other hand, is a branch of feminism that focuses upon both the public and private spheres of a woman's life and argues that liberation can only be achieved by working to end both the economic and cultural sources of women's oppression. The social feminist viewpoint represents the human face of the left movement. Social feminists differ from their redical colleagues in their emphasis on patriarchy as the root cause of all other societal inequalities in relation to women, and they reject the notion that 'men are the enemy' and favour unity between men and women in the struggle between classes, though they continue to focus on female subordination and the ways and means of alleviating it. Thus, social feminism represents the welfare stream of radical feminism. One of the ways to achieve this goal is through women empowerment. Since the 1970s feminist writings have shed considerable light on the subjects of women's empowerment. The term empowerment has a contextual connotation. Empowerment gives to the expression to emergence of creativity, self-reliance, the blossoming of thoughts drawing on both external and internal sources to secure self-determination, a productive self-improvement and individualistic identity and personality for the women. The faminst writers have identified two broad strategies of empowerment 'power wihin' and 'power with'. 'Power within' refers to affirmation of the individual. It may come out through education getting the skills and expertise. 'Power with' refers to collective action, the organization and mobilization of self-confident individuals.

Feminist Geography

Human geographers have variously defined geography as a spatial science yet a definition of space universally acceptable to all of them remained elusive. From the earliest times until the 1970s geographers understood space as physically real, measurable and definable—a grid within which objects were located and events occurred. For spatial scientists, space is a key variable in shaping the organisation of human activity. Feminist geographers have focused the ways in which space is gendered. They believe that under patriarchy, ground rules determine how men and women can use space. They examine the ways in which gender relations are expressed in space. The lives of women and men differ because the lives of women revolve around private space—the realm of house and family, while men's lives revolve around public space—the space of money, power and institutions. This does not simply make women's lives different from men's lives, but the relegation of women to private space and reproductive labour greatly limits the opportunities available to them and makes them subservient to and dependent on men. Men have long dominated women, and like those of other dominated groups the voices of women have remained silent. Most of what we know has come from the perception of those who do the domination. History books are full of the exploits of men. Men are leaders and statesmen; they are the winners of war and builders of empires, the philosophers and thinkers, founders of religion, the explorers and discoverers, the scientists and inventors. The musicians, artists, painters and authors too have mostly been men. In short, men have always dominated public life. Women are totally missing from the scene of activity and therefore, also from the pages of history, making history the story of men—in the literal sense of the word. Women, who constitute one half of mankind, are relegated to the background. What is true of history is true of every other academic discipline. Since, men founded them all, and their tradition was, at least till recently, carried forward almost exclusively by men, the female worldview was either neglected totally or greatly marginalised.

Feminist geography is an approach in human geography which applies the theories, methods and critiques of feminism to the study of the human environment, society and geographical space. Its principal focus of study is geographic differences in gender relations and gender equality. In addition to its analysis of the real world, it reflects on the geographical nature of the female experience. In this respect feminist geography is part of a broader post-modern approach to the subject which is not primarily concerned with the development of conceptual theory in itself but rather focuses on the real experiences of individuals and groups in their own localities, upon the geographies that they live in within their own communities. However, feminist geography is not limited to the local scale. One example of a global topic of feminist geography research is the worldwide migration of women from the Third World to perform domestic labour and various other jobs. The term feminist geography subsumes perspectives on feminist politics and the theories that explore gender relations in society. The interest in the study of feminism dates back to the mid-1970s. It drew inspiration from the women's movements in the West in the 1960s as part of the radicalization of societal perspectives. As the Women and Geography Study Group of Institute of British Geographers put it, "In common with other approaches in geography which are critical of mainstream work", feminist geography tries to "analyse and understand why women remain in subordinate position" in society and the professions.[4] Seen in this light, feminist geography may be defined as "a geography which explicitely takes into account the socially created gender structure of society; and in which a commitment, both towards the alleviation of gender inequality in the short term and towards its removal, through social change towards real equality, in the long term, is expressed." A feature common to all feminist argument is that the differences in social positions of men and women systematically work to the advantage of men so that women and men in fact have unequal power, opportunities, and social status. Feminist geography aims at exploring the ways in which current practices in society might be changed in order to release women from the state of subordination. In addition to societal studies, Feminist Geography also critiques Human Geography and other academic disciplines, arguing that academic structures have been traditionally characterized by a patriarchal perspective, and that contemporary studies which do not confront the nature of previous work reinforce the masculine bias of academic study.

It was not until the late 1980s that feminist geography became well established as a sub-discipline of geography—several years after it was established in other social sciences. Up until then, the research agenda being set by men, geography examined only issues that were of interest of men, vowing masculine rationality to derive 'universally applicable' laws. What passed as human geography was really male geography and the geography of the world was really geography of the male world. The male viewpoint thought it appropriate to explore only the public spaces, which were dominated by men. Private spaces, the realm of women on account of their association with reproductive labour, held little interest for male geographers. Feminist geographers developed a more 'progressive' and 'engaged' form of geographical enquiry, bringing a new theoretical agenda to the subject. They undertook a reinterpretation of the uses and structures of specific places, reappraised methodological issues in human geography and presented a critique of underlying assumptions within geographical theory. Feminist geography does not concern itself only with women's issues, nor is it geography of female world. It takes into account all the processes that contribute to the constitution and reproduction of gender. Feminist geographers have tried to understand the spatial dimensions of gendered division of labour and its effects on women's well-being by asking 'who works where?' rather than 'where work was taken place?'

One of the major contributions of feminist geographers has been to the study of space. Feminist geographers have insisted that geography holds a series of unstated assumptions about men and women do, and that the discipline concentrates on spaces, places and landscapes that it sees as men's. They contend that a complete and comprehensive understanding of human cannot be achieved without exploring private space where a large part of our daily lives are lived. Private space is the domain of domestic and reproductive labour, which is as important as public space, the domain of productive labour. Traditional andro-centric geography has chosen to focus on the latter, to the exclusion of the former. The association between men and labour markets, social and political institutions was taken for granted, the absence of women from them envinced no interest. Feminist geographers have not only expanded the reach of geographical enquiry, they have challenged conventional explanations and altered certain assumptions made about human society[5]. They have focused on the exclusion of women from public spaces of waged employment and their segregation into a particular range of occupations identified as feminine as well as their absence of the portals of power and positions of decision-making. It was neither the lack of ability nor the desire that kept women out of these spaces. Rather, it was the gender relations that are fundamental part of social structure that contrived to effect this exclusion. Ground rules laid down by patriarchy construct social, economic and political structures which influence the ways in which men and women form their own gender identities. Just as social roles are gendered, patriarchy also constructs certain spaces as feminine as others as masculine.

NOTES

1. 'Feminism', Wikipedia, *The Free Encyclopedia.*
2. Patriarchy is a social system in which the role of the male as the primary authority figure is central to social organization, and where fathers hold authority over women, children, and property. It implies the institutions of male rule and privilege, and is dependent on female subordination.
3. MacMillan, *Encyclopedia of Sex and Gender,* 1104.
4. Dikshit, R.D. (1997), 263-69.
5. Tewari, 2010, 33.

REFERENCES

Dikshit, R.D. (1997), *Geographical Thought: A Contextual History of Ideas,* Prentice Hall of India.

Gender, *Place and Culture—A Journal of Feminist Geography.* Routledge.

Johnston, R.J. (2004), *Geography and Geographers: Anglo-American Human Geography Since 1945,* 6th Ed., Edward Arnold, London.

MacMillan, *Encyclopedia of Sex and Gender.*

McDowell, Linda (1993), "Space, place and gender relations" in *Progress in Human Geography,* 17(2).

Peet, R. (1998), *Modern Geographical Thought,* Blackwell.

Rose, Gillian (1993), *Feminism and Geography: The Limits of Geographical Knowledge,* University of Minnesota Press.

Stanford, *Encyclopedia of Philosophy.*

Tewari, Preeti (2010), *Division of Work and Gender Space in Delhi,* Ph.D. research work submitted at the Department of Geography, University of Delhi, pp. 1-55.

Wikipedia, *The Free Encyclopedia.*

CHAPTER-14

Positivism

Positivism is the belief that the scientific method is the best approach to discover knowledge about things or humans. In other words, all true knowledge is scientific. More narrowly, the term designates the philosophical and religious doctrines elaborated by Auguste Comte[1], who held that human thought had passed inevitably through a theological stage into a metaphysical stage and was passing into a positive, or scientific, stage. As a philosophical system or method, Positivism denies the validity of metaphysical speculations, and maintains that the data of sense experience are the only object and the supreme criterion of human knowledge. In the original Comtean usage, the term meant the use of scientific methods to uncover the laws according to which both physical and human events occur. The basic principle of Positivism is that all factual knowledge is based on the "positive" information gained from observable experience, and that any ideas beyond this realm of demonstrable fact are metaphysical. Only analytic statements are allowed to be known as true through reason alone. For instance, 'roses are flowers' is analytic, whilst 'roses are fragrant' is synthetic and requires evidence. Positivism seeks empirical regularities. It is based on the view that in the social as well as natural sciences, data derived from sensory experience, and logical and mathematical treatments of such data, are together the exclusive source of all authentic knowledge. Obtaining and verifying data that can be received from the senses is known as empirical evidence. This view holds that society operates according to laws like the physical world. Introspective and intuitional attempts to gain knowledge are rejected. One extended form of Positivism is Logical Positivism, promoted by the 'Vienna Circle'[2]. Logical Positivism places particular emphasis on sense experience and observation and attempted to eradicate metaphysics and synthetic statements. Logical Positivists include early Wittgenstein, Bertrand Russell and Alfred Whitehead and Rudolph Carnap. Logical positivism attempted to introduce the methodology and precision of mathematics and the natural sciences into the field of philosophy.

Positivism claims to be a fundamentally non-metaphysical philosophy, modelled on empirical sciences and providing them with a methodology. The basic theme of positivism is that in scientific discipline, knowledge is based on the facts that can be observed and the relationships among these facts. The science should concern itself only with empirical questions, i.e. how things are in reality. The real world is the world that can be sensed actually (and is not mind-based). The reliable knowledge can come only from basic observation of actual conditions. The philosophy of positivism is anti-idealism. Since, the positivists believe that the moral norms like attitudes, beliefs, prejudices, customs, traditions, taste, aesthetics, etc. cannot be tested, measured or investigated, such aspects should be kept out of scientific disciplines. The six tenets of Positivism are: (1) *Naturalism*: the principles of the natural sciences should be used for social science; (2) *Phenomenalism*: only observable phenomena provide valid information; (3) *Nominalism*: words of scientific value have fixed and single meanings and the existence of a word does not imply the existence of what it describes; (4) *Atomism*: things can be studied by reducing them to their smallest parts (and the whole is the sum of the parts); (5) *Scientific laws*: the goal of science is to create generalised laws (which are useful for such as prediction); and (6) *Facts and values*: facts are to sought and values have no meaning for science.

In its strongest original formulation, positivism could be thought of as a set of five principles: (1) The unity of the scientific method—i.e., the logic of inquiry is the same across all sciences (social and natural); (2) The aim of science is to explain and predict; (3) Scientific knowledge is testable. Research can be proved only by empirical means, not arguments alone; (4) Science does not equal common sense. Researchers must be careful not to let common sense bias their research; and (5) Science should be as value-neutral as possible. The ultimate goal of science is to produce knowledge, regardless of any politics, morals, or values held by those involved in the research. Science should be judged by logic.

History

Also referred to as 'empiricism' or 'comtism', positivism is actually the philosophical system of Auguste Comte, a French Social Philosopher. It recognizes only positive facts and observable phenomena and rejects metaphysics and theism[3]. As an approach to the philosophy of science, positivism was first systematically theorized by Comte, who saw the scientific method as replacing metaphysics in the history of thought, and who observed the circular dependence of theory and observation in science. Positivism emerged in response to the inability of speculative philosophy (e.g. Classical German Idealism) to solve philosophical problems which had arisen as a result of scientific development. Positivists went to an opposite extreme and rejected theoretical speculation as a means of obtaining knowledge. Positivism declared false and senseless all problems, concepts and propositions of traditional philosophy on being, substances, causes, etc., that could not be solved or verified by experience due to a high degree of abstract nature. Although the positivist approach has been a "recurrent theme in the history of western thought from the Ancient Greeks to the present day", the concept was developed in the early 19th century by the philosopher and founding sociologist,

Auguste Comte. Positivism took shape as a distinct trend in the 1830s. During its history of more than a century, positivism has evolved steadily towards expressing more clearly and carrying to a logical conclusion its inherent tendency towards subjective idealism. Historically, the concept of positivism has emerged after French Revolution. It originated out of the French Enlightenment. At the time of Comte, science was having a huge impact and was steadily replacing religion as the key authority for knowledge about what was true or false. When something is pronounced scientific then it is generally held to be verifiable. The roots of Positivism lie particularly with Empiricism, which works only with observable facts, seeing that beyond this is the realm of logic and mathematics. The positivism movement broke a range of taboos and religious beliefs against empirical investigations. Especially in the latter half of 19th century, positivism had a considerable influence on the natural and social sciences.

Auguste Comte first described the epistemological perspective of positivism in *"Cours de Philosophie Positive"* (*The Course in Positive Philosophy*), a series of texts published between 1830 and 1842. These texts were followed by the 1844 work, *A General View of Positivism* (published in French 1848, English in 1865). The first three volumes of the *Course* dealt chiefly with the physical sciences already in existence (mathematics, astronomy, physics, chemistry, biology), whereas, the latter two emphasized the inevitable coming of social science. Observing the circular dependence of theory and observation in science, and classifying the sciences in this way, Comte may be regarded as the first philosopher of science in the modern sense of the term. Comte believed that: "all scientific knowledge must be based on direct experience of immediate reality, since direct observation is the surest guarantee that the knowledge acquired is scientific." Comte had put forward his ideas in the form of private lectures in his apartment at Paris. Interestingly, the lectures were attended among others by Humboldt (then nearly 60 years in age). In fact, Comte's lectures were published in the form of *"Cours de Philosophie Positive"*.

Percepts of Comte's Philosophy of Positivism

- The direct experience of reality should be complemented by *"la certilude"*, i.e. the unity of scientific method. This implies that the different branches of knowledge are distinguished by their object of study (i.e. subject matter) and not their method. In other words, sciences differ from one another in 'what they study' rather than 'how they study'.
- Positivist view of science incorporated the principle of *"l'utile"*, which means that all scientific knowledge must serve some useful purpose—it should be utilizable, it should be a means to an end, and a tool for social engineering.
- The concept of unity of scientific method requires *"le précis"*, i.e. a common scientific goal for formulating testable theories. This implies that there is no place for subjective value judgements in scientific enquiry, since being based on ethical assertions, value judgements are not products of scientific observations, and are as such, not verifiable.
- The next percept is *"le relative"*, which means that scientific knowledge is

essentially unfinished and relative, because knowledge keeps progressing by gradual unification of theories that in turn enhance man's awareness of social laws.

Comte offered an account of social evolution, proposing that society undergoes three phases in its quest for the truth according to a general 'law of three stages'. The idea bears some similarity to Marx's view that human society would progress towards a communist peak. Both Comte and Marx intended to develop secular-scientific ideologies in the wake of European secularisation. Comte's stages were: (1) the *theological,* (2) the *metaphysical,* and (3) the *positive.* The theological phase of man was based on whole-hearted belief in all things with reference to God. God, Comte says, had reigned supreme over human existence pre-Enlightenment. Humanity's place in society was governed by its association with the divine presences and with the church. The theological phase deals with humankind's accepting the doctrines of the church (or place of worship) rather than relying on its rational powers to explore basic questions about existence. It dealt with the restrictions put in place by the religious organization at the time and the total acceptance of any "fact" adduced for society to believe. Comte describes the metaphysical phase of humanity as the time since the Enlightenment, a time steeped in logical rationalism, to the time right after the French Revolution. This second phase states that the universal rights of humanity are most important. The central idea is that humanity is invested with certain rights that must be respected. In this phase, democracies and dictators rose and fell in attempts to maintain the innate rights of humanity. The final stage of the trilogy of Comte's universal law is the scientific, or positive, stage. The central idea of this phase is that individual rights are more important than the rule of any one person. Comte stated that the idea of humanity's ability to govern itself makes this stage innately different from the rest. There is no higher power governing the masses and the intrigue of any one person can achieve anything based on that individual's free will and authority. The third principle is most important in the positive stage. Comte calls these three phases the universal rule in relation to society and its development. Neither the second nor the third phase can be reached without the completion and understanding of the preceding stage. All stages must be completed in progress. Comte believed that the appreciation of the past and the ability to build on it towards the future was key in transitioning from the theological and metaphysical phases. The idea of progress was central to Comte's new science, sociology. Sociology would "lead to the historical consideration of every science" because "the history of one science, including pure political history, would make no sense unless it was attached to the study of the general progress of all of humanity". As Comte would say: "from science comes prediction; from prediction comes action." It is a philosophy of human intellectual development that culminated in science. The irony of this series of phases is that though Comte attempted to prove that human development has to go through these three stages, it seems that the positivist stage is far from becoming a realization. This is due to two truths. The positivist phase requires having a complete understanding of the universe and world around us and requires that society should never know if it is in this positivist phase.

At the end of the 19th century, positivism went through a crisis caused by the radical breakdown of many concepts in physics at the turn of the century, as well as by progress in the natural sciences, which cancelled out or reduced the importance of many of the "synthetic" generalizations viewed by positivism as eternal and unquestionable attainments of science. The crisis in positivism was promoted by the intensive development of research in psychology, which forced scientists to analyze "ultimate" philosophical questions of knowledge—the very questions that positivism had always done its utmost to avoid. In addition, the crisis was fostered by the failure of all attempts to find the objective basis for the positivist system of values in mechanistic and metaphysical sociology. (The positivist criterion of what is scientific makes it impossible to introduce the consideration of values into scientific research and to deduce "what should be" from "what is".) As a result of all of these developments, it became necessary to reassert the question of philosophy's role in the sciences. Transformed, positivism entered a new, second stage of its evolution with the appearance of Machism (empirio-criticism). The Machist trend has been further developed in the current, or third stage in the evolution of positivism-neopositivism, which emerged during the 1920s. Retreating from the attempt to solve fundamental philosophical problems, neopositivism concentrates on concrete logical and methodological research on language or immediate experience.

Thus, historically, there are three stages in the development of positivism. The exponents of the first were Comte, E. Littré and P. Laffitte in France, J.S. Mill and Herbert Spencer in England. Alongside the problems of the theory of knowledge (Comte) and logic (Mill), the main place in the first Positivism was assigned to sociology (Comte's idea of transforming society on the basis of science, Spencer's organic theory of society). The rise of the second stage in Positivism-empirio-criticism dates back to the 1870s-1890s and is associated with Ernst Mach and Avenarius, who renounced even formal recognition of objective real objects, which was a feature of early Positivism. In Machism, the problems of cognition were interpreted from the viewpoint of extreme psychologism, which was merging with subjectivism. The rise and formation of the latest Positivism, or neo-positivism, is linked up with the activity of the Vienna Circle (O. Neurath, Carnap, Schlick, Frank and others). The main place in the third positivism is taken by the philosophical problems of language, symbolic logic, the structure of scientific investigations, and others. Having renounced psychologism, the exponents of the third positivism took the course of reconciling the logic of science with mathematics, the course of formalisation of epistemological problems.

Logical Positivism

Logical positivism was an important development in early 20th century Vienna, led by Moritz Schlick[4]. It held that not just that propositions were true or false, but that statements which could not be verified (shown to be either true or false) were meaningless. Thus statements like "There is a God" or "There is no God" are not true or false, but meaningless. This is the original view of positivism that had dominated scientific thinking for almost a century. As a further development of this philosophy in 1920s there emerged 'Logical Positivism' (or Logical Empiricism), a new idea—a body of principles developed by a group of philosophers at the University of Vienna (Austria)—the Vienna Circle. The logical positivists accepted the basic scheme of Comte but codified its principles for the conduct of scientific enquiry in order that it could lead to well-formulated generalizations and theories. The logical positivists were against everything that cannot be verified empirically and investigated by scientific method. As a set of scientific principles, logical positivism is concerned with acquisition of knowledge in the form of general statements obtained through accepted scientific procedures of observation and analysis, which can be used in manipulating phenomena with a view to bringing about desired results. Logical positivism is a school of philosophy that combines empiricism, the idea that observational evidence is indispensable for knowledge of the world, with a version of rationalism, the idea that our knowledge includes a component that is not derived from observation. Until the 1950s, logical positivism was the leading school in the philosophy of science. Logical positivism used formal logic to underpin an empiricist account of our knowledge of the world. Philosophers such as Rudolf Carnap and Hans Reichenbach, along with other members of the Vienna Circle, claimed that the truths of logic and mathematics were tautologies, and those of science were verifiable empirical claims. These two constituted the entire universe of meaningful judgements; anything else was nonsense. The claims of ethics and aesthetics were subjective preferences. Theology and other metaphysics were pseudo-statements, neither true nor false, simply meaningless nonsense. Karl Popper's insistence upon the role of falsification in the philosophy of science was a reaction to the logical positivists. With the rise of Adolf Hitler and National Socialism in Germany and Austria, some members of the Vienna and Berlin Circles fled Germany, mainly to Britain and the USA, which helped to reinforce the dominance of logical positivism and analytic philosophy in the Anglophone world. Logical positivists typically considered philosophy as having a very limited function. For them, philosophy is concerned with the organization of thoughts, rather than having distinct topics of its own. The positivists adopted the principle of verificationism, according to which every meaningful statement is either analytic or is capable of being verified by experience. This caused the logical positivists to reject many traditional problems of philosophy, especially those of metaphysics or ontology, as meaningless.

Basic Tenets

Although the logical positivists held a wide range of views on many matters, they were all interested in science and skeptical of theology and metaphysics. Early, most logical positivists proposed that all knowledge is based on logical inference from simple "protocol sentences" grounded in observable facts. Many logical positivists endorsed forms of materialism, metaphysical naturalism, and empiricism. Perhaps the view for which the logical positivists are best known is the verifiability criterion of meaning, or verificationism. In one of its earlier and stronger formulations, this is the doctrine that a proposition is "cognitively meaningful" only if there is a finite procedure for conclusively determining its truth. An intended consequence of this opinion, for most logical positivists, is that metaphysical, theological, and ethical statements fail this criterion, and so are not cognitively meaningful. They distinguished cognitive from other varieties of meaningfulness (e.g. emotive, expressive, figurative), and most authors concede that the non-cognitive statements of the history of philosophy possess some other kind of meaningfulness. The positive characterization of cognitive meaningfulness varies from author to author. It has been described as the property of having a truth value, corresponding to a possible state of affairs, naming a proposition, or being intelligible or understandable in the sense in which scientific statements are intelligible or understandable. Another characteristic feature of logical positivism is the commitment to "Unified Science"; that is, the development of a common language or, in Neurath's phrase, a "universal slang" in which all scientific propositions can be expressed.

Influence

Logical positivism spread throughout almost the entire western world. It was disseminated throughout the European continent. It was spread to Britain by the influence of A.J. Ayer. And later, it was brought to American Universities by members of the Vienna Circle after they fled Europe and settled in the United States during and after Second World War. Logical positivism was essential to the development of early analytic philosophy. The term subsequently came to be almost interchangeable with "analytic philosophy" during the first half of the twentieth century. Logical positivism was immensely influential in the philosophy of language and represented the dominant philosophy of science between First World War and the Cold War. Key tenets of logical positivism, including its atomistic philosophy of science, the verifiability principle, and the fact-value distinction, came under attack after the Second World War by philosophers such as Nelson Goodman, Quine, J.L. Austin, and Peter Strawson.

Thus, the logical positivism—a descendant of Comte's basic thesis but an independent movement—sprang up in Vienna and grew to become one of the dominant schools in Anglo-American philosophy and the analytic tradition in the early 20th century. Logical positivists (or 'neopositivists') rejected metaphysical speculation and attempted to reduce statements and propositions to pure logic. Critiques of this approach by philosophers such as Karl Popper, Willard Van Orman Quine and Thomas Kuhn have been highly influential, and led to the development of post-positivism.

Post-Positivism

In its broadest sense, positivism is a rejection of metaphysics. It is a position that holds that the goal of knowledge is simply to describe the phenomena that we experience. The purpose of science is simply to stick to what we can observe and measure. Knowledge of anything beyond that, a positivist would hold, is impossible. In a positivist view of the world, science was seen as the way to get at truth, to understand the world well enough so that we might predict and control it. The world and the universe were deterministic—they operated by laws of cause and effect that we could discern if we applied the unique approach of the scientific method. The positivist believed in *empiricism*—the idea that observation and measurement was the core of the scientific endeavour. The key approach of the scientific method is the experiment, the attempt to discern natural laws through direct manipulation and observation. Things have changed in our views of science since the middle part of the 20th century. The most important has been our shift away from positivism into what we term *post-positivism*.

Post-positivism is a wholesale rejection of the central tenets of positivism. A post-positivist might begin by recognizing that the way scientists think and work and the way we think in our everyday life are not distinctly different. Scientific reasoning and common sense reasoning are essentially the same process. There is no difference in kind between the two, only a difference in degree. One of the most common forms of post-positivism is a philosophy called *critical realism*. A critical realist believes that there is a reality independent of our thinking about it that science can study. Positivists were also realists. The difference is that the post-positivist critical realist recognizes that all observation is fallible and has error and that all theory is revisable. In other words, the critical realist is *critical* of our ability to know reality with certainty. Where the positivist believed that the goal of science was to uncover the truth, the post-positivist critical realist believes that *the goal of science is to hold steadfastly to the goal of getting it right about reality, even though we can never achieve that goal*! Because all measurement is fallible, the post-positivist emphasizes the importance of multiple measures and observations, each of which may possess different types of error, and the need to use *triangulation* across these multiple errorful sources to try to get a better bead on what's happening in reality. The post-positivist also believes that all observations are theory-laden and that scientists (and everyone else, for that matter) are inherently biased by their cultural experiences, world views, and so on. This is not cause to give up in despair, however. Just because I have my world view based on my experiences and you have yours doesn't mean that we can't hope to translate from each other's experiences or understand each other. That is, post-positivism rejects the *relativist* idea of the *incommensurability* of different perspectives, the idea that we can never understand each other because we come from different experiences and cultures. Most post-positivists are *constructivists* who believe that we each construct our view of the world based on our perceptions of it. Because perception and observation is fallible, our constructions must be imperfect. Post-positivists reject the idea that any individual can see the world perfectly as it really is. We are all biased and all of our observations are affected (theory-laden). Our best hope for achieving objectivity is to triangulate across multiple fallible perspectives! Thus, objectivity is not the characteristic of an individual, it is inherently a social phenomenon. It is what multiple individuals are trying to achieve when they criticize each other's work. We never achieve objectivity perfectly, but we can approach it. The best way for us to improve the objectivity of what we do is to do it within the context of a broader contentious community of truth-seekers (including other scientists) who criticize each other's work. The theories that survive such intense scrutiny are a bit like the species that survive in the evolutionary struggle. (This is sometimes called the *natural selection theory of knowledge* and holds that ideas have 'survival value' and that knowledge evolves through a process of variation, selection and retention.) They have adaptive value and are probably as close as our species can come to being objective and understanding reality.

At the turn of the 20th century, the first wave of German sociologists formally introduced methodological anti-positivism, proposing that research should concentrate on human cultural norms, values, symbols, and social processes viewed from a subjective perspective. This first wave of German sociologists, included Max Weber and Georg Simmel, who rejected the doctrine, thus founding the antipositivist tradition. "Antipositivism" is based on the belief that natural and human sciences are ontologically and epistemologically distinct. The extent of anti-positivist criticism has also become broad, with many philosophies broadly rejecting the scientifically based social epistemology and other ones only seeking to amend it to reflect 20th century developments in the philosophy of science. In contemporary social science, strong accounts of positivism have long since fallen out of favour. Practitioners of positivism today acknowledge in far greater detail observer bias and structural limitations. Modern positivists generally eschew metaphysical concerns in favour of methodological debates concerning clarity, replicability, reliability and validity. This positivism is generally equated with "quantitative research" and thus carries no explicit theoretical or philosophical commitments.

Criticism

Historically, positivism has been criticized for its universalism, i.e. for contending that all "processes are reducible to physiological, physical or chemical events," "social processes are reducible to relationships between and actions of individuals," and that "biological organisms are reducible to physical systems."

Max Horkheimer[5] criticized the classic formulation of positivism on two grounds. First, he claimed that it falsely represented human social action. The first criticism argued that positivism systematically failed to appreciate the extent to which the so-called social facts it yielded did not exist 'out there', in the objective world, but were themselves a product of socially and historically mediated human consciousness. Positivism ignored the role of the 'observer' in the constitution of social reality and thereby failed to consider the historical and social conditions affecting the representation of social ideas. Positivism falsely represented the object of study by reifying social reality as existing objectively and independently and labour actually produced those conditions. Secondly, he argued, representation of social reality produced by positivism was inherently and artificially conservative, helping to support the *status quo*, rather than challenging it. This character may also explains the popularity of positivism in certain political circles. Horkheimer argued, in contrast, that critical theory possessed a reflexive element lacking in the positivistic traditional theory.

Since the time of his writing, critiques of positivism, especially from philosophy of science, have led to the development of post-positivism. Positivism has also come under fire on religious and philosophical grounds, whose proponents assert that truth begins in sense experience, but does not end there. Positivism fails to prove that there are not abstract ideas, laws, and principles, beyond particular observable facts and relationships and necessary principles, or that we cannot know them. Nor does it prove that material and corporeal things constitute the whole order of existing beings, and that our knowledge is limited to them. According to positivism, our abstract concepts or general ideas are mere collective representations of the experimental order—for example, the idea of "man" is a kind of blended image of all the men observed in our experience. This runs contrary to a Platonic or Christian ideal, where an idea can be abstracted from any concrete determination, and may be applied identically to an indefinite number of objects of the same class. From the idea's perspective, the latter is more precise as collective images are more or less confused, become more so as the collection represented increases; an idea by definition remains always clear.

Positivism asserts that sense experiences are the only object of human knowledge, but does not prove its assertion. It is true that all our knowledge has its starting point in sense experience, but it is not proved that knowledge stops there. Positivism fails to demonstrate that, above particular facts and contingent relations, there are not abstract notions, general laws universal and necessary principles, or that we cannot know them. Nor does it prove that material and corporeal things constitute the whole order of existing beings, and that our knowledge is limited to them. Concrete beings and individual relations are not only perceptible by our senses, but they have also their causes and laws of existence and constitution; they are intelligible. These causes and laws pass beyond the particularness and contingency of individual facts, and are elements as fundamentally real as the individual facts which they produce and control. They cannot be perceived by our senses, but why can they not be explained by our intelligence? Again, immaterial beings cannot be perceived by sense experience, it is true, but their existence is not contradictory to our intelligence, and, if their existence is required as a cause and a condition of the actual existence of material things, they certainly exist. We can infer their existence and know something of their nature. They cannot indeed be known in the same way as material things, but this is no reason for declaring them unknowable to our intelligence.

NOTES

1. Isidore Auguste Marie François Xavier Comte (1798-1857), better known as Auguste Comte, was a French philosopher. He was a founder of the discipline of sociology and of the doctrine of positivism. He may be regarded as the first philosopher of science in the modern sense of the term.
2. The Vienna Circle was a group of philosophers of the Vienna University in the 1920s. Members of the Vienna Circle had a common attitude towards philosophy, characterized by two main beliefs: first, experience is the only source of knowledge; second, logical analysis performed with the help of symbolic logic is the preferred method for solving philosophical problems.

3. *Metaphysics*: philosophy of mind; *theism*: belief in the existence of god.
4. Friedrich Albert Moritz Schlick (1882-1936) was a German philosopher, physicist and the founding father of logical positivism and the Vienna Circle.
5. Max Horkheimer (1895-1973) was a German-Jewish philosopher-sociologist, famous for his work in critical theory as a member of the 'Frankfurt School' of social research. His most important works include *The Eclipse of Reason* (1947) and, in collaboration with Theodor Adorno, *The Dialectic of Enlightenment* (1947). Through the Frankfurt School, Horkheimer planned, supported and made other significant works possible.

REFERENCES

Ayer, Alfred Jules (1959), *Logical Positivism*. Glencoe, Ill: Free Press.

Catholic Encyclopaedia.

Giddens, Anthony (1974), *Positivism and Sociology*. Heinemann. London.

Hanfling, Oswald (1981), *Logical Positivism*. Oxford: B. Blackwell.

Harvey, M.E. and B.P. Holly (1989), *Themes in Geographic Thought*, Rawat Publications, Jaipur and New Delhi.

Mill, John Stuart. *August Comte and Positivism*. web-books.com.

Mises, Richard von. (1951), *Positivism: A study in human understanding*. Harvard University Press. Cambridge, Mass.

Pickering, Mary (1993), *Auguste Comte: An intellectual biography*. Cambridge University Press, Cambridge.

The Dictionary of Philosophy, Progress Publishers.

Wikipedia, *The Free Encyclopedia*.

CHAPTER-15

Pragmatism

Pragmatism is a philosophical movement that includes those who claim that an ideology or proposition is true if it works satisfactorily, that the meaning of a proposition is to be found in the practical consequences of accepting it, and that unpractical ideas are to be rejected. Pragmatism is a modified form of Positivism and believes that all what is observed may or may not be of significance. It is a doctrine that evaluates any assertion solely by its practical consequence and its bearing on human interests; giving a matter of fact treatment to things. It portrays life as a continuous process of experience, experiment and evaluation by which beliefs are continually being reconstructed; such reconstruction is a social process, whereby individuals learn and behave in the context of the beliefs of those with whom they interact. Pragmatism deals with both theoretical as well as practical solutions. The doctrine is especially suggested for social sciences. Because it emphasizes human experiences, experimental enquiry and truth as criteria for evaluating consequences or for finding practical solutions for human problems. Pragmatism is action-oriented, user-oriented (i.e. subject-based; not object-based as positivism). Pragmatism holds that what is important about knowledge is that it solves certain problems that are constrained both by the world and by human purposes. The place of knowledge in human activity is to resolve the problems that arise in conflicts between belief and action. The pragmatists are typically committed to the use of the experimental method in all forms of enquiry. Most of the thinkers who describe themselves as pragmatists point to some connection with practical consequences or real effects as vital components of both meaning and truth.

Basic Theme

Pragmatism means 'to rationalize or dealing with matters according to their practical significance'. It is a practical approach to problems and affairs marked by the doctrines that the meaning of conceptions is to be sought in their practical bearings, that the function of thought is to guide action, and that truth is pre-eminently to be tested by the practical consequences of belief. Pragmatism, as a tendency in philosophy, signifies the insistence on usefulness as a test of truth. In its negative phase, it opposes what it styles the formalism or rationalism of Intellectualistic philosophy. That is, it objects to the view that concepts, judgements, and reasoning processes are representative of reality and the processes of reality. It considers them to be merely symbols, hypotheses and *schemáta* devised by man to facilitate or render possible the use, or experience, of reality. This use, or experience, is the true test of real existence. In its positive phase, therefore, Pragmatism sets up as the standard of truth some non-rational test, such as action, satisfaction of needs, realization in conduct, the possibility of being lived, and judges reality by this norm to the exclusion of all others. Pragmatism is Individualistic. Despite the disclaimers of some of its exponents, it sets up the Protagorean principle, "Man is the measure of all things". For if Pragmatism means anything, it means that human consequences, "consequences to you and me", are the test of the meaning and truth of our concepts, judgements, and reasonings. Pragmatism is Nominalistic. It denies the validity of content of universal concepts, and scornfully rejects the mere possibility of universal, all-including or even many-including, reality. It is, by implication, Sensistic. For in describing the functional value of concepts it restricts that function to immediate or remote sense-experience. It is, in a sense, Anarchistic. Discarding Intellectualistic logic, it discards principles, and has no substitute for them except individual experience.

Origins

Pragmatism was a philosophical tradition that originated in the United States around 1870. The most important of the 'classical pragmatists' were Charles Sanders Peirce (1839-1914), William James (1842-1910) and John Dewey (1859-1952). The influence of pragmatism declined during the first two-thirds of the 20th century, but it has undergone a revival since 1970s with philosophers being increasingly willing to use the writings and ideas of classical pragmatists, and also a number of other thinkers, developing philosophical views that represent later stages of the pragmatist tradition. The core of pragmatism was the 'pragmatist maxim', a rule for clarifying the contents of hypotheses by tracing their 'practical consequences'. In the work of Peirce and James, the most influential application of the pragmatist maxim was to the concept of truth. But the pragmatists have also tended to share a distinctive epistemological outlook, a fallibilist anti-Cartesian[1] approach to the norms that govern inquiry. Pragmatism is but a new name for old ways of thinking.

In a paper contributed to the "*Popular Science Monthly*" in 1878 entitled "How to make our Ideas clear", C.S. Peirce first used the word Pragmatism to designate a principle, *Pragmatist Maxim*, put forward by him as a rule to guide the scientist and the mathematician. The principle is that the meaning of any conception in the mind is the practical effect it will have in action. "Consider what effects which might conceivably have practical bearings we consider the object of our conception to have. Then our conception of these effects is the whole of our conception of the object." This rule remained unnoticed for twenty years, until it was taken up by Professor William James in his address delivered at the University of California in 1898. "Pragmatism", according to James, "is a temper of mind, an attitude; it is also a theory of the nature of ideas and truth; and finally, it is a theory about reality"[2]. As he uses the word, therefore, it designates—(a) an attitude of mind towards philosophy, (b) an epistemology, and (c) a metaphysics. The attitude which he calls Pragmatism he defines as follows: "The whole function of philosophy ought to be to find out what definite difference it will make to you and me, at definite instants of our lives, if this world-formula or that world-formula be the true one". Next in importance to James is Professor John Dewey, who in a number of his articles and lectures, defends the doctrine known variously as Instrumentalism, or Immediate Empiricism. According to Dewey, we are constantly acquiring new items of knowledge which are at first unrelated to the previous contents of the mind; or, in moments of reflection, we discover that there is some contradiction among the items of knowledge already acquired. This condition causes a strain or tension, the removal of which gives satisfaction to the thinker. An idea is "a plan of action", which we use to relieve the strain; if it performs that function successfully, that is, satisfactorily, it is true. The adjustment is not, however, one-sided. Both the old truths in the mind and the new truth that has just entered the mind must be modified before we can have satisfaction. Thus there is no static truth, much less absolute truth; there are truths, and these are constantly being made true. This is the view which, under the names Personalism and Humanism, has been emphasized by Professor F. S. Schiller, the foremost of the English exponents of Pragmatism. "Humanism", and "Studies in Humanism" are the titles of his principal works. Pragmatism, Schiller thinks, "is in reality only the application of Humanism to the theory of knowledge", and Humanism is the doctrine that there is no absolute truth, but only truths, which are constantly being made true by the mind working on the data of experience.

Pragmatist Maxim

The pragmatist maxim is a distinctive rule or method for becoming reflectively clear about the contents of concepts and hypotheses: we clarify a hypothesis by identifying its practical consequences. Peirce's illustrative example urges that what we mean by calling something *hard* is that 'it will not be scratched by many other substances.' Unless there are cases where something's being hard makes a difference to what we experience and what it is rational for us to do, the concept is empty. The principle has a verificationist character: 'our idea of anything *is* our idea of its sensible effects' but the use of the phrase 'practical consequences' suggests that these are to be understood as having implications for what we will or should do. We become clearer about the concept *hard*, for example, by identifying how there can be conceivable circumstances in which we have desires that would call for different patterns of action if some object were hard to from those it would call for if the object were not hard. If one wants to break a window by throwing something through it, then he would need an object which is hard, not one which is soft. It is important that, as Peirce hints here, the consequences we are concerned with are *general* ones: we are to look for the laws that govern the behaviour of hard things and for laws that show how such modes of behaviour on the part of things can make a difference to what it is rational for us to do. Peirce insisted that his maxim was a *logical* principle and it was defended as an important component of the method of science, his favoured method for carrying out inquiries. This is reflected in the applications of the maxim that we find in his writings. First, he used it to clarify hard concepts that had a role in scientific reasoning: concepts like *probability*, *truth*, and *reality*. It also had a role in scientific testing. The pragmatist clarification of a scientific hypothesis, for example, provides us with just the information we need for testing it empirically. Pragmatism, described by Peirce as a 'laboratory philosophy', shows us how we test theories by carrying out experiments (performing rational actions) in the expectation that if the hypothesis is not true, then the experiment will fail to have some predetermined sensible effect. In later work, Peirce insisted that the maxim revealed *all* the information that was need for theory testing and evaluation. The pragmatist clarification revealed all the information we would need for testing hypotheses and theories empirically.

As is evident from the pragmatist maxim, pragmatism is a form of empiricism. Our ability to think about external things and to steadily improve our understanding of them rests upon our experience. However, the pragmatists all adopted accounts of experience and perception that were radically different from the views of earlier modern philosophers such as David Hume and Descartes[3]. The established view linked experience to what is sometimes called 'the given': we are the passive recipients of atomistic, determinate and singular sensory contents, the kinds of things that are sometimes called sense data. Experience provides the material for knowledge and conceptualization, but it does not itself have a content that is informed by concepts, practical needs, or anything else non-sensory. Our only contact with the external world is through receiving such experiences that, we suppose, are caused by external things; but since these sensory inputs are our only source of knowledge of the external world, we have no direct sensory awareness of external things. It is no surprise that this way of thinking about experience can easily lead to skepticism about the external world.

In different ways, Peirce, James, and Dewey all argued that experience is far richer than the tradition had supposed. They thought that experience was 'full of inference'. Experience is a process through which we interact with our surroundings, obtaining information that helps us to meet our needs. What we experience is shaped by our habits of expectation and there is no basis for extracting from this complex process the kind of 'thin given' beloved of sense datum theorists. We experience all sorts of objects, events and processes.

CENTRAL PRAGMATIST TENETS

Primacy of Practice

The pragmatist proceeds from the basic premise that the human capability of theorizing is integral to intelligent practice. Theory and practice are not separate spheres; rather, theories and distinctions are tools or maps for finding our way in the world. As John Dewey put it, there is no question of theory *versus* practice but rather of intelligent practice *versus* uninformed practice. Theory is an abstraction from direct experience and ultimately must return to inform experience in turn. Thus an organism navigating his or her environment is the grounds for pragmatist inquiry.

Anti-reification[4] of Concepts and Theories

Dewey has criticized what he called "the philosophical fallacy": philosophers often take categories (such as the mental and the physical) for granted because they don't realize that these are merely nominal concepts that were invented to help solve specific problems. This causes metaphysical and conceptual confusion.

Naturalism and Anti-Cartesianism

From the outset, pragmatists wanted to reform philosophy and bring it more in line with the scientific method as they understood it. They argued that idealist and realist philosophy had a tendency to present human knowledge as something beyond what science could grasp. These philosophies then resorted either to a phenomenology inspired by Kant or to correspondence theories of knowledge and truth. Pragmatists criticized the former for its a priorism, and the latter because it takes correspondence as an unanalyzable fact. Pragmatism instead tries to explain, psychologically and biologically, how the relation between knower and known 'works' in the world. C.S. Peirce denied that introspection and intuition (staple philosophical tools at least since Descartes) were valid methods for philosophical investigation. He argued that intuition could lead to faulty reasoning, e.g. when we reason intuitively about infinity. Furthermore, introspection does not give privileged access to knowledge about the mind - the self is a concept that is derived from our interaction with the external world and not the other way around.[5]

Reconciliation of Anti-skepticism and Fallibilism

Hilary Putnam has suggested that the reconciliation of anti-skepticism and fallibilism is the central goal of American pragmatism. Although all human knowledge is partial, with no ability to take a 'God's-eye-view,' this does not necessitate a globalized skeptical attitude, a radical philosophical skepticism (as distinguished from that which is called scientific skepticism). Peirce insisted that—(1) in reasoning, there is the presupposition, and at least the hope, that truth and the real are discoverable and would be discovered, sooner or later but still inevitably, by investigation taken far enough, and (2) contrary to Descartes' famous and influential methodology, doubt cannot be feigned or created by verbal *fiat* so as to motivate fruitful inquiry, and much less can philosophy begin in universal doubt. Doubt, like belief, requires justification. Inquiry is then the rationally self-controlled process of attempting to return to a settled state of belief about the matter. Note that anti-skepticism is a reaction to modern academic skepticism in the wake of Descartes. The pragmatist insistence that all knowledge is tentative is actually quite congenial to the older skeptical tradition.

Pragmatist Theory of Truth and Epistemology

The epistemology of early pragmatism was heavily influenced by Charles Darwin. Pragmatism was not the first to apply evolution to theories of knowledge: Schopenhauer[6] advocated a *biological idealism* as what's useful to an organism to believe might differ wildly from what is true. Here knowledge and action are portrayed as two separate spheres with an absolute or transcendental truth above and beyond any sort of inquiry organisms use to cope with life. Pragmatism challenges this idealism by providing an "ecological" account of knowledge: inquiry is how organisms can get a grip on their environment. *Real* and *true* are functional labels in inquiry and cannot be understood outside of this context. It is not *realist* in a traditionally robust sense of realism (what Hilary Putnam would later call metaphysical realism), but it is realist in how it acknowledges an external world which must be dealt with. Many of James' best-turned phrases—*truth's cash value* and *the true is only the expedient in our way of thinking*[7]—were taken out of context and caricatured in contemporary literature as representing the view where any idea with practical utility is true. William James wrote:

> "It is high time to urge the use of a little imagination in philosophy. The unwillingness of some of our critics to read any but the silliest of possible meanings into our statements is as discreditable to their imaginations as anything I know in recent philosophic history. Schiller says the truth is that which 'works'. Thereupon he is treated as one who limits verification to the lowest material utilities. Dewey says truth is what gives 'satisfaction'! He is treated as one who believes in calling everything true which, if it were true, would be pleasant."[8]

LEGACY AND CONTEMPORARY RELEVANCE

In the 20th century, the movements of logical positivism and ordinary language philosophy have similarities with pragmatism. Like pragmatism, logical positivism provides a verification criterion of meaning that is supposed to rid us of nonsense metaphysics. However, logical positivism doesn't stress action like pragmatism does. Furthermore, the pragmatists rarely used their maxim of meaning to rule out all metaphysics as nonsense. Usually, pragmatism was put forth to correct metaphysical doctrines or to construct empirically verifiable ones rather than to provide a wholesale rejection. Ordinary language philosophy is closer to pragmatism than other philosophy of language because of its nominalist character and because it takes the broader functioning of language in an environment as its focus instead of investigating abstract relations between *language* and *world*. Behaviourism and functionalism in psychology and sociology also have ties to pragmatism, which is not surprising considering that James and Dewey were both scholars of psychology.

Influence of Pragmatism in Social Sciences

Symbolic interactionism, a major perspective within sociological social psychology, was derived from pragmatism in the early 20th century. Increasing attention is being given to pragmatist epistemology in other branches of the social sciences, which have struggled with divisive debates over the status of social scientific knowledge.

Influence of Pragmatism in Public Administration

The classical pragmatism of John Dewey, William James and Charles Sanders Peirce has influenced research in the field of Public Administration. Scholars claim classical pragmatism had a profound influence on the origin of the field of Public Administration.[9] At the most basic level, public administrators are responsible for making programmes "work" in a pluralistic, problems-oriented environment. Public administrators are also responsible for the day-to-day work with citizens. Dewey's participatory democracy can be applied in this environment. Dewey and James' notion of theory as a tool, helps administrators craft theories to resolve policy and administrative problems. Further, the birth of American public administration coincides closely with the period of greatest influence of the classical pragmatists.

Pragmatism and Feminism

Since the mid-1990s, feminist philosophers have re-discovered classical pragmatism as a source of feminist theories. Their works explore the historic and philosophic links between feminism and pragmatism. The connection between pragmatism and feminism took so long to be rediscovered because pragmatism itself was eclipsed by logical positivism during the middle decades of the 20th century. As a result it was lost from feminine discourse. The very features of pragmatism that led to its decline are the characteristics that feminists now consider its greatest strength. These are "persistent and early criticisms of positivist interpretations of scientific methodology; disclosure of value dimension of factual claims"; viewing aesthetics as informing everyday experience; subordinating logical analysis to political, cultural and social issues; linking the dominant discourses with domination; "realigning theory with praxis; and resisting the turn to epistemology and instead emphasizing concrete experience". These feminist philosophers point to Jane Addams[10] as a founder of classical pragmatism. In addition, the ideas of Dewey, Mead and James are consistent with many feminist tenets. Jane Addams, John Dewey and George Herbert Mead developed their philosophies as all three became friends, influenced each other and were engaged in the women's rights causes.

Pragmatism and Urbanism

One application of pragmatism that is being developed, is the one between pragmatism and urbanism/urban transformation. A pragmatic approach to urban transformation values and evaluates the consequences of a design, rather than only considering the initial intentions. According to the pragmatic maxim, an object or conception can only be fully understood through its practical consequences. In an urban context this signifies how the implementation (and its effects) of a concept or design alters the overall understanding of the concept. Richard Rorty mentions that "a sea change" is occurring in recent philosophical thought— "a change so profound that we may not recognize that it is occurring." While the world that the movement is rooted in has had many changes, as a frame to perceive the world, pragmatism also has experienced different levels of modifications. Those changes are very relevant to the development of cities and basic themes, such as anti-foundationalism, fallibilism, community as inquirers, questioning the sharp distinction between theory and practice, pluralism and democracy, of pragmatism can be applied to the urbanism even more strongly.

NOTES

1. Cartesianism is the name given to the philosophical doctrine of René Descartes (1596-1650), a French philosopher and a major figure in 17th-century rationalism. Cartesians view the mind as being wholly separate from the corporeal body. Sensation and the perception of reality are thought to be the source of untruth and illusions, with the only reliable truths to be had in the existence of a metaphysical mind. Such a mind can perhaps interact with a physical body, but it does not exist in the body, nor even in the same physical plane as the body. In general the Cartesian satamevjaita divides the world into three areas of existence: that inhabited by the physical body (matter), that inhabited by the mind, and that inhabited by God.
2. *Journal of Phil.*, V, 85.
3. Smith 1978: chapter 3.
4. Reification may refer to Concretization, or making less abstract or less generalized.
5. De Waal 2005, 7-10.
6. Arthur Schopenhauer (1788-1860) was a German philosopher known for his pessimism and philosophical clarity. At age 25, he published his doctoral dissertation, *On the Fourfold Root of the Principle of Sufficient Reason*, which examined the four separate manifestations of reason in the phenomenal world. Schopenhauer's most influential work, *The World as Will and Representation*, claimed that the world is fundamentally what humans recognize in themselves as their will. His analysis of will led him to the conclusion that emotional, physical, and sexual desires can never be fully satisfied. The corollary of this is an ultimately painful human condition. Consequently, he considered that a lifestyle of negating desires, similar to the ascetic teachings of Vedanta, Buddhism and the Church Fathers of early Christianity, was the only way to attain liberation.
7. James 1907, 200-202.
8. James 1907, 90.
9. 'Pragmatism', Wikipedia, *The Free Encyclopedia*.
10. Jane Addams (1860-1935) can be labelled the first woman *"public philosopher"* in United States history.

REFERENCES

Daniel Dennett (1998), *Post-modernism and Truth.*

Goodman, R. (ed) (1995), *Pragmatism,* London: Routledge.

______(ed) (2005), *Pragmatism: Critical Concepts in Philosophy* (four volumes covering: volume one: Pragmatism's first decade; volume two: metaphysics and epistemology; volume three: moral and political issues; volume four: neopragmatism and aesthetics). London: Routledge.

Haack, S. (ed) (2006). *Pragmatism, Old and New,* Amherst NY: Prometheus.

Internet Encyclopaedia of Philosophy.

James, W. (1907), *A New Name for Some Old Ways of Thinking: Popular Lectures on Philosophy,* Longmans; Harvard University Press, 1975.

Menand, L. (ed) (1998), *Pragmatism,* New York: Random House.

Misak, C.J. (ed) (1999), *Pragmatism,* Calgary: University of Calgary Press.

Putnam, H. (1994), *Pragmatism,* Oxford: Blackwell.

psychology.wikia.com/wiki/Pragmatism.

Stanford Encyclopaedia of Philosophy.

Thayer, H.S. (ed) (1982),. *Pragmatism: The Classic* Writings, Hackett.

The Catholic Encyclopedia.

Wikipedia, *The Free Encyclopedia.*

CHAPTER-16

Phenomenology

In its central use "phenomenology" names a movement in 20th century philosophy. Phenomenology utilizes a distinctive method to study the structural features of experience and of things as experienced. Topics discussed within the phenomenological tradition include the nature of intentionality, perception, time-consciousness, self-consciousness, awareness of the body and consciousness of others. Although elements of this 20th century movement can be found in earlier philosophers—such as David Hume, Immanuel Kant and Franz Brentano—phenomenology as a philosophical movement really began with the work of Edmund Husserl. Following Husserl, phenomenology was adapted, broadened and extended by, amongst others, Martin Heidegger, Jean-Paul Sartre, Maurice Merleau-Ponty, Emmanuel Levinas and Jacques Derrida. The domain of phenomenology is the range of experiences including perception, imagination, thought, emotion, desire, volition, and action. Classical phenomenologists practiced three distinguishable methods—(1) Describing a type of experience just as we find it in our own (past) experience. Thus, Husserl and Merleau-Ponty spoke of pure description of lived experience; (2) Interpreting a type of experience by relating it to relevant features of context. In this vein, Heidegger and his followers spoke of hermeneutics, the art of interpretation in context, especially social and linguistic context; and (3) Analysing the form of a type of experience. In the end, all the classical phenomenologists practiced analysis of experience, factoring out notable features for further elaboration. These traditional methods have been ramified in recent decades, expanding the methods available to phenomenology.

Basic Theme

The discipline of phenomenology (from Greek: phainómenon "that which appears"; and lógos "study") may be defined initially as the study of structures of experience, or consciousness. Literally, phenomenology is the study of "phenomena": appearances of things, or things as they appear in our experience, or the ways we experience things, thus the meanings things have in our experience. Phenomenology is the study of structures of consciousness as experienced from the first-person point of view[1]. It is a philosophy of enquiry based on the premise that reality consists of objects and events as they are perceived by human consciousness and not of anything independent of human consciousness. The phenomenologist is concerned with providing a description of the life-world as it is actually experienced[2]. This concern avoids the abstraction involved in both positivism and idealism. However, in seeking to describe life-worlds there is strong element of subjectivity and this subjectivity is inevitable. The positivist fails to recognize the difference between objective and behavioural environment. To the phenomenologist, all cognitive experiences possess an intentional structure through which objects are made to mean something specific to an individual or group[3]. To the phenomenologist there is no objective world independent of man's existence and "all knowledge proceeds from the world of experience and cannot be independent of the world". This means that the phenomenologists describe, rather than explain. The student of phenomenology has no presuppositions; he does not use his own constructs in seeking to comprehend those of his subject. The subjectivity of phenomenologist research is that of the subject being studied, and not the student.

Phenomenology, in Husserl's conception, is primarily concerned with the systematic reflection on and study of the structures of consciousness and the phenomena that appear in acts of consciousness. This phenomenological ontology can be clearly differentiated from the Cartesian method of analysis which sees the world as objects, sets of objects, and objects acting and reacting upon one another. In its most basic form, phenomenology attempts to create conditions for the objective study of topics usually regarded as subjective: consciousness and the content of conscious experiences such as judgements, perceptions, and emotions. Although phenomenology seeks to be scientific, it does not attempt to study consciousness from the perspective of clinical psychology or neurology. Instead, it seeks through systematic reflection to determine the essential properties and structures of experience.

Techniques

Phenomenology operates rather differently from conventional social science[4]. Phenomenology is a theoretical orientation, but it does not generate deductions from propositions that can be empirically tested. It operates more on a meta-sociological level, demonstrating its premises through descriptive analyses of the procedures of self-situational, and social constitution. Through its demonstrations, audiences apprehend the means by which phenomena, originating in human consciousness, come to be experienced as features of the world. Phenomenological tools include the use of introspective and Verstehen methods to offer a detailed description of how consciousness itself operates[5]. Introspection requires the phenomenologist to use his or her own subjective process as a resource for study, while Verstehen requires an empathic effort to move into the mind of the other[6]. Not only are introspection and

Verstehen tools of phenomenological analysis, but they are procedures used by ordinary individuals to carry out their projects. Thus, the phenomenologist as analyst might study himself or herself as an ordinary subject dissecting his or her own self-consciousness and action schemes[7]. In this technique, an analytic attitude toward the role of consciousness in designing everyday life is developed.

Phenomenological concerns are frequently researched using qualitative methods[8]. Phenomenological researchers frequently undertake analyses of small groups, social situations, and organizations using face-to-face techniques of participant observation[9]. Ethnographic research frequently utilizes phenomenological tools. Intensive interviewing to uncover the subject's orientations or his or her "life world" is also widely practiced[10]. Qualitative tools are used in phenomenological research either to yield insight into the microdynamics of particular spheres of human life for its own sake or to exhibit the constitutive activity of human consciousness[11].

Origin

Phenomenology is a broad philosophical movement founded in the early years of 20th century by Edmund Husserl, expanded upon by a circle of his followers at the Universities of Göttingen and Munich in Germany. It then spread to France, the United States and elsewhere. In that movement, the discipline of phenomenology was prized as the proper foundation of all philosophy—as opposed, say, to ethics or metaphysics or epistemology. Phenomenology as a discipline has been central to the tradition of continental European philosophy, while philosophy of mind has evolved in the Austro-Anglo-American tradition of analytic philosophy throughout the 20th century. The fundamental character of our mental activity is pursued in overlapping ways within these two traditions. The Oxford English Dictionary presents the following definitions of Phenomenology: (a) the science of phenomena as distinct from being (ontology); (b) that division of any science which describes and classifies its phenomena. In its root meaning, then, phenomenology is the study of phenomena: literally, appearances as opposed to reality. However, the discipline of phenomenology did not blossom until the 20th century. Originally, in the 18th century, "phenomenology" meant the theory of appearances fundamental to empirical knowledge, especially sensory appearances. The Latin term "Phenomenologia" was introduced by Christoph Friedrich Oetinger in 1736. Subsequently, the German term "Phänomenologia" was used by Johann Heinrich Lambert. Immanuel Kant used the term occasionally in various writings, as did Johann Gottlieb Fichte. In 1807, G.W.F. Hegel wrote a book titled *Phänomenologie des Geistes* (usually translated as Phenomenology of Spirit). By 1889 Franz Brentano used the term to characterize what he called "descriptive psychology". From there Edmund Husserl took up the term for his new science of consciousness and the rest is history.

Phenomenology as we know it was launched by Edmund Husserl in his *Logical Investigations* (1900-01)[12]. Indeed, phenomenology and the modern concept of intentionality emerged hand-in-hand in Husserl's *Logical Investigations*. Intentionality (often described as "aboutness"), is the notion that consciousness is always consciousness of something. The object of consciousness is called the intentional object, and this object is constituted for consciousness in many different ways, through, for instance, perception, memory, retention and protention, signification, etc. Throughout these different intentionalities, though they have different structures and different ways of being "about" the object, an object is still constituted as the identical object; consciousness is directed at the same intentional object in direct perception as it is in the immediately following retention of this object and the eventual remembering of it. It was Husserl's adoption of this term (circa 1900) that propelled it into becoming the designation of a philosophical school. As envisioned by Husserl, phenomenology is a method of philosophical inquiry that rejects the rationalist bias that has dominated Western thought since Plato in favour of a method of reflective attentiveness that discloses the individual's "lived experience". Sometimes depicted as the "science of experience," the phenomenological method is rooted in intentionality, Husserl's theory of consciousness. Intentionality represents an alternative to the representational theory of consciousness, which holds that reality cannot be grasped directly because it is available only through perceptions of reality that are representations of it in the mind. Hence, the phenomenological method relies on the description of phenomena as they are given to consciousness, in their immediacy. The phenomenological method serves to momentarily erase the world of speculation by returning the subject to his or her primordial experience of the matter, whether, the object of inquiry is a feeling, an idea, or a perception. According to Husserl the suspension of belief in what we ordinarily take for granted or infer by conjecture diminishes the power of what we customarily embrace as objective reality.

Phenomenology came into its own with Husserl, much as epistemology came into its own with Descartes, and ontology or metaphysics came into its own with Aristotle on the heels of Plato. Yet phenomenology has been practiced, with or without the name, for many centuries. When Hindu and Buddhist philosophers reflected on states of consciousness achieved in a variety of meditative states, they were practicing phenomenology. When Descartes, Hume, and Kant characterized states of perception, thought, and imagination, they were practicing phenomenology. When Brentano classified varieties of mental phenomena (defined by the directedness of consciousness), he was practicing phenomenology. When William James appraised kinds of mental activity in the stream of consciousness (including their embodiment and their dependence on habit), he too was practicing phenomenology. And when recent analytic philosophers of mind have addressed issues of consciousness and intentionality, they have often been practicing phenomenology. Still, the discipline of phenomenology, its roots tracing back through the centuries, came to full flower in Husserl.

Husserl's work was followed by a flurry of phenomenological writing in the first half of 20th century. The diversity of traditional phenomenology is apparent in the *Encyclopaedia of Phenomenology*, which features seven types of phenomenology— (1) Transcendental constitutive phenomenology studies how objects are constituted in pure or transcendental consciousness, setting aside questions of any relation to the natural world around us, (2) Naturalistic constitutive phenomenology studies how consciousness constitutes or takes things in the world of nature, assuming with the natural attitude that consciousness is part of nature, (3) Existential phenomenology studies concrete human existence, including our experience of free choice or action in concrete situations, (4) Generative historicist phenomenology studies how meaning, as found in our experience, is generated in historical processes of collective experience over time, (5) Genetic phenomenology studies the genesis of meanings of things within one's own stream of experience, (6) Hermeneutical phenomenology studies interpretive structures of experience, how we understand and engage things around us in our human world, including ourselves and others, and (7) Realistic phenomenology studies the structure of consciousness and intentionality, assuming it occurs in a real world that is largely external to consciousness and not somehow brought into being by consciousness. The most famous of the classical phenomenologists were Husserl, Heidegger, Sartre, and Merleau-Ponty. In these four thinkers we find different conceptions of phenomenology, different methods, and different results.

Theory

Phenomenology commences with an analysis of the natural attitude. This is understood as the way ordinary individuals participate in the world, taking its existence for granted, assuming its objectivity, and undertaking action projects as if they were predetermined. Language, culture, and common sense are experienced in the natural attitude as objective features of an external world that are learned by actors in the course of their lives. Human beings are open to patterned social experience and strive towards meaningful involvement in a knowable world. They are characterized by a typifying mode of consciousness tending to classify sense data. In phenomenological terms humans experience the world in terms of typifications: Children are exposed to the common sounds and sights of their environments, including their own bodies, people, animals, vehicles, and so on. They come to apprehend the categorical identity and typified meanings of each in terms of conventional linguistic forms. In a similar manner, children learn the formulas for doing common activities. These practical means of doing are called recipes for action. Typifications and recipes, once internalized, tend to settle beneath the level of full awareness, that is, become sedimented, as do layers of rock. Thus, in the natural attitude, the foundations of actors' knowledge of meaning and action are obscured to the actors themselves. Actors assume that knowledge is objective and all people reason in a like manner. Each actor assumes that every other actor knows what he or she knows of this world: All believe that they share common sense. However, each person's biography is unique, and each develops a relatively distinct stock of typifications and recipes. Therefore, interpretations may diverge. Everyday social interaction is replete with ways in which actors create feelings that common sense is shared, that mutual understanding is occurring, and that everything is all right. Phenomenology emphasizes that humans live within an intersubjective world, yet they at best approximate shared realities. While a paramount reality is commonly experienced in this manner, particular realities or finite provinces of meaning are also constructed and experienced by diverse cultural, social, or occupational groupings. For phenomenology, all human consciousness is practical—

it is always of something. Actors intend projects into the world; they act in order to implement goals based on their typifications and recipes, their stock of knowledge at hand. Consciousness as an intentional process is composed of thinking, perceiving, feeling, remembering, imagining, and anticipating directed towards the world. The objects of consciousness, these intentional acts, are the sources of all social realities that are, in turn, the materials of common sense. Thus, typifications derived from common sense are internalized, becoming the tools that individual consciousness uses to constitute a lifeworld, the unified arena of human awareness and action. Common sense serves as an ever-present resource to assure actors that the reality that is projected from human subjectivity is an objective reality. Since all actors are involved in this intentional work, they sustain the collaborative effort to reify their projections and thereby reinforce the very frameworks that provide the construction tools.

PHENOMENOLOGICAL TERMINOLOGY

Intentionality

Intentionality refers to the notion that consciousness is always the consciousness of something. The word itself should not be confused with the "ordinary" use of the word intentional, but should rather be taken as playing on the etymological roots of the word. Originally, intention referred to a "stretching out". Intentionality is often summed up as "aboutness". Whether this something that consciousness is about is in direct perception or in fantasy is inconsequential to the concept of intentionality itself; whatever consciousness is directed at, that is what consciousness is consciousness of. This means that the object of consciousness doesn't have to be a physical object apprehended in perception: it can just as well be a fantasy or a memory. Consequently, these "structures" of consciousness, i.e. perception, memory, fantasy, etc., are called intentionalities. The cardinal principle of phenomenology, the term *intentionality* originated with the Scholastics in the medieval period and was resurrected by Brentano who in turn influenced Husserl's conception of phenomenology, who refined the term and made it the cornerstone of his theory of consciousness. The term should not be confused with "intention" or the psychoanalytic conception of unconscious "motive" or "gain".

Intuition

Intuition in phenomenology refers to those cases where the intentional object is directly present to the intentionality at play; if the intention is "filled" by the direct apprehension of the object, you have an intuited object. Having a cup of coffee in front of you, for instance, seeing it, feeling it, or even imagining it—these are all filled intentions, and the object is then intuited. The same goes for the apprehension of mathematical formulae or a number. If you do not have the object as referred to directly, the object is not intuited, but still intended, but then emptily. Examples of empty intentions can be signitive intentions—intentions that only imply or refer to their objects.

Evidence

In everyday language, we use the word evidence to signify a special sort of relation between a state of affairs and a proposition: State A is evidence for the proposition "A is true." In phenomenology, however, the concept of evidence is meant to signify the "subjective achievement of truth." This is not an attempt to reduce the objective sort of evidence to subjective "opinion," but rather an attempt to describe the structure of having something present in intuition with the addition of having it present as intelligible: "Evidence is the successful presentation of an intelligible object, the successful presentation of something whose truth becomes manifest in the evidencing itself."

Noesis and Noema

In Husserl's phenomenology, which is quite common, this pair of terms, derived from the Greek nous (mind), designate respectively the real content, noesis, and the ideal content, noema, of an intentional act (an act of consciousness). The Noesis is the part of the act that gives it a particular sense or character (as in judging or perceiving something, loving or hating it, accepting or rejecting it, and so on). This is real in the sense that it is actually part of what takes place in the consciousness (or psyche) of the subject of the act. The Noesis is always correlated with a Noema; for Husserl, the full Noema is a complex ideal structure comprising at least a noematic sense and a noematic core.

Empathy and Intersubjectivity

In phenomenology, empathy refers to the experience of one's own body as another. While we often identify others with their physical bodies, this type of phenomenology requires that we focus on the subjectivity of the other, as well as our intersubjective engagement with them. In Husserl's original account, this was done by a sort of apperception built on the experiences of your own lived-body. The lived-body is your own body as experienced by yourself, as yourself. Your body is also experienced as a duality, both as object (you can touch your own hand) and as your own subjectivity (you experience being touched). The experience of your own body as your own subjectivity is then applied to the experience of another's body, which, through apperception, is constituted as another subjectivity. You can thus recognise the Other's intentions, emotions, etc. This experience of empathy is important in the phenomenological account of intersubjectivity. In phenomenology, intersubjectivity constitutes objectivity (i.e. what you experience as objective is experienced as being intersubjectively available—available to all other subjects. This does not imply that objectivity is reduced to subjectivity nor does it imply a relativist position, cf. for instance intersubjective verifiability). In the experience of intersubjectivity, one also experiences oneself as being a subject among other subjects, and one experiences oneself as existing objectively for these Others; one experiences oneself as the noema of Others' noeses, or as a subject in another's empathic experience. As such, one experiences oneself as objectively existing subjectivity. Intersubjectivity is also a part in the constitution of one's lifeworld, especially as "homeworld".

Lifeworld

The lifeworld (German: Lebenswelt) is the "world" each one of us lives in. One could call it the "background" or "horizon" of all experience, and it is that on which each object stands out as itself (as different) and with the meaning it can only hold for us. The lifeworld is both personal and intersubjective (it is then called a "homeworld"), and, as such, it does not enclose each one of us in a *solus ipse*.

PHENOMENOLOGY AND GEOGRAPHY

Phenomenology has to do with beginnings, with phenomena that are first experienced and are only subsequently formulated as concepts. Geography as a formal body of knowledge presupposes our geographical experiences of the world. In other words, geography has an experiential or phenomenological foundation. Concepts of space, landscape, city, region, have meaning for us because we can refer to them to our direct experiences of these phenomena. We live in a world of buildings, streets, sunshine and rainfall and other people with all their sufferings and joys, and we know intersubjectively the meanings of these things and events. This pre-intellectual world, or life-world, we experience not as a set of objects somehow apart from us and fixed in time and space, but as a set of meaningful and dynamic relations. A recurrent theme in the phenomenological writings of the geographers is that Phenomenology leads to self-awareness and a heightened sense of responsibility for the environments we live in and the ideas we express about them. Phenomenology helps people to appreciate their worlds and lives more fully and also to begin to see what modern science, technology and consumerism have wrought in the name of efficiency and satisfaction. So it might be suggested that once the insight of grasping the world from the perspective of one's own experience has been achieved then necessarily one must do geography phenomenologically[13].

NOTES

1. *Stanford, Encyclopaedia of Philosophy.*
2. Buttimer, 1976.
3. Dikshit, 1997, 150.
4. Darroch and Silvers, 1982.
5. Hitzler and Keller, 1989.
6. Helle 1991; Truzzi, 1974.
7. Bleicher, 1982.
8. Bogdan and Taylor, 1975; Denzin and Lincoln 1994, 1998.
9. Bruyn, 1966.
10. Costelloe, 1996; Grekova, 1996.
11. Langsdorf, 1995.
12. Myron Orleans, *Encyclopedia of Sociology.*
13. Harvey and Holly, 1989, 109-114.

REFERENCES

Bleicher, Josef (1982), *The Hermeneutic Imagination: Outline of a Positive Critique of Scientism and Sociology*. London: Routledge and Kegan Paul.

Bogdan, Robert, and Steven, J. Taylor (1975), *Introduction to Qualitative Research Methods: A Phenomenological Approach to the Social Sciences*. New York: Wiley.

Bruyn, Severn, T. (1966), *The Human Perspective in Sociology: The Methodology of Participant Observation*. Englewood Cliffs, N.J.: Prentice-Hall.

Buttimer, Anne (1976), "Grasping the Dynamism of Lifeworld", *Annals of the Association of American Geographers*, Volume 66, Issue 2, pp. 277-292, June 1976.

Cerbone, David R. (2006), *Understanding Phenomenology*. Chesham: Acumen.

Costelloe, Timothy M. (1996), "Between the Subject and Sociology: Alfred Schutz's Phenomenology of the Life-World", *Human Studies*, 19:247-266.

Darroch, Vivian, and Ronald J. Silvers (eds.) (1982), *Interpretive Human Studies: An Introduction to Phenomenological Research*. Washington, D.C.: University Press.

David Stewart and Algis Mickunas (1990), *Exploring Phenomenology: A Guide to the Field and its Literature*, Athens: Ohio University Press.

Dikshit, R.D. (1997), *Geographical Thought: A Contextual History of Ideas*, Prentice Hall of India.

Harvey, M.E. and B.P. Holly (1989), *Themes in Geographic Thought*, Rawat Publications, New Delhi.

Helle, Horst J., ed. (1991), *Verstehen and Pragmatism: Essays in Interpretative Sociology*. Frankfurt, Federal Republic Germany: Peter Lang.

Herbert Spiegelberg (1983), *The Phenomenological Movement: A Historical Introduction*, 3rd ed. The Hague: Martinus Nijhoff.

Hitzler, Ronald, and Reiner Keller (1989), "On Sociological and Common-Sense Verstehen.", *Current Sociology*, 37, pp. 91-101.

Husserl, Edmund. (1999) (1907), *The Idea of Phenomenology*. Trans. Lee Hardy. Dordrecht: Kluwer.

Internet *Encyclopaedia of Philosophy*.

Langsdorf, Lenore (1995), "Treating Method and Form as Phenomena: An Appreciation of Garfinkel's Phenomenology of Social Action", *Human Studies*, 18:177-188.

Michael Hammond, Jane Howarth, and Russell Kent (1995), *Understanding Phenomenology*, Oxford: Blackwell.

Moran, D. (2000), *Introduction to Phenomenology*, Oxford: Routledge.

Myron Orleans, *Encyclopedia of Sociology*.

Peet, R. (1998), *Modern Geographical Thought*, Blackwell.

Richard, M. Zaner (1970), *The Way of Phenomenology*, Indianapolis: Pegasus.

Robert Sokolowski (2000), *Introduction to Phenomenology*, Cambridge: Cambridge University Press.

Stanford, *Encyclopaedia of Philosophy*.

Truzzi, Marcello (1974), *Verstehen: Subjective Understanding in the Social Sciences*. Reading, Mass.: Addison-Wesley.

Wikipedia, *The Free Encyclopaedia*.

CHAPTER-17

Existentialism

Closely associated with phenomenology is the philosophy of existentialism. It is another anthropocentric approach used by social scientists. It emphasises that the explanation should begin from space occupied by man or society, and then the other relationships should be explored. The central concept is all sorts of explanations should begin from the existential space of mankind. The view supports the conception that man is responsible for making his own space or the living environment. Both, phenomenology and existentialism look at the environment from an experimental perspective[1]. The basic difference is that phenomenology believes in "being before essence". The process of defining oneself (creating an essence) involves creating an environment. Thus environments can be read as biographies, as representations of man creating himself[2]; whereas, 'existential' means 'concerned with existence', and 'especially with human existence'. In this sense, the 'existentialism' may be defined as the philosophical theory emphasizing existence of the individual person as free and responsible agent determining his own development. Here, the basic idea is "existence comes before essence", i.e. existence comes before anything, any other reality. For the social scientists, the existence of man should be the starting point. The scale of existential space may range from an individual to group of individuals to society and ultimately the entire mankind. Existentialism considers a study that pursues meaning in existence and seeks value for the existing individual. Existentialism, unlike other fields of philosophy, does not treat the individual as a concept, and values individual subjectivity over objectivity. As a result, questions regarding the meaning of life and subjective experience are seen as being of paramount importance, above all other scientific and philosophical pursuits. Existentialism tends to view human beings as subjects in an indifferent, objective, often ambiguous, and "absurd" universe in which meaning is not provided by the natural order, but rather can be created, however, provisionally and unstably, by human beings' actions and interpretations.

Basic Theme

Existentialism is a philosophy that is centred upon the analysis of existence and of the way humans find themselves existing in the world. The notion is that humans exist first and then each individual spends a lifetime changing their essence or nature. In simpler terms, existentialism is a philosophy concerned with finding self and the meaning of life through free will, choice, and personal responsibility. The belief is that people are searching to find out who and what they are throughout life as they make choices based on their experiences, beliefs, and outlook. And personal choices become unique without the necessity of an objective form of truth. An existentialist believes that a person should be forced to choose and be responsible without the help of laws, ethnic rules, or traditions. Existentialism emphasizes action, freedom, and decision as fundamental to human existence and is fundamentally opposed to the rationalist tradition and to positivism. That is, they argue against definitions of human beings as primarily rational. Rather, existentialists look at where people find meaning. Existentialism asserts that people actually make decisions based on the meaning to them rather than rationally.

On the existential view, to understand what a human being is it is not enough to know all the truths that natural science—including the science of psychology—could tell us. The dualist who holds that human beings are composed of independent substances—"mind" and "body"—is no better off in this regard than is the physicalist, who holds that human existence can be adequately explained in terms of the fundamental physical constituents of the universe. Existentialism does not deny the validity of the basic categories of physics, biology, psychology, and the other sciences (categories such as matter, causality, force, function, organism, development, motivation, and so on). It claims only that human beings cannot be fully understood in terms of them. Nor can such an understanding be gained by supplementing our scientific picture with a moral one. Categories of moral theory such as intention, blame, responsibility, character, duty, virtue, and the like do capture important aspects of the human condition, but neither moral thinking (governed by the norms of the good and the right) nor scientific thinking (governed by the norm of truth) suffices. "Existentialism", therefore, may be defined as the philosophical theory which holds that a further set of categories, governed by the norm of authenticity, is necessary to grasp human existence.

Origin

The term "existentialism" was actually coined by the French philosopher Gabriel Marcel in the mid-1940s. It was adopted by Jean-Paul Sartre[3]. Existentialism began in the mid-19th century as a reaction against then—dominant systematic philosophies, with Søren Kierkegaard and Friedrich Nietzsche, came to be seen as precursors of the movement. Kierkegaard and Nietzsche were the first philosophers considered fundamental to the existentialist movement, though neither used the term "existentialism". They focused on subjective human experience rather than the objective truths of mathematics and science, which they believed were too detached or observational to truly get at the human experience. Opposed to Hegelianism and Kantianism, Kierkegaard posited that it is the individual who is solely responsible for giving meaning to life and for living life passionately and sincerely. Existentialism became popular in the years following Second World War and influenced a range of disciplines besides philosophy, including theology, drama, art, literature, and psychology. Existentialistic ideas came out of a time in society when there was a deep sense of despair following the Great Depression and Second World War. There was a spirit of optimism in society that was destroyed by First World War and its mid-century calamities. This despair has been articulated by existentialist philosophers well into the 1970s and continues on to this day as a popular way of thinking and reasoning (with the freedom to choose one's preferred moral belief system and lifestyle).

Like empiricism, 'existentialism' is a term that belongs to intellectual history. The term was explicitly adopted by Jean-Paul Sartre, and through the wide dissemination of the post-war literary and philosophical output of Sartre and his associates—notably Simone de Beauvoir, Maurice Merleau-Ponty, and Albert Camus—existentialism became identified with a cultural movement that flourished in Europe in the 1940s and 1950s. Among the major philosophers identified as existentialists were Karl Jaspers, Martin Heidegger, and Martin Buber in Germany, Jean Wahl and Gabriel Marcel in France, the Spaniards José Ortega y Gasset and Miguel de Unamuno, and the Russians Nikolai Berdyaev and Lev Shestov. In the first decades of 20th century, a number of philosophers and writers explored existentialist ideas. Following the Second World War, existentialism became a well-known and significant philosophical and cultural movement.

CONCEPTS

Existence Precedes Essence

A central proposition of existentialism is that existence precedes essence, which means that the actual life of the individual is what constitutes what could be called his or her "essence" instead of there being a predetermined essence that defines what it is to be a human. Thus, human beings—through their own consciousness—create their own values and determine a meaning to their life. Although, it was Sartre who explicitly coined the phrase, similar notions can be found in the thought of many existentialist philosophers from Kierkegaard to Heidegger. Sartre writes it in his work Existentialism is a Humanism: "man first of all exists, encounters himself, surges up in the world—and defines himself afterwards." Sartre's slogan—"existence precedes essence"—may serve to introduce what is most distinctive of existentialism, namely the idea that no general, non-formal account of what it means to be human can be given, since that meaning is decided in and through existing itself. Existence is "self-making-in-a-situation"[4]. Traditionally, philosophers have connected the concept of existence with that of essence in such a way that the former signifies merely the instantiation of the latter. If "essence" designates what a thing is and "existence" that it is, it follows that what is intelligible about any given thing, what can be thought about it, will belong to its essence. It is from essence in this sense—say, human being as rational animal or imago Dei—that ancient philosophy drew its prescriptions for an individual's way of life, its estimation of the meaning and value of existence. Having an essence meant that human beings could be placed within a larger whole, a kosmos that provided the standard for human flourishing. Modern philosophy retained this framework even as it abandoned the idea of a "natural place" for man in the face of the scientific picture of an infinite, labyrinthine universe. The fundamental contribution of existential thought lies in the idea that one's identity is constituted neither by nature nor by culture, since to "exist" is precisely to constitute such an identity. It is in light of this idea that key existential notions such as facticity, transcendence (project), alienation, and authenticity must be understood.

Facticity and Transcendence

Facticity includes all those properties that third-person investigation can establish about an inddividual: natural properties such as weight, height, and skin colour; social facts such as race, class, and nationality; psychological properties such as his web of belief, desires, and character traits; historical facts such as his past actions, family background, and his broader historical milieu; and so on. Transcendence refers to to one's attitude toward himself characteristic of his practical engagement in the world, the agent's perspective. An agent is oriented by the task at hand as something to be brought about through its own will or agency. Such orientation does not take itself as a theme but loses itself in what is to be done. Thereby, things present themselves not as indifferent givens, facts, but as meaningful: salient, expedient, obstructive, and so on. To speak of "transcendence" here is to indicate that the agent "goes beyond" what simply is towards what can be: the factual—including the agent's own properties—always emerges in light of the possible, where the possible is not a function of anonymous forces (third-person or logical possibility) but a function of the agent's choice and decision. Because existence is co-constituted by facticity and transcendence, the self cannot be conceived as a Cartesian ego but is embodied being-in-the-world, a self-making in situation.

Alienation

The anti-Cartesian view of the self as in situation yields the familiar existential theme of the "alienated" self, the estrangement of the self both from the world and from itself. This experience, basic to existential thought, contrasts most sharply with the ancient notion of a kosmos in which human beings have a well-ordered place, and it connects existential thought tightly to the modern experience of a meaningless universe. An individual's very engagement in the world alienates him from his authentic possibility. This theme is brought out most clearly by Heidegger: the anti-Cartesian idea that the self is defined first of all by its practical engagement entails that this self is not properly individual but rather indisinguishable from anyone else (das Man) who engages in such practices: such a "they-self" does what "one" does. Thus, if who I am is defined through existing, this "who" is normally pre-defined by what is average, by the roles available to me in my culture, and so on. The "I" that gets defined is thereby "anonymous," or "anyone"; self-making is largely a function of not distinguishing myself from others. If there is nevertheless good sense in talking of the singularity of my existence, it will not be something with which one starts but something that gets achieved in recovering oneself from alienation or lostness in the "crowd". If the normative is first of all the normal, however, it might seem that talk about a norm for the singularity of existence, a standard for thinking about what is my ownmost just as I myself, would be incoherent. It is here that the idea of "authenticity" must come into focus.

Authenticity

The theme of authentic existence is common to many existentialist thinkers. It is often taken to mean that one has to "find oneself" and then live in accordance with this self. What is meant by authenticity is that in acting, one should act as oneself, not as one acts or as one's genes or any other essence requires. The authentic act is one that is in accordance with one's freedom. Existentialism arises with the collapse of the idea that philosophy can provide substantive norms for existing, ones that specify particular ways of life. Nevertheless, there remains the distinction between what I do "as" myself and as "anyone", so in this sense existing is something at which I can succeed or fail. Authenticity—in German, Eigentlichkeit—names that attitude in which I engage in my projects as my own (eigen). What this means can perhaps be brought out by considering moral evaluations. In keeping my promise I act in accord with duty; and if I keep it because it is my duty, I also act morally (according to Kant) because I am acting for the sake of duty. But existentially there is still a further evaluation to be made. My moral act is inauthentic if, in keeping my promise for the sake of duty, I do so because that is what "one" does (what "moral people" do). But I can do the same thing authentically if, in keeping my promise for the sake of duty, acting this way is something I choose as my own, something to which, apart from its social sanction, I commit myself. Similarly, doing the right thing from a fixed and stable character—which virtue ethics considers a condition of the good—is not beyond the reach of existential evaluation: such character may simply be a product of my tendency to "do what one does," including feeling "the right way" about things and betaking myself in appropriate ways as one is expected to do. Thus the norm of authenticity refers to a kind of "transparency" with regard to my situation, a recognition that I am a being who can be responsible for who I am. In choosing in light of this norm I can be said to recover myself from alienation, from my absorption in the anonymous "one-self" that characterizes me in my everyday engagement in the world. Authenticity thus indicates a certain kind of integrity—not that of a pre-given whole, an identity waiting to be discovered, but that of a project to which I can either commit myself (and thus "become" what it entails) or else simply occupy for a time, inauthentically drifting in and out of various affairs. Thus existentialism's focus on authenticity leads to a distinctive stance towards ethics and value-theory generally. The possibility of authenticity is a mark of my freedom, and it is through freedom that existentialism approaches questions of value, leading to many of its most recognizable doctrines.

Criticism

Many critics argue Sartre's philosophy is contradictory. Specifically, they argue that Sartre makes metaphysical arguments despite his claiming that his philosophical views ignore metaphysics. Herbert Marcuse criticized Being and Nothingness (1943) by Jean-Paul Sartre for projecting anxiety and meaninglessness onto the nature of existence itself: "In so far as Existentialism is a philosophical doctrine, it remains an idealistic doctrine: it hypostatizes specific historical conditions of human existence into ontological and metaphysical characteristics. Existentialism thus becomes part of the very ideology which it attacks, and its radicalism is illusory." Heidegger criticizes Sartre's existentialism: Existentialism says existence precedes essence. In this statement he is taking existentia and essentia according to their metaphysical meaning, which, from Plato's time on, has said that essentia precedes existentia. Sartre reverses this statement. But the reversal of a metaphysical statement remains a metaphysical statement. With it, he stays with metaphysics, in oblivion of the truth of Being.

Existentialism Today

As a cultural movement, existentialism belongs to the past. As a philosophical inquiry that introduced a new norm, authenticity, for understanding what it means to be human—a norm tied to distinctive, post-Cartesian concept of the self as practical, embodied, being-in-the-world—existentialism has continued to play an important role in contemporary thought, in both the continental and analytic traditions. The Society for Phenomenology and Existential Philosophy, as well as societies devoted to Heidegger, Sartre, Merleau-Ponty, Jaspers, Beauvoir, and other existential philosophers, provide a forum for ongoing work—both of a historical, scholarly nature and of more systematic focus—that derives from classical existentialism, often bringing it into confrontation with more recent movements such as structuralism, deconstruction, hermeneutics, and feminism. In the area of gender studies Judith Butler (1990) draws importantly on existential sources, as does Lewis Gordon (1995) in the area of race theory. Interest in a narrative conception of self-identity—for instance, in the work of Charles Taylor (1999), Paul Ricoeur, David Carr (1986), or Charles Guignon—has its roots in the existential revision of Hegelian notions of temporality and its critique of rationalism. Hubert Dreyfus (1979) developed an influential criticism of the Artificial Intelligence programme drawing essentially upon the existentialist idea, found especially in Heidegger and Merleau-Ponty, that the human world, the world of meaning, must be understood first of all as a function of our embodied practices and cannot be represented as a logically structured system of representations.

NOTES

1. Dikshit, 1997, 151.
2. Johnston, 1983, 155.
3. Jean-Paul Charles Aymard Sartre (1905-1980) was a French existentialist philosopher, playwright, novelist, screenwriter, political activist, biographer, and literary critic. He was one of the key figures in the philosophy of existentialism, and one of the leading figures in 20th century French philosophy and Marxism. His work has also influenced sociology, critical theory, post-colonial theory, and literary studies, and continues to influence these disciplines. Sartre has also been noted for his relationship with the prominent feminist theorist Simone de Beauvoir.
4. Fackenheim, 1961: 37.

REFERENCES

Appignanesi, R. and O. Zarate (2001), *Introducing Existentialism*, Cambridge, UK: Icon. ISBN 1-84046-266-3.

Collins, J. (1952), *The Existentialists: A Critical Study*, Chicago: Henry Regnery Company.

Dikshit, R.D. (1997), *Geographical Thought: A Contextual History of Ideas*, Prentice Hall of India.

Flynn, T. (2006), *Existentialism: A Very Short Introduction*, Oxford: Oxford University Press.

Guignon, C., and D. Pereboom (eds.), *Existentialism: Basic Writings*, Indianapolis: Hackett.

Harvey, M.E. and B.P. Holly (1989), *Themes in Geographic Thought*, Rawat Publications, Jaipur and New Delhi.

Johnston, R.J. (1983), *Philosophy and Human Geography*, Edward Arnold, London.

Luper, Steven (ed.) (2000), *An Introduction to Existential Thought*. Mountain View, California: Mayfield. ISBN 0-7674-0587-0.

McBride, W. (ed.) (1997), *The Development and Meaning of Twentieth Century Existentialism*, New York: Garland Publishers.

MacIntyre, A. (1967), "Existentialism," in *The Encyclopedia of Philosophy*, Vol. III. Ed. Paul Edwards. New York: Macmillan Publishing Company.

Marcel, G. (1968), *The Philosophy of Existentialism*, New York: Citadel Press.

Marino, Gordon (ed.) (2004), *Basic Writings of Existentialism*. New York: Modern Library. ISBN 0-375-75989-1.

Peet, R. (1998), *Modern Geographical Thought*, Blackwell.

Reynolds, J. (2006), *Understanding Existentialism*. London: Acumen.

Sartre, Jean-Paul (1945), *Existentialism and Humanism*.

Solomon, R. (ed.), (1974), *Existentialism*, New York: Random House.

Stanford *Encyclopaedia of Philosophy*.

Wikipedia, *The Free Encyclopaedia*.

CHAPTER-18

Idealism

In philosophy, idealism is the group of philosophies which assert that reality is fundamentally mental, mentally constructed, or otherwise immaterial. Epistemologically, idealism manifests as a skepticism about the possibility of knowing any mind-independent thing. In a sociological sense, idealism emphasizes how human ideas—especially beliefs and values—shape society. As an ontological doctrine, idealism goes further, asserting that all entities are composed of mind or spirit. Idealism thus rejects physicalist and dualist theories that fail to ascribe priority to the mind. An extreme version of this idealism can exist in the philosophical notion of solipsism.[1]

Roots

Religious and philosophical thought privileging the immaterial or supernatural over the material and natural is ubiquitous and ancient. However, the earliest extant arguments that the world of experience is grounded in the mental derive from India and Greece. The Hindu idealists in India and the Greek Neoplatonists gave pantheistic arguments for an all-pervading consciousness as the ground or true nature of reality. In contrast, the Yogācāra school which arose within Mahayana Buddhism in India in the 4th century CE, based its "mind-only" idealism to a greater extent on phenomenological analyses of personal experience. This turn towards the subjective anticipated empiricists such as George Berkeley, who revived idealism in 18th century Europe by employing skeptical arguments against materialism.

Beginning with Immanuel Kant, German idealists such as G.W.F. Hegel, Johann Gottlieb Fichte, Friedrich Wilhelm Joseph Schelling, and Arthur Schopenhauer dominated 19th century philosophy. This tradition, which emphasized the mental or "ideal" character of all phenomena, birthed idealistic and subjectivist schools ranging from British idealism to phenomenalism to existentialism. The historical influence of this branch of idealism remains central even to the schools that rejected its metaphysical assumptions, such as—marxism, pragmatism, and positivism.

Idealism is a term used to describe a wide variety of philosophical positions. One can distinguish two general senses: A Platonic sense, and a modern sense. Idealism in the Platonic sense involves the claim that ideal things occupy a metaphysically privileged position in the universe. Idealism in the modern sense centres around the claim that at least large portions of reality (in particular, the experienced physical world) are metaphysically based in something mental (minds and their ideas or representations). Such a view stands in stark opposition with "materialist" views of reality, which claim that mental entities and properties are somehow based or grounded in non-mental, material entities and properties, of the sort with which physics is concerned (there are positions between the two extremes, such as dualism).

Though both types of idealism are first and foremost metaphysical positions, their proponents have typically tried to motivate them using epistemological considerations. Plato's concern with the ideal realm appears to have been largely motivated by questions concerning knowledge. Epistemological arguments play a central role in the defences of modern idealism presented by the two most prominent idealists in modern Western philosophy: George Berkeley and Immanuel Kant.

Modern Idealism

Kant defined "genuine" idealism as consisting in the assertion that, "there are none but thinking beings; all other things which we believe are perceived in intuitions are nothing but representations in the thinking beings, to which no object external to them corresponds". His view, which he described as "transcendental" or "critical" idealism (4:293-94), did *not* involve the claim that all non-mental things must exist in representations. The distinction Kant aimed to draw can be turned into a useful general point. It is clearest to understand the term "idealism" in a relative sense and an absolute sense. In the relative sense, a philosopher is an idealist about a certain sort of entity or property, where this simply means that she believes that the existence and nature of that entity or property ultimately reduces to facts about minds and their representations. Given this, certain forms of idealism should be generally accepted—for instance, we might be idealists about a certain fictional character. Kant, then, was an idealist about a certain set of properties (including space and time), but not about others (for instance, the property of being able to affect other entities). The absolute sense of "idealism," then, is relative idealism about *all* entities and properties. This is then a much stronger position, and one that cannot be conclusively argued for one entity or property at a time.

Inspired by the work of the French philosopher and theologian Nicolas Malebranche, the Irish Bishop George Berkeley believed that philosophical positions that posited absolutely non-mental entities in the universe (in particular, Cartesian material substance) were responsible for the spread of atheism and skepticism across Europe in the seventeenth and eighteenth centuries. According to a philosophical picture such as that advanced by John Locke, material substance was the crucial aspect of the physical world, and was responsible for causing representations in the mind. It could not, however, be directly perceived, and could only be known indirectly through the representations it caused. But if material substance was at the core of physical reality and could not be directly known, then, Berkeley believed, it was inevitable that people would come to doubt whether it existed, and thereby come to question the reality of the world of everyday objects. Worse, in his view, this view described a universe that seemed capable of operating independently of God. Were people to become convinced of such a picture, it was inevitable that they would come to wonder if they had any reason for believing in God at all. On the other hand, if people believed—(1) that all that existed were minds and their representations, (2) that the world of everyday objects was simply composed of representations, and (3) that most of their representations were directly caused by God, then the source of those temptations towards skepticism and atheism would dry up. In his two major works, the *Principles of Human Knowledge* (1710) and *Three Dialogues Between Hylas and Philonous* (1713), Berkeley presented two general arguments for his idealism: The first based on the differing representations we have of supposedly unchanging objects, and the second based on the very conceivability of something non-mental. The first general argument might be schematized as follows: Our perceptions of objects change with changes in us (e.g. objects appear different shapes from different perspective angles), but, on the view that there exists some non-mental material substance, the underlying substance needn't change with (e.g.) changes in our position. Yet there is no non-arbitrary way of determining which of those changing perceptions is *correct,* in the sense of revealing the true nature of the object. Because those perceptions are often incompatible, they cannot all reveal the nature of the object, but since they are all on par, the only reasonable conclusion is that none of them do. But that, Berkeley claimed, is obviously absurd; *of course* human perceptions say something about the nature of the object. That's why people use their perception in the first place. Given this, he thought that the only reasonable alternative was to identify the object with one's perceptions of it, thereby allowing one direct epistemic access to it (this relied on the uncontroversial assumption that people have direct access to their perceptions). The first argument, however, is not nearly strong enough to establish absolute idealism, which was Berkeley's aim. It leaves open the possibility that the objects people perceive have an unknown reality, as well as the possibility that there might be unperceivable and non-mental objects. To rule out those possibilities, Berkeley presented another line of argument. Accepting a strong form of empiricism, Berkeley claimed that the only understanding of "existence" one can have must be one derived from his experiences. Human experiences, however, are all of one's own mind and one's own representations. But in that case, the only *meaning* that existence can have is "to have a representation or be a representation." Material substance, however, was supposed to be something that was neither a representation nor a possessor of representations. The conclusion is that "material substance exists" is in fact a contradiction.

Berkeley's second argument (presented above) relied heavily on the claim that all of one's meaningful thoughts must be based in direct experience. While this thought has appealed to some philosophers (perhaps most notably, in the twentieth century, theological positivists), it strikes most people as highly problematic. For instance, people seem to be able to think thoughts with universal and necessary content (for instance, all events have a cause), even though experience alone seems insufficient to yield ideas of universality or necessity. Motivated by just such thoughts, Kant rejected the strong empiricist assumptions that underlay Berkeley's most radical arguments. Nevertheless, in his *Critique of Pure Reason,* he advanced arguments for forms of relative idealism about almost all qualities of objects, including their spatiality, temporality, and all sensible qualities. With respect to space and time, Kant believed that some form of idealism was required to explain the vast store of *a priori* knowledge people have concerning the spatial and temporal properties of objects (the clearest example being geometry). After all, people have experienced only a minute fraction of what exists, so they are hardly in a place to draw any inductive inference to such a conclusion. The only way one could explain this bulk of necessary, universal knowledge, Kant believed, was if space and time only existed as representations in the mind that one *imposes* on objects she encounters. Nevertheless, Kant was clear that this does not mean that the objects people encounter only *exist* in their representations. The objects exist on their own—it is rather a certain set of their properties that are ideal. They almost certainly have other properties beyond those people encounter, and those properties needn't have any relation to anything mental. Kant often puts this distinction in terms of a contrast between "things as they appear to us" and "things as they are in themselves." By emphasizing ignorance of how things are in themselves, Kant hoped to rule out the possibility that natural science (which has to do only with things as they appear) could disprove the existence of freedom of the will or the existence of God.

Basic Theme

Idealism is an approach to philosophical enquiry which asserts that direct and immediate knowledge can only be had through ideas or mental pictures. Objects that are the basis of these ideas can only be known indirectly or mediately. Idealism is often contrasted with materialism[2]. Idealism is based on the philosophy that "reality is mind-based or mind-dependent". All spatial expressions of human beings are the manifestations of man's thinking or thinking of society. Activity of mind is the foundation of human existence and knowledge. There is no real world that can be known independently of mind—the main point that separates social sciences from the natural sciences. For example, when people grow crops, build houses, exploit resources, clear forests, lay down roads, grow trees in deserts, dig canals or make any other use of space—their actions are the results of rational thought (logical thinking of mind); all these activities, which alter the natural landscape, are mind based. Idealism, therefore, is a system of philosophy that regards reality as residing in or constituted by the mind. To the idealist, all actions result from rational thought ensuing from a theoretical construct present at the back of the decision-makers' mind. Such theories are generally the part of socio-cultural milieu and may include religions, beliefs, myths and traditions. They represent the order which man has himself created through his social organization. Idealism holds that what we refer to and perceive as the external world is in some way an artifice of the mind. The idealist is concerned with what the actor believes, not why he believes it, since according to the idealist 'all knowledge is ultimately based on an individual's objective experience of the world, and comprises mental constructs and ideas'.

UNDERSTANDING IDEALISM

Contents

What idealism is may be clarified by approaching it through its basic doctrines and principles. Six basic conceptions are discussed here in order to distinguish idealistic philosophy:

(1) The Union of Individuality and Universality

Abstract universals, such as "canineness," which express the common nature or essence that the members of a class (e.g. individual dogs or wolves) share with one another, are acknowledged by many philosophers. Many idealists, however, emphasize the concept of a concrete universal, one that is also a concrete reality, such as "humankind" or "literature," which can be imagined as gatherable into one specific thing. As opposed to the fixed, formal, abstract universal, the concrete universal is essentially dynamic, organic, and developing. Thus, universality and individuality merge.

(2) The Contrast between Contemporaneity and Eternity

While most philosophers tend to focus on matters of contemporary concern, idealists always seek a much wider perspective that embraces epochs and eras in the broad sweep of history. In the words of the 17th century rationalist philosopher Benedict de Spinoza, they strive to view the contemporary world "under the aspect of eternity." Thus, in spite of the extensive formative influence of culture, idealists claim that their philosophy transcends the parochialism of a particular culture; and idealisms are found, in fact, in all of the major cultures of the world.

(3) The Doctrine of Internal Relations and the Coherence Theory of Truth

It seems natural to suppose, as non-idealists usually do, that the consideration of two things in their relatedness to one another can have no effect on the things themselves—i.e. that a relation is something in addition to the things or terms related and is thus external. On this basis, truth would be defined as a relation of correspondence between a proposition and a state of affairs. The idealist believes, however, that reality is more subtle than this. The relationship between a mineral deposit and the business cycle, for example, is an internal one: the deposit of an ore changes when prices render it profitable to mine the mineral. Similarly, it is part of the essence of a brick that it is related to a wall or pavement. Thus, terms and relations logically determine one another. Ultimate reality is, therefore, a system of judgements or propositions, and truth is defined in terms of the coherence of these propositions with one another to form a harmonious whole. Thus, a successful spy is judged either a hero or a villain only in relation to a total system of international relations, an accepted philosophy of history, and the moral judgements involved. There are, therefore, degrees of reality and degrees of truth within a system of truth cohering by internal relations, and the truth of a judgement reflects its place in this system.

(4) The Dialectical Method

Idealism seeks to overcome contradictions by penetrating into the overall coherent system of truth and continually creating new knowledge to be integrated with earlier discoveries. Idealism is thus friendly to all quests for truth, whether in the natural or behavioural sciences or in art, religion, and philosophy. It seeks the truth in every positive judgement and in its contradictory as well. Thus, it uses the dialectical method of reasoning to remove the contradictions characteristic of human knowledge. Such removal leads to a new synthetic judgement that incorporates in a higher truth the degree of truth that was present in each of the two lower judgements.

(5) The Centrality of Mind in Knowledge and Being

Idealism is not reductive, as are opposing philosophies that identify mind with matter and reduce the higher level of reality to the protons and electrons of mathematical physics. On the contrary, idealism defends the principle that the lower is explained by the higher—specifically, that matter can be explained by mind but that mind cannot be explained by matter. The word spirit can be substituted for "mind" or even placed above it; and at one time "spiritualism" was used, especially in Europe, as a synonym for idealism.

(6) The Transmutation of Evil into Good

Nearly all idealists accept the principle that the evils with which humankind has to deal may become ingredients in a larger whole that overcomes them. The American Hegelian Josiah Royce held that the larger whole is the Absolute Mind, which keeps evils under control as a person might hold a viper under the sole of his boot. Along with this doctrine of the sublimation or transmutation of evil, Royce incorporated into his metaphysics a point from the irrationalism of Arthur Schopenhauer, itself a voluntaristic form of idealism, viz. that "the world is my idea". Schopenhauer, however, was probably the only idealist who defended the converse principle that good is transmuted into evil.

DEFINITIONS

Idealism is a term with several related meanings. It comes via *idea* from the Greek *idein*, meaning "to see". The term entered the English language in 18th century. In ordinary use, as when speaking of Woodrow Wilson's political idealism, it generally suggests the priority of ideals, principles, values, and goals over concrete realities. Idealists are understood to represent the world as it might or should be, unlike pragmatists, who focus on the world as it presently is. In the arts, similarly, idealism affirms imagination and attempts to realize a mental conception of beauty, a standard of perfection, in opposition to aesthetic naturalism and realism.

Any philosophy that assigns crucial importance to the ideal or spiritual realm in its account of human existence may be termed "idealist". Metaphysical idealism is an ontological doctrine that holds that reality itself is incorporeal or experiential at its core. Beyond this, idealists disagree on which aspects of the mental are more basic. Platonic idealism affirms that abstractions are more basic to reality than the things we perceive, while subjective idealists and phenomenalists tend to privilege sensory experience over abstract reasoning. Epistemological idealism is the weaker view that reality can only be known through ideas, that only psychological experience can be apprehended by the mind. Subjective idealists like George Berkeley are anti-realists in terms of a mind-independent world, whereas transcendental idealists like Immanuel Kant are strong skeptics of such a world, affirming epistemological and not metaphysical idealism. Thus Kant defines *idealism* as "the assertion that we can never be certain whether all of our putative outer experience is not mere imagining". As a rule, transcendental idealists like Kant affirm idealism's epistemic side without committing themselves to whether reality is *ultimately* mental; objective idealists like Plato affirm reality's metaphysical basis in the mental or abstract without restricting their epistemology to ordinary experience; and subjective idealists like Berkeley affirm both metaphysical and epistemological idealism.[3]

Classical Idealism

Monistic idealism holds that consciousness, not matter, is the ground of all being. It is monist because it holds that there is only one type of thing in the universe and idealist because it holds that one thing to be consciousness. Anaxagoras (480 B.C.) introduced the term "*Nous*" ("Mind") and taught that "all things" were created by Mind, that Mind held the cosmos together and gave human beings a connection to the cosmos or a pathway to the divine. Many religious philosophies are specifically idealist. The belief that beings with knowledge (God/s, angels and spirits) preceded insentient matter seems to suggest that an experiencing subject is a necessary reality. Hindu idealism, for instance, is central to Vedanta philosophy.[4] Christian theologians have held idealist views, often based on Neoplatonism, despite the influence of Aristotelian scholasticism from the 12th century onward. Several modern religious movements, for example the organizations within the New Thought Movement and the Unity Church, may be said to have a particularly idealist orientation. The theology of

Christian Science includes a form of idealism: it teaches that all that truly exists is God and God's ideas; that the world as it appears to the senses is a distortion of the underlying spiritual reality, a distortion that may be corrected (both conceptually and in terms of human experience) through a reorientation (spiritualization) of thought. Wang Yangming, a Ming Chinese neo-Confucian philosopher, held that objects do not exist entirely apart from the mind because the mind shapes them. It is not the world that shapes the mind but the mind that gives reason to the world, so the mind alone is the source of all reason, having an inner light, an innate moral goodness and understanding of what is good.

Platonism and Neoplatonism

Plato's theory of forms or "ideas" describes ideal forms (for example the platonic solids in geometry or abstracts like Goodness and Justice), as universals existing independently of any particular instance. Arne Grøn calls this doctrine "the classic example of a metaphysical idealism as a *transcendent* idealism", while Simone Klein calls Plato "the earliest representative of metaphysical objective idealism".[5] Nevertheless, Plato holds that matter is real, though transitory and imperfect, and is perceived by our body and its senses and given existence by the eternal ideas that are perceived directly by our rational soul. Plato was, therefore, a metaphysical and epistemological dualist, an outlook that modern idealism has striven to avoid: Plato's thought cannot, therefore, be counted as idealist in the modern sense.

With the neoplatonist Plotinus, wrote Nathaniel Alfred Boll: "there even appears, probably for the first time in Western philosophy, *idealism* that had long been current in the East even at that time, for it taught...that the soul has made the world by stepping from eternity into time...". Similarly, in regard to passages from the Enneads, "The only space or place of the world is the soul" and "Time must not be assumed to exist outside the soul". Ludwig Noiré wrote: "For the first time in Western philosophy we find idealism proper in Plotinus. However, Plotinus does not address whether we know external objects, unlike Schopenhauer and other modern philosophers."

Subjective Idealism

Subjective Idealism (immaterialism or phenomenalism) describes a relationship between experience and the world in which objects are no more than collections or "bundles" of sense data in the perceiver. Proponents include Berkeley, Bishop of Cloyne, an Irish philosopher who advanced a theory he called immaterialism, later referred to as "subjective idealism", contending that individuals can only know sensations and ideas of objects directly, not abstractions such as "matter", and that ideas also depend upon being perceived for their very existence—*esse est percipi*; "to be is to be perceived".

Transcendental Idealism

Transcendental idealism, founded by Immanuel Kant in the 18th century, maintains that the mind shapes the world we perceive into the form of space-and-time. "If I remove the thinking subject, the whole material world must at once vanish because it is nothing but a phenomenal appearance in the sensibility of ourselves as a subject, and a manner or species of representation". Kant focused on ideas drawn from British philosophers such as Locke, Berkeley and Hume but distinguished his transcendental or critical idealism from previous varieties. Kant distinguished things as they appear to an observer and things in themselves, "that is, things considered without regard to whether and how they may be given to us". We cannot approach the *noumenon*, the "thing in Itself" (German: *Ding an sich*) without our own mental world. He added that the mind is not a blank slate, *tabula rasa* but rather comes equipped with categories for organizing our sense impressions.

Pluralistic Idealism

Pluralistic idealism such as that of Gottfried Leibniz takes the view that there are many individual minds that together underlie the existence of the observed world and make possible the existence of the physical universe. The pluralistic idealism does not assume the existence of a single ultimate mental reality.

Idealism in the Philosophy of Science

Idealist notions took a strong hold among physicists of the early 20th century confronted with the paradoxes of quantum physics and the theory of relativity. In *The Grammar of Science*, Preface to the 2nd Edition, 1900, Karl Pearson wrote, "There are many signs that a sound idealism is surely replacing, as a basis for natural philosophy, the crude materialism of the older physicists." This book influenced Einstein's regard for the importance of the observer in scientific measurements. Sir Arthur Eddington, a British astrophysicist of the early 20th century, wrote in his book *The Nature of the Physical World*, "The stuff of the world is mind-stuff": "The mind-stuff of the world is, of course, something more general than our individual conscious minds.... The mind-stuff is not spread in space and time; these are part of the cyclic scheme ultimately derived out of it.... It is necessary to keep reminding ourselves that all knowledge of our environment from which the world of physics is constructed, has entered in the form of messages transmitted along the nerves to the seat of consciousness.... Consciousness is not sharply defined, but fades into subconsciousness; and beyond that we must postulate something indefinite but yet continuous with our mental nature.... It is difficult for the matter-of-fact physicist to accept the view that the substratum of everything is of mental character. But no one can deny that mind is the first and most direct thing in our experience, and all else is remote inference."[6]

FORMS

Idealism, in philosophy, any view that stresses the central role of the ideal or the spiritual in the interpretation of experience. It may hold that the world or reality exists essentially as spirit or consciousness, that abstractions and laws are more fundamental in reality than sensory things, or, at least, that whatever exists is known in dimensions that are chiefly mental—through and as ideas. Thus, the two basic forms of idealism are *metaphysical idealism*, which asserts the ideality of reality, and *epistemological idealism*, which holds that in the knowledge process the mind can grasp only the psychic or that its objects are conditioned by their perceptibility. In its metaphysics, idealism is thus directly opposed to materialism, the view that the basic substance of the world is matter and that it is known primarily through and as material forms and processes. In its epistemology, it is opposed to realism, which holds that in human knowledge objects are grasped and seen as they really are—in their existence outside and independently of the mind.

As a philosophy often expressed in bold and expansive syntheses, idealism is also opposed to various restrictive forms of thought: to skepticism, with occasional exceptions as in the work of the modern British Hegelian F.H. Bradley; to positivism, which stresses observable facts and relations as opposed to ultimates and therefore, spurns the speculative "pretensions" of every metaphysics; and often to atheism, since the idealist commonly extrapolates the concept of mind to embrace an infinite Mind. The essential orientation of idealism can be sensed through some of its typical tenets: "Truth is the whole, or the Absolute"; "to be is to be perceived"; "reality reveals its ultimate nature more faithfully in its highest qualities (mental) than in its lowest (material)"; "the Ego is both subject and object."

CRITICISM

The most natural response to idealism is that it violates some tenet of common sense. Berkeley was well aware of this, and spent much of his *Three Dialogues* attempted to argue to the contrary.

Yet a sustained philosophical attack on idealism was made (largely in response to Hegelian idealism) by the British philosopher G.E. Moore in the early 20th century (Bertrand Russell made a parallel attack). Moore directly attacked that essential assumption of idealism, that what people are directly aware of are their representations. Instead, Moore proposed that people should understand the objects of their thoughts to be *propositions,* where propositions can be understood as states of affairs constituted by genuinely non-mental objects in the world. Such a picture has become the dominant one in contemporary analytic philosophy, and idealism is not often counted as a viably philosophical position. Nevertheless, defenders of idealism may well note that Moore's alternative picture is no more self-evident than the picture it meant to replace, so that the matter is far from settled.

NOTES

1. 'Idealism', Last updated 2 days ago.

2. Wikipedia, *The Free Encyclopaedia.*
3. In philosophy, materialism is that form of physicalism which holds that the only thing that can truly be said to *exist* is matter; that fundamentally, all things are composed of *material* and all phenomena are the result of material interactions. Science uses a working assumption, sometimes known as methodological naturalism, that observable events in nature are explained only by natural causes without assuming the existence or non-existence of the supernatural. In terms of singular explanations of the phenomenal reality, materialism stands in sharp contrast to *idealism.*
4. Mark Kulstad and Laurence Carlin, "Leibniz's Philosophy of Mind", *Stanford Encyclopedia of Philosophy.*
5. Arne Grøn, "Idealism". *Encyclopedia of Science and Religion.*
6. *Op. cit.*
7. A.S. Eddington, *The Nature of the Physical World,* 276-81.

REFERENCES

Encyclopedia Britannica.

Encyclopedia of Science and Religion.

Harvey, M.E. and B.P. Holly (1989), *Themes in Geographic Thought,* Rawat Publications, Jaipur and New Delhi.

Kant, Immanuel (1963), *Critique of Pure Reason.* Trans. N.K. Smith. London: MacMillan and Co.

New World Encyclopedia.

Sprigge, T.L.S. (1998). "Idealism." In E. Craig, ed. *Routledge Encyclopedia of Philosophy.* London: Routledge.

Stanford *Encyclopedia of Philosophy.*

The Encyclopedia of Philosophy, vol. 3, "Idealism," New York, 1967.

Wikipedia, *The Free Encyclopaedia.*

CHAPTER-19

Functionalism

Functionalism is a practical and utilitarian approach based on the concept of interaction. The diversity of function has resulted in diversity of meanings. However, taken in the simplest sense, it is a viewpoint that investigates functional linkages with emphasis on the goals, the needs and links between role and actor. In other words, functionalism is concerned with functions (occupations) and the analysis of the functions of a society. It is a perspective which views the world as a set of differentiated and interdependent systems, whose collective actions are instances of repeatable and predictable regularities in which form and function can be assumed to be related, and which explains these form-function relations in terms of their role in maintaining the continuity of the systems[1]. The distinction between Pragmatism and Functionalism is weak. Still the latter viewpoint is very popular among geographers. Taken as a viewpoint, it investigates the functional linkages between elements and is concerned mainly with the function of the object or phenomenon, rather than its aesthetic consideration. The proponents believe in two types of functions, viz. 'manifest' (visible) and 'latent' (hidden/unseen). The importance of the centres like Mumbai, Tatanagar and Gulmurg, for instance, can be explained in terms of their functions—as a chief port, an iron and steel manufacturing centre, and a centre of tourism, respectively. Then each of them has its manifest and latent function. For example, manufacturing of iron and steel is the manifest function of Tatanagar (Bihar, India) but its latent function is the exchange of ideas and social meeting of people with each other. Following functionalism, the human societies should be examined holistically in an interrelated system's framework. However, the functionalists are less interested in the history of a society, but more concerned with social interaction. They attempt to find the interrelationships between the compounds of social structure. Recently, the functional approach has become very popular in geography.

Functionalism is the most familiar view among philosophers of mind and cognitive science. It is a theory about the nature of mental states (metaphysics). Accordingly, mental states are identified by what they do rather than by what they are made of. In particular, the original motivation for functionalism comes from the comparison of minds with computers. But that is only an analogy. Functionalism is a theory of the mind in contemporary philosophy, developed largely as an alternative to both the identity theory of mind and behaviourism.

Basic Theme

Functionalism is the doctrine believing that mental state of a particular type does not depend on its internal constitution, but rather on the way it functions, or the role it plays, in the system of which it is a part. The main arguments for functionalism depend on showing that it is superior to its primary competitors: identity theory and behaviourism. Contrasted with behaviourism, functionalism retains the traditional idea that mental states are internal states of thinking creatures. Contrasted with identity theory, functionalism introduces the idea that mental states are multiply realized. Objectors to functionalism generally charge that it classifies too many things as having mental states, or at least more states than psychologists usually accept. The effectiveness of the arguments for and against functionalism depends in part on the particular variety in question, and whether it is a stronger or weaker version of the theory. More precisely, functionalist theories take the identity of a mental state to be determined by its causal relations to sensory stimulations, other mental states and behaviour. Its core idea is that mental states (beliefs, desires, being in pain, etc.) are constituted solely by their functional role—that is, they are causal relations to other mental states, sensory inputs, and behavioural outputs.[2] Functionalism is a theoretical level between the physical implementation and behavioural output.[3] Therefore, it is different from its predecessors of Cartesian dualism (advocating independent mental and physical substances) and Skinnerian behaviourism and physicalism (declaring only physical substances) because it is only concerned with the effective functions of the brain, through its organization or its 'software programmes'. Since, mental states are identified by a functional role, they are said to be realized on multiple levels; in other words, they are able to be manifested in various systems, even perhaps computers, so long as the system performs the appropriate functions. While computers are physical devices with electronic substrate that perform computations on inputs to give outputs, so brains are physical devices with neural substrate that perform computations on inputs which produce behaviours.

Roots

This doctrine is rooted in Aristotle's conception of the soul, and has antecedents in Hobbes' conception of the mind as a "calculating machine", but it has become fully articulated (and popularly endorsed) only in the last third period of the 20th century. Although functionalism attained its greatest prominence as a theory of mental states in the last third period of the 20th century, it has antecedents in both modern and ancient philosophy, as well as in early theories of computation and artificial intelligence. The earliest view that can be considered an ancestor of functionalism is Aristotle's theory of the soul (350 B.C.). In contrast to Plato's claim that the soul can exist apart from the body, Aristotle argued that the (human) soul is the *form* of a natural, organized human body—the set of powers or capacities that enable it to express its "essential whatness", which for Aristotle is a matter of fulfilling the function or purpose that defines it as the kind of thing it is. Just as the form of an axe is whatever enables it to cut, and the form of an eye is whatever enables it to see, the (human) soul is to be identified with whichever powers and capacities enable a natural, organized human body to fulfil its defining function, which, according to Aristotle, is to survive and flourish as a living, acting, perceiving and reasoning being. So, Aristotle argues, the soul is inseparable from the body, and comprises whichever capacities are required for a body to live, perceive, reason and act.

In a seminal paper[4], A.M. Turing proposed that the question, "Can machines think?" can be replaced by the question, "Is it theoretically possible for a finite state digital computer, provided with a large but finite table of instructions, or programme, to provide responses to questions that would fool an unknowing interrogator into thinking it is a human being?" Now, in deference to its author, this question is most often expressed as "Is it theoretically possible for a finite state digital computer (appropriately programmed) to pass the Turing Test?" Turing identifies thoughts with states of a system defined solely by their roles in producing further internal states and verbal outputs, a view that has much in common with contemporary functionalist theories. Indeed, Turing's work was explicitly invoked by many theorists during the beginning stages of 20th century functionalism and was the avowed inspiration for a class of theories, the "machine state" theories are most firmly associated with Hilary Putnam (1960, 1967)[5] that had an important role in the early development of the doctrine.

Other important recent antecedents of functionalism are the behaviourist theories that emerged in the early-to-mid 20th century. The development of two important strains of functionalism, "psycho-functionalism" and "analytical" functionalism are related, respectively, to empirical and logical behaviourism. As an empirical psychological theory, behaviourism holds that the behaviour of humans (and other animals) can be explained by appealing solely to behavioural dispositions, that is, to the law like tendencies of organisms to behave in certain ways, given certain environmental stimulations. Behavioural dispositions, unlike thoughts, feelings and other internal states that can be directly observed only by introspection, are objectively observable and are indisputably part of the natural world. Logical behaviourism, in contrast to behaviourism as a psychological theory, is a thesis about the meanings of our mental state terms or concepts. According to logical behaviourism, all statements about mental states and processes are equivalent in meaning to statements about behavioural dispositions.

TYPES OF FUNCTIONALISM

The functionalist theories belong to one of three major strains—machine state functionalism, psycho-functionalism and analytic functionalism.

Machine State Functionalism

The broad position of "functionalism" can be articulated in many different varieties. The first formulation of a functionalist theory of mind was put forth by Hilary Putnam.[6] This formulation, which is now called machine-state functionalism, or just machine functionalism, was inspired by the analogies which Putnam and others noted between the mind and the theoretical machines or computers capable of computing any given algorithm which were developed by Alan Turing (called Universal Turing machines). The early functionalist theories of Putnam (1960, 1967) can be seen as a response to the difficulties facing behaviourism as a scientific psychological theory, and as an endorsement of the (new) computational theories of mind which were becoming increasingly significant rivals to it. According to Putnam's *machine state functionalism*, any creature with a mind can be regarded as a Turing machine (an idealized finite state digital computer), whose operation can be fully specified by a set of instructions (a 'machine table' or programme) each having the form:

> If the machine is in state S_i, and receives input I_j, it will go into state S_k and produce output O_l (for a finite number of states, inputs and outputs).

A machine table of this sort describes the operation of a *deterministic* automation, but most machine state functionalists take the proper model for the mind to be that of a *probabilistic* automation: one in which the programme specifies, for each state and set of inputs, the *probability* with which the machine will enter some subsequent state and produce some particular output. On either model, however, the mental states of a creature are to be identified with such "machine table states" ($S_1, ..., S_n$). These states are not mere behavioural dispositions, since they are specified in terms of their relations not only to inputs and outputs, but also to the state of the machine at the time. For example, if *believing it will rain* is regarded as a machine state, it will not be regarded as a disposition to take one's umbrella after looking at the weather report, but rather as a disposition to take one's umbrella if one looks at the weather report and is in the state of wanting to stay dry. So machine state functionalism can avoid what many have thought to be a fatal difficulty for behaviourism. In addition, machines of this sort provide at least a simple model of how internal states whose effects on output occur by means of mechanical processes can be viewed as *representations* (though the question of *what*, exactly, they represent has been an ongoing topic of discussion). Finally, machine table states are not tied to any particular physical (or other) realization; the same programme, after all, can be run on different sorts of computer hardware. It's easy to see, therefore, why Turing machines provided a fruitful model for early functionalist theories. But because machine table states are total states of a system, the early functionalist equation of mental states with machine table states faded in importance as a model for the functional characterization of the complex of distinct internal states that can be simultaneously realized in a human (or other) subject. Nonetheless, the idea that internal states can be fully described in terms of their relations to input, output, and No: *one another*, and can figure in law like descriptions, and predictions, of a system's output, was a rich and important idea that is retained by contemporary functionalist theories.

Psycho-Functionalism

A second form of functionalism is based on the rejection of behaviourist theories in psychology and their replacement with empirical cognitive models of the mind. The fundamental idea of psycho-functionalism is that psychology is an irreducibly complex science and that the terms that we use to describe the entities of mind in our best psychological theories cannot be redefined in terms of simple behavioural dispositions, and further, that such a redefinition would not be desirable were it achievable. Psycho-functionalists view psychology as employing the same sorts of irreducibly teleological or purposive explanations as the biological sciences. Thus, for example, the function or role of the heart is to pump blood, that of the kidney is to filter it and to maintain certain chemical balances and so on—this is what accounts for the purposes of scientific explanation and taxonomy. There may be an infinite variety of physical realizations for all of the mechanisms, but what is important is only their role in the overall biological theory. In an analogous manner, the role of mental states, such as belief and desire, is determined by the functional or causal role that is designated for them within our best *scientific* psychological theory.

This strain of functionalism derives primarily from reflection upon the goals and methodology of "cognitive" psychological theories. In contrast to the behaviourist's insistence that the laws of psychology appeal only to behavioural dispositions, cognitive psychologists argue that the best empirical theories of behaviour take it to be the result of a complex of mental states and processes, introduced and individuated in terms of the roles they play in producing the behaviour to be explained. For example, a psychologist may begin to construct a theory of memory by postulating the existence of "memory trace" decay, a process whose occurrence or absence is responsible for effects such as memory loss and retention, and which is affected by stress or emotion in certain distinctive ways.[7] On a theory of this sort, what makes some neural process an instance of memory trace decay is a matter of how it functions, or the role it plays, in a cognitive system; its neural or chemical properties are relevant only in so far as they enable that process to do what trace decay is hypothesized to do; and similarly for all mental states and processes invoked by cognitive psychological theories. Cognitive psychology, that is, is intended by its proponents to be a 'higher-level' science like biology: just as, in biology, physically disparate entities can all be hearts as long as they function to circulate blood in a living organism, and physically disparate entities can all be eyes as long as they enable an organism to see, disparate physical structures or processes can be instances of memory trace decay—or more familiar phenomena such as thoughts, sensations and desires—as long as they play the roles described by the relevant cognitive theory. Psycho-functionalism, therefore, can be seen as straightforwardly adopting the methodology of cognitive psychology in its characterization of mental states and processes as entities defined by their role in a cognitive psychological theory. All versions of functionalism, however, can be regarded as characterizing mental states in terms of their roles in *some* psychological theory or other. What is distinctive about psycho-functionalism is its claim that mental states and processes are just those entities, with just those properties, postulated by the best *scientific* explanation of human behaviour. This means, first, that the form of the theory can diverge from the "machine table" specifications of machine state functionalism. It also means that the information used in the functional characterization of mental states and processes needn't be restricted to what is considered common knowledge or common sense, but can include information available only by careful laboratory observation and experimentation. For example, a psycho-functional theory might be able to distinguish phenomena such as depression from sadness or listlessness even though the distinctive causes and effects of these syndromes are difficult to untangle solely by consulting intuitions or appealing to common sense. And psycho-functional theories will not include characterizations of mental states for which there is no scientific evidence, such as Buyerto's regret or hysteria, even if the existence and efficacy of such states is something that common sense affirms.

Analytic Functionalism

There is another important strain of functionalism, "analytic" functionalism that takes there to be reason to restrict the defining theory not just to generalizations sufficiently close to those that "the folk" take to hold between mental states, environmental stimulations, and behaviour, but rather to *a priori* information about these relations. This is because, for analytic functionalists, there are equally important goals that require strictly a priori characterizations of mental states. This form of functionalism is concerned with the meanings of theoretical terms in general. This view is most closely associated with David Lewis[8] and is often referred to as conceptual functionalism. The basic idea of analytic functionalism is that theoretical terms are implicitly defined by the theories in whose formulation they occur and not by intrinsic properties of the phonemes they comprise. In the case of ordinary language terms, such as "belief", "desire", or "hunger", the idea is that such terms get their meanings from our commonsense "folk psychological" theories about them, but that such conceptualizations are not sufficient to withstand the rigor imposed by materialistic theories of reality and causality. Such terms are subject to conceptual analyses which take something like the following form:

Mental state M is the state that is pre-conceived by P and causes Q.

For example, the state of *pain* is caused by sitting on a tack and causes loud cries, and higher order mental states of anger and resentment directed at the careless person who left a tack lying around. These sorts of functional definitions in terms of causal roles are claimed to be *analytic* and *a priori* truths about the sub mental states and the (largely fictitious) propositional attitudes they describe. Hence, its proponents are known as *analytic* or *conceptual* functionalists. The essential difference between analytic and psycho-functionalism is that the latter emphasizes the importance of laboratory observation and experimentation in the determination of which mental state terms and concepts are genuine and which functional identifications may be considered to be genuinely contingent and *a posteriori* identities. The former, on the other hand, claims that such identities are necessary and not subject to empirical scientific investigation.

Like the logical behaviourism from which it emerged, the goal of analytic functionalism is to provide "topic-neutral" translations, or analyses, of our ordinary mental state terms or concepts. Analytic functionalism, of course, has richer resources than logical behaviourism for such translations, since it permits reference to the causal relations that a mental state has to stimulations, behaviour, *and other mental states.* True enough, analytic functionalists can acknowledge, terms like 'pain', 'thought', and 'desire' are not equivalent to any descriptions expressed in the language of physics, chemistry, or neurophysiology. But if there are *functional* descriptions that preserve the meanings of these terms, then a creature's mental states can be identified simply by determining which of that creature's internal states and processes play the relevant functional roles. And since the capacity to play these roles is merely a matter of having certain causal relations to stimulations, behaviour, and one another, the possession of these properties is compatible with a materialistic theory of the mind.[9]

OBJECTIONS

There have been many objections to functionalism, however, that apply to all versions of the theory.

Functionalism and Holism

One difficulty for every version of the theory is that functional characterization is *holistic.* Functionalists hold that mental states are to be characterized in terms of their roles in a psychological theory—be it common sense, scientific, or something in between—but all such theories incorporate information about a large number and variety of mental states. Thus if pain is interdefined with certain highly articulated beliefs and desires, then animals who don't have internal states that play the roles of our articulated beliefs and desires can't share our pains, and humans without the capacity to feel pain can't share certain of our beliefs and desires. In addition, differences in the ways people reason, the ways their beliefs are fixed, or the ways their desires affect their beliefs—due either to cultural or individual idiosyncrasies—might make it impossible for them to share the same mental states. These are regarded as serious worries for all versions of functionalism.

Functionalism and Mental Causation

Another worry for functionalism is the "causal exclusion problem", the worry about whether role functionalism can account for what we take to be the causal efficacy of our mental states. For example, if pain is realized in me by some neural state-type, then in so far as there are purely physical law—like generalizations linking states of that type with pain behaviour; one can give a complete causal explanation of my behaviour by citing the occurrence of that neural state (and the properties by virtue of which it figures in those laws). And thus, some have argued, the higher-level role properties of that state—its being a pain—are causally irrelevant. There have been a number of different responses to this problem. Some suggest that it arises from an overly restrictive account of causation, in which a cause must "generate" or "produce" its effect, a view which would count the macroscopic properties of other special sciences as causally irrelevant as well. Instead, some argue, causation should be regarded as a special sort of counterfactual dependence between states of certain types, or as a special sort of regularity that holds between them[10]. If this is correct, then functional role properties (along with the other macroscopic properties of the special sciences) could count as causally efficacious.

Alternatively, some argue that causation is best regarded as a relation between types of events that must be invoked to provide sufficiently general explanations of behaviour. Though many who are moved by the exclusion problem maintain that there is a difference between generalizations that are truly causal and those that contribute in some other (merely epistemic) way to our understanding of the world, theorists who advocate this response to the problem charge that this objection, once again, depends on a restrictive view of causation that would rule out too much.

Functionalism and Introspective Belief

Another important question concerns the beliefs that we have about our own "occurrent" (as opposed to dispositional) mental states such as thoughts, sensations, and perceptions. We seem to have immediately available, non-inferential beliefs about these states, and the question is how this is to be explained if mental states are identical with functional properties. The answer depends on what one takes these introspective beliefs to involve. Broadly speaking, there are two dominant views of the matter. One popular account of introspection—the "inner sense" model on which introspection is taken to be a kind of "internal scanning" of the contents of one's mind has been taken to be unfriendly to functionalism, on the grounds that it's hard to see how the objects of such scanning could be second-order relational properties of one's neural states. Some theorists, however, have maintained that functionalism can accommodate the special features of introspective belief on the "inner sense" model, since it would be only one of many domains in which it's plausible to think that we have immediate, non-inferential knowledge of causal or dispositional properties. Another account of introspection, is that the immediacy of introspective belief follows from the fact that occurrent mental states and our introspective beliefs about them are functionally interdefined. For example, one satisfies the definition of being in pain only if one is in a state that tends to cause (in creatures with the requisite concepts who are considering the question) the belief that one is in pain, and one believes that one is in pain only if one is in a state that plays the belief role, and is caused directly by the pain itself. On this account of introspection, the immediacy and non-inferential nature of introspective belief is not merely compatible with functionalism, but required by it.

Functionalism and the Norms of Reason

Yet another objection to functionalism raises the question of whether any "theory" of the mind that invokes beliefs, desires and other intentional states could ever be, or even approximate, an empirical theory. Whereas, even analytic functionalists hold that mental states are implicitly defined in terms of their (causal or probabilistic) roles in *producing* behaviour, these critics take mental states, or at least intentional states, to be implicitly defined in terms of their roles in *rationalizing* or *making sense of* behaviour. This is a different enterprise, they claim, since rationalization, unlike causal explanation, requires showing how an individual's beliefs, desires, and behaviour conform, or at least approximate, to certain *a priori norms* or *ideals* of theoretical and practical reasoning—prescriptions about how we *should* reason, or what, given our beliefs and desires, we *ought* to do. Thus the defining ("constitutive") normative or rational relations among intentional states expressed by these principles cannot be expected to correspond to empirical relations among our internal states, sensory stimulations and behaviour, since they comprise a kind of explanation that has sources of evidence and standards for correctness that are different from those of empirical theories. One can't, that is, extract facts from values.

Thus, although attributions of mental states can in some sense explain behaviour, by permitting an observer to 'interpret' it as making sense, they should not be expected to denote entities that figure in empirical laws. This is not to say, these theorists stress, that there are no causes, or empirical laws of, behaviour. These, however, will be expressible only in the vocabularies of the neurosciences, or other lower-level sciences, and not as relations among beliefs, desires and behaviour. Functionalists have replied to these worries in different ways. Many just deny the intuition behind the objection, and maintain that even the strictest conceptual analyses of our intentional terms and concepts purport to define them in terms of their *bonafide* causal roles, and that any norms they reflect are explanatory rather than prescriptive. They argue, that is, that if these generalizations are idealizations, they are the sort of idealizations that occur in any scientific theory: just as Boyle's Law depicts the relations between the temperature, pressure, and volume of a gas under certain ideal experimental conditions, our *a priori* theory of the mind consists of descriptions of what normal humans *would* do under (physically specifiable) ideal conditions, not prescriptions as to what they *should*, or are *rationally required*, to do.

While functionalism has its advantages, there have been several arguments against it, claiming that it is an insufficient account of the mind. An important part of some accounts of functionalism is the idea of multiple realizability. Since, according to standard functionalist theories, mental states are the corresponding functional role, mental states can be sufficiently explained without taking into account the underlying physical medium (e.g. the brain, neurons, etc.) that realizes such states; one need only take into account the higher-level functions in the cognitive system. Since, mental states are not limited to a particular medium, they can be realized in multiple ways, including, theoretically, within non-biological systems, such as computers. In other words, a silicon-based machine could, in principle, have the same sort of mental life that a human being has, provided that its cognitive system realized the proper functional roles. Thus, mental states are individuated much like a valve; a valve can be made of plastic or metal or whatever material, as long as it performs the proper function (say, controlling the flow of liquid through a tube by blocking and unblocking its pathway).

However, there have been some functionalist theories that combine with the identity theory of mind, which deny multiple realizability. Such Functional Specification Theories (FSTs), as they are called, were most notably developed by David Lewis and David Malet Armstrong.[11] According to FSTs, mental states are the particular "realizers" of the functional role, not the functional role itself. The mental state of belief, for example, just is whatever brain or neurological process that realizes the appropriate belief function. Thus, unlike standard versions of functionalism (often called "Functional State Identity Theories", FSTs do not allow for the multiple realizability of mental states, because the fact that mental states are realized by brain states is essential. What often drives this view is the belief that if we were to encounter an alien race with a cognitive system composed of significantly different material from humans' (e.g., silicon-based) but performed the same functions as human mental states (e.g. they tend to yell "Yowzas!" when poked with sharp objects, etc.) then we would say that their type of mental state is perhaps similar to ours, but too different to say it's the same. For some, this may be a disadvantage to FSTs. Indeed, one of Hilary Putnam's arguments for his version of functionalism relied on the intuition that such alien creatures would have the same mental states as humans do, and that the multiple realizability of standard functionalism makes it a better theory of mind.

THE FUTURE

In the last part of the 20th century, functionalism stood as the dominant theory of mental states. Like behaviourism, functionalism takes mental states out of the realm of the "private" or subjective, and gives them status as entities open to scientific investigation. But, in contrast to behaviourism, functionalism's characterization of mental states in terms of their roles in the production of behaviour grants them the causal efficacy that common sense takes them to have. And in permitting mental states to be multiply realized, functionalism offers an account of mental states that is compatible with materialism, without limiting the class of those with minds to creatures with brains like ours. The sophistication of functionalist theories has increased since their introduction, but so has the sophistication of the objections to functionalism, especially to functionalist accounts of mental causation, introspective knowledge, and the qualitative character of experiential states. For those unconvinced of the plausibility of dualism, however, and unwilling to restrict mental states to creatures physically like ourselves, the initial attractions of functionalism remain. The primary challenge for future functionalists, therefore, will be to meet these objections to the doctrine, either by articulating a functionalist theory in increasingly convincing detail, or by showing how the intuitions that fuel these objections can be explained away.

NOTES

1. Husain, M. (1995), 369-70.
2. Block, Ned. (1996).
3. Marr, D. (1982).
4. Turing, A.M. (1950).
5. Putnam, H. (1960); Putnam, H. (1967a and b).
6. Putnam, Hilary, (1960 and 1967).
7. Fodor, J. (1968).
8. David Kellogg Lewis (September 28, 1941—October 14, 2001) was an American philosopher.
9. Lewis, D. (1966).
10. Melnyk, Andrew. (2003 a and b).
11. Armstrong, D.M. (1968).

REFERENCES

Armstrong, D.M. (1968), *A Materialistic Theory of the Mind*. London: RKP.

Block, Ned. (1980a), "Introduction: What Is Functionalism?" in *Readings in Philosophy of Psychology*. Cambridge, MA: Harvard University Press.

Block, Ned. (1996), "What is functionalism?" in *The Encyclopedia of Philosophy Supplement*, Macmillan.

Fodor, J. (1968), *Psychological Explanation*. New York: Random House.

Harvey, M.E. and B.P. Holly (1989), *Themes in Geographic Thought*, Rawat Publications, Jaipur and New Delhi.

Husain, M. (1995), *Evolution of Geographical Thought*, 3rd Revised Edition, Rawat Publications, Jaipur.

Internet Encyclocpaedia of Philosophy.

Levin, Janet. (2004), "Functionalism", *The Stanford Encyclopedia of Philosophy* (Fall 2004 Edition), E. Zalta (ed.).

Lewis, D. (1966), "An Argument for the Identity Theory". *Journal of Philosophy*, 63: 17-25.

Marr, D. (1982), *Vision: A Computational Approach*. San Francisco: Freeman and Co.

Melnyk, Andrew (2003a), *A Physicalist Manifesto: Thoroughly Modern Materialism*. Cambridge: Cambridge University Press.

Melnyk, Andrew (2003b), "Some Evidence for Physicalism." In *Physicalism and Mental Causation: The Metaphysics of Mind and Action*, eds. Sven Walter and Heinz-Dieter Heckmann, 155-172.

Putnam, H. (1960) "Minds and Machines", in S. Hook (ed.), *Dimensions of Mind*, New York: University of New York Press, pp. 148-80.
Putnam, H. (1967a) "The Mental Life of some Machines", in Hector-Neri Castañeda (ed.), *Intentionality, Minds and Perception*, Detroit: Wayne State University Press, pp. 177-200.
Putnam, H. (1967b) "The Nature of Mental States", in Captain, W.H. and Merrill, D.D. (eds.), *Art, Mind and Religion*, Pittsburgh: University of Pittsburgh Press, pp. 37-48.
Turing, A.M. (1950). Computing machinery and intelligence. *Mind*, 59, 433-460.
Wikipedia, *The Free Encyclopedia.*

CHAPTER-20

Realism

Realism is commonly defined as a concern for fact or reality and a rejection of the impractical and visionary. As a word in common use, however, realism is employed to suggest a wide variety of meanings. Broadly speaking, realism denotes two distinct sets of philosophical theories, one regarding the nature of universal concepts and the other dealing with knowledge of objects in the world. Accordingly, a distinction may be made between the 'Direct' (or Naïve) Realism and the 'Philosophical' Realism. Indeed, the realism is a very interesting and different philosophy. In whatever form the view of realism is taken, it is certainly a mix of both idealism and positivism. It is the view regarding things in their true nature and dealing with them as they are. Realism says that facts (whether hidden or seen) speak themselves. Accordingly, the realities are neither totally mind-based (like idealism) nor solely directly observable (like positivism).

In its most general form realism asserts that objects in the external world exist independently of what is thought about them. The most straightforward of such theories is usually known as "Naïve Realism". It contends that in perception humans are made directly aware of objects and their attributes and thus have immediate access to the external world. Thus, Direct or Naïve Realism is a theory of perception that claims that the senses provide us with direct awareness of the external world. Perception is an exemplar of direct contact with something. Examples of indirect perception might be seeing something in a photograph, or hearing a recording of a voice. Most people, until they start thinking philosophically, are naïve realists. This theory is also known as "common sense realism". It holds that the view of the world that we derive from our senses is to be taken at face value. There are objects out there in the world, and those objects have the properties that they appear to us to have. For instance, if one has an experience as of a large apple tree, then that's because there's a large apple tree in front of him. If the apples on the tree appear to him to be red, then that's because there are objects in front of him—apples—that have the property of redness. In other words, this form of realism claims that the world is pretty much as it appears to our senses. All objects are composed of matter; they occupy space, and have properties such as size, shape, texture, smell, taste and colour. These properties are perceived directly. So, when we look at and touch things we see and feel those things themselves and so perceive them as they really are. The naïve or direct realism is distinguished from 'representative' realism, the view that we cannot perceive the external world directly; or 'critical realism', the philosophy of perception concerned with the accuracy of human sense-data; or the 'transcendental realism', the concept implying that individuals have a perfect understanding of the limitations of their own minds.[1]

The Philosophical Realism, on the other hand, is the belief that reality exists independently of observers. Also referred to as Metaphysical Realism it refers to the thesis that general properties, technically known as universals, have a form of reality that is in a certain sense independent of the things that possess them. This view of realism believes that a particular phenomenon is only appearance which shall disappear in due course of time[2]. For example, a specific mountain, like the Himalayas, does not exist; it will wear down to the ocean floor over geological time. Contrary to this, the general and universal term 'mountain' is rigid and fixed. Philosophical realism is also referred to as *Platonic realism* or *Scholastic realism*. Plato was a Greek philosopher, who lived during 427-347 B.C. He was a student of Socrates and the teacher of Aristotle. The rational theory that the forms we see, touch, taste and smell in time and space do not exist and are not knowable with certainty by our senses was first put by Platonic-Socratic writers[3]. Platonic realism describes a philosophy articulated by Plato, positing the existence of universals. In its simplest sense, according to philosophical realism the theory of knowledge that maintains that "universals" (general concepts representing the common elements belonging to individuals of the same genus or species) have a separate existence apart from individual objects. Plato's insistence that there is a realm of universals above the material universe as real as individual objects themselves had a great influence on medieval thought. The philosophical realists tend to believe that whatever we believe now is only an approximation of reality and that every new observation brings us closer to understanding reality. They tend to embrace what they believe is actually real, despite how unattractive reality itself may be. Such realists arrive at their understanding of reality through critical thinking.

SCIENTIFIC REALISM

Scientific realism is, at the most general level, the view that the world described by science is the real world, as it is, independent of what we might take it to be. Within philosophy of science, it is often framed as an answer to the question "how is the success of science to be explained?" The debate over what the success of science involves centres primarily on the status of unobservable entities apparently talked about by scientific theories. Generally, those who are scientific realists assert that one can make reliable claims about unobservables (viz., that they have the same ontological status) as observables, as opposed to instrumentalism. Scientific realism involves two basic positions. First, it is a set of claims about the features of an ideal scientific theory; an ideal theory is the sort of theory science aims to produce. Second, it is the commitment that science will eventually produce theories very much like an ideal theory and that science has done pretty well thus far in some domains. It is important to note that one might be a scientific realist regarding some sciences while not being a realist regarding others. For example, one might hold realist attitudes towards physics, chemistry and biology, and not towards economics, psychology and sociology.

ROOTS

Realism is a widely used term in the arts. In literature, it came into being as a response to Romanticism[4]. While Romanticism focused on the inner, spiritual side of human nature, and was skewed towards the exceptional and sublime, Realism focused on the mundane, the everyday. Realism focused on the ideology of objective reality and revolted against exaggerated emotionalism of Romanticism. It was more "democratic" in orientation, interested in the life of the majority, not the elite. As an artistic strategy, it was an attempt to focus literature on the objective, the concrete, the physical and social milieu was depicted in painstaking detail to convey the ethos of the society. Characters were portrayed in their social setting, which shaped their actions and their choices. Realism is often referred to as an attempt to portray things "as they are".[5]

Literary Realism began as a cultural movement with its roots in France, where it was a very popular art form, not only in France but the in rest of Europe as well, from the mid to late 1800s. It was aided with the introduction of photography—a new visual source that created a desire for people to produce things that looked "objectively real". It became popular in America largely in the early twentieth century.

In the visual arts it refers to a style of depiction that attempts to portray subjects as they appear in everyday life, without embellishment or interpretation. The term is also used to describe works of art which, in revealing a truth, may emphasize the ugly or sordid. Realist artists focused on that side of "reality" which had often been excluded in Romantic art, the unflattering truth of the underside of elite culture.

Realism is a literary technique practiced by many schools of writing. This style focuses particularly on the representation of middle-class life and is a reaction against Romanticism. According to William Harmon and Hugh Holman, "Where romanticists transcend the immediate to find the ideal, and naturalists plumb the actual or superficial to find the scientific laws that control its actions, realists centre their attention to a remarkable degree on the immediate, the here and now, the specific action, and the verifiable consequence". In American literature, the term "realism" encompasses the period of time from the Civil War to the turn of the century, during which William Dean Howells, Rebecca Harding Davis, Henry James, Mark Twain, and others wrote fiction devoted to accurate representation and exploration of American lives in various contexts.

Nineteenth Century Realism

Realism was a response to both Neo-classicism and Romanticism, and for the entire group, history had no artistic relevance or importance.[2] Gustave Courbet, the leader of the realism movement, defined Realism as a "human conclusion which awakened the very forces of man against paganism, Greco-Roman art, the Renaissance, Catholicism, and the gods and demigods, in short against the conventional ideal." The Realists, who were influenced by the Dutch and Flemish naturalists of the 17th century, were dedicated wholeheartedly to an establishment founded on justice for the working class, the ordinary citizens of society. In fact, all the artists, politicians, economists, and critics congregated in the Andler Keller, a type of restaurant serving food at all hours, which eventually became known as the temple of Realism. In 1863, after being shunned by Count Nieuwerkerke at the Universal Exposition of 1855, Courbet and friends organized a Salon de Refusés. This was an exhibition that included the works of those who are now recognized as the premier painters of the period. Astoundingly, two of the greatest Realist masters, Daumier and Courbet, were actually forced to serve prison sentences as a result of their involvement in the rebellion against uniformity.

With the arrival of photography, the world of visual arts would be altered significantly. The idea of photography itself was not new, and some artists had even employed some form of it. The concept of photography revolved around light passing through a small aperture as it registers the image of its subjects upon any surface which it may strike. The *camera obscura* was used by artists throughout the ages and specialized particularly by Vermeer. Daguerreotypes soon became popular by the hundreds of thousands. The first photo portrait was made by Samuel F.B. Morse, inventor of the telegraph. The possibilities were enormous, but for many artists, a point of concern. With the invention of photography, the art of portraiture would become almost non-existent. By 1858, photography was an assured fact, and photographers were able to prove at last how living beings really look in motion, to the great discomfiture of artists in the classic tradition with their contrived poses. In other words, photographs capture the essence of the action, the movement as it is, and there is absolutely no doubt in the veracity or accuracy of the photograph. This fits in perfectly with the realists because their sole focus is to portray the world, as it is, and not in a blown-up, romantic manner.

THE REALIST DOCTRINE

"The doctrine of the Realists, like those of the Neo-classicists and the Romanticists could not be maintained for long in their original purity. Too many aspects of natural human feeling and imagination were excluded. But the immense historic value of Realism lay in its insistence on the priority of vision over either abstract principles of form and composition or emotional and narrative content. The Realist emphasis on the here and now was instrumental in the formation of Impressionism. It would reappear minimally in the twentieth century, and at times, was even more fanatical.

NOTES

1. 'Realism', Last updated 4 days ago Wikipedia, *The Free Encyclopaedia.*
2. Husain, 1995, 377-80.
3. Harvey and Holly, 1989, 148-62.
4. Romanticism (or the Romantic Era/Period) was an artistic, literary, and intellectual movement that originated in Europe toward the end of the 18th century and in most areas was at its peak in the approximate period from 1800 to 1840. Partly a reaction to the Industrial Revolution, it was also a revolt against aristocratic social and political norms of the Age of Enlightenment and a reaction against the scientific rationalization of nature. It was embodied most strongly in the visual arts, music, and literature, but had a major impact on historiography, education and the natural sciences.
5. 'Realism', *New World Encyclopaedia.*

REFERENCES

Harvey, M.E. and B.P. Holly (1989), *Themes in Geographic Thought*, Rawat Publications, Jaipur and New Delhi.

Husain, M. (1995), *Evolution of Geographical Thought*, Third Revised Edition, Rawat Publications, Jaipur.

New World Encyclopaedia.

Peet, R. (1998), *Modern Geographical Thought*, Blackwell.

Wikipedia, *The Free Encyclopaedia.*

SECTION-V
FUTURE OF GEOGRAPHY

CHAPTER-21

Scope of Geography

Geography, as a branch of learning, is focused on understanding the relationship between man and nature. The edifice of geography as a discipline is built on the experiences of successive generations of mankind in trying to comprehend the world, of which their homeland forms a but small part. The thrust for geographical knowledge is as old as human curiosity. There is a natural urge in man to gain knowledge about the lands and peoples lying beyond his own territory. As human beings we are concerned with what m an knows about himself, the earth on which he lives and the universe in which he exists. Geography has been defined differently at different times, depending upon the demands made on it. For instance, in the early stage of its growth it was concerned with 'what is where' and fixing absolute locations. After the Age of Discovery it was required to or anize the mass of information about different lands and people. It was followed by the areal differentiation view, and so on. But, according to a recent definition of geography its overriding problem is concerned with a 'full understanding of vast system on the earth's surface comprising man and the natural environment'[1]. The field of geographical study appears to be so broad that it merits the remark—'it is fit subject for encyclopaedia, but far too inclusive for a 'scientific discipline'. But there is no reason for despair. Geography has a domain of its own—the terrestrial space, which makes it distinct from other disciplines. The broad emphasis is on the spatial arrangement of phenomena and their interrelationships, and all those processes that are responsible for spatial patterns are taken into account. Geography shares the man-environment system with many earth, biological and social sciences. While focusing its attention on spatial patterns and processes geography performs a function that no other discipline does, i.e. integrating the data from a number of disciplines and conceptualing the total unity of the earth, a holistic view which no other discipline can achieve by itself, and which is the ultimate objective of man's quest. Geography is the backbone of human knowledge and is, therefore, the most important science. The Greek Scholar, Eratosthenes, first used the word 'geography' in the 3rd century B.C. It is derived from Greek 'geo' (the earth) and 'graphe' (description), so that geography as a discipline is focused on the description of the world of man on the earth's surface. As a discipline, "geography is concerned to provide accurate, orderly and rational description and interpretation of the variable character of the Earth's surface"[2]. By the phrase 'earth's surface', the geographers imply the zone extending as far down below the surface as man has been able to penetrate and as far high above the surface as man normally goes. Since man's reach above as well as below the earth's surface is relative to the level of his technological progress, the thickness of this zone of study has been progressively increasing.

In the contemporary period, the prestige of geography as a useful branch of knowledge has been further enhanced owing to the valuable service rendered by geographers in the context of inter-disciplinary work in the study of nature and society. The geographers' locational perspective and their cartographic expertise have proved particularly useful. This has led to liberal funding for geographic research and education. Though the word is originally derived from the ancient Greek "geographikos", meaning 'the description of the earth', the geographical scope extends far beyond describing the earth. Today, the geographers' ambit includes spatial analysis, the creation of theoretical models involving extensive use of mathematics, as an essential component. If one wishes to study the subject, he should be interested in the earth, its produce and its people and their activities. Mathematical ability would be an advantage, as would basic dwelling skills. Besides, a geographer's work calls for practice and meticulousness to handle detailed tasks and analyze data. As geography has a number of sister-disciplines, it also provides assistance in the understanding of various other disciplines. In fact, geography serves all those disciplines that serve it. For instance, geography and history have always gone hand in hand, the common thing between the two being 'description', i.e. in terms of space for the former and in terms of time for the latter. More recently we find that geography has become the backbone of the fields like Remote Sensing and Geographic Information System (GIS).

Beside its academic growth, the professionalization in geography is also increasing day by day. Penetrating in so many spheres, geography opens job avenues in numerous areas, and the discipline is becoming more and more job-oriented. A graduate or post-graduate in geography, for instance, has a lot of scope being absorbed in the fields of Education, Administration, Research, Cartography, Town and Country Planning, Surveying and Civil Services. The job prospects are also open in the departments of Remote-Sensing, GIS, Demography, Defence, Meteorology, Soil Science, Forestry, Industry, Agriculture, Environmental Studies and Travel and Tourism. The students of geography can opt for teaching at school and college level, and take up professions as TGTs (Trained Graduate Teachers) or PGTs (Post-Graduate Teachers) in the school level and as Lecturers, Readers or Professors at university level. B.Ed. or M.Ed. is desirable at school level, whereas, at university level it is M.Phil. and Ph.D. In the field of administration, the jobs may be taken through UPSC (Union Public Service

Commission), All India Central Services, SPSC (State Public Service Commission), SSC (Staff Selection Commission) and Block Development Offices.

Remote Sensing and GIS is the recently emerging area of job prospects for the geographers. This field also opens a booming career options for the students of geography those who wish to be a breed apart in the corporate world. The geoinformatic[3] institutes like SIG[4] aim to train and empower the students with analytical decision-making and management skills, besides enabling them to offer complex geoinformation solutions in infrastructure development, natural resources monitoring and urban development wherein the data are analysed and used in areas like disaster management, environment, logistics and asset management, to name a few. A number of Institutes offer courses of different types and durations in this field[5]. Growing interest in Geographic Information and the success of GIS technologies have stimulated the development of new application fields, technologies, hardware and software to meet a very wide range of requests from both the scientific research community and the environmental policy-making community. It is possible to find solutions to critical environmental, social, economic, anthropogenic and other situations; in other words, all kinds of situations related to the territorial management, or governance. Introducing geo-knowledge in traditional e-government can improve land-use governance. It can help citizens to perceive the state of the environment and accept constraints on land use and planning strategies more readily. It naturally implies the availability and usability of good Geographic Information System (GIS). *Geo-Governance* is an initiative of the University of Potsdam's focus research areas of Earth Sciences and Politics, Administration and Management. The initiative aims to better integrate the results of current research on global problems such as natural hazards, climate change and sustainable resource into policy decision-making. *Geo-Governance* bridges the gap between natural and social sciences. The interdisciplinary focus lies on incorporating Earth and environmental expertise into political and administrative decision-making processes. This provides a common reference point, which is currently being addressed in some initial research projects and a variety of pilot models of training and continuing education.

Computer technology is now overwhelming the world and computers can handle vast amount of information in comparison to the traditional skills. Therefore, Computer Assisted Cartography (CAC) is emerging as another field of profession. Particularly the young generation of geographers has a considerable degree of mastery over this technology. A number of organizations are offering professional courses in this field. Besides, the above, there are a score of other institutes offering different professional courses where the students of geography may be absorbed. These courses are related to the fields of Demography, Environment, Forestry, Planning, Social Work and Tourism[6].

In spite of its inter-dependence cutting across on so many sister disciplines, geography in itself is an independent field of study. It is the science of place, i.e. the study of the surface of the earth, the location and distribution of its physical and cultural features, the areal patterns or places that they form, and the interrelation of these features as they affect humans. Geography is a synoptic science that uses the same elements as the other sciences but in a different context. It integrates data spatially, making elaborate use of maps as its special tool. Geography may be studied by way of several interrelated approaches, i.e. systematically, regionally, descriptively, and analytically. Not only the definition, purpose and methodology of geography is changing, but its scope and subject matter is also widening with time.

NOTES

1. Ackerman, *et al.*, 1965, 1.
2. Hartshorne, 1959.
3. Symbiosis Institute of Geoinformatics, SIMS Campus, 2nd Floor, Range Hills Corner, Kirkee Cantt., Pune.
4. Geoinformatics is a science which develops and uses information science infrastructure to address the problems of geosciences and related branches of engineering. The three main tasks of geoinformatics are: (1) development and management of databases of geodata; (2) analysis and modelling of geodata; and (3) development and integration of computer tools and software for the first two tasks. Geoinformatics is related to geocomputation and to the development and use of geographic information systems.
5. Rana, 2008, 37-43.
6. Rana, *op. cit.*

REFERENCES

Ackerman, E.A. *et al.*, (1965), *The Science of Geography*, National Academy of Science Publication, Washington D.C.

Hartshorne, R. (1959), *Perspective on the Nature of Geography*, Rand McNally and Co., Chicago.

Rana, L. (2008), *Geographical Thought: A Systematic Record of Evolution*, Concept Publishing Company, New Delhi.

Bibliography

Abler, Adams and Gould (1971), *Spatial Organization: The Geographers' View of the World*, Engelwood Cliffs, New Jersey, Prentice Hall.

Ackerman, A.E. (1963), "Where is a Research Frontier", *Annals of Association of American Geographers*, Vol. 53, pp. 429-440.

Ackerman, E.A. *et al.* (1965), *The Science of Geography*, National Academy of Science Publication, Washington D.C.

Ackoff, R.L. (1962), *Scientific Method: Optimising Applied Research Decisions*, New York.

Adhikari, S. (1992), *Fundamentals of Geographical Thought*, Chaitanya Publishing House, Allahabad, India.

Ambrose, P., Ed. (1970), *Analytical Human Geography*, Longmans.

Ancient History, www.infoplease.com.

Appignanesi, R. and O. Zarate (2001), *Introducing Existentialism*, Cambridge, UK: Icon. ISBN 1-84046-266-3.

Armstrong, D.M. (1968), *A Materialistic Theory of the Mind*. London: RKP.

Artmann, Benno (1999), *Euclid: The Creation of Mathematics*. New York: Springer. ISBN 0387984232.

Avineri, Shlomo (1968), *The Social and Political Thought of Karl Marx*. Cambridge University Press.

Ayer, Alfred Jules (1959), *Logical Positivism*. Glencoe, Ill: Free Press.

Baird, F.E., and Kaufmann, W.A. (2008), *Philosophic classics: From Plato to Derrida*. Upper Saddle River, N.J: Pearson/Prentice Hall.

Ball, W.W. Rouse (1960) [1908], *A Short Account of the History of Mathematics* (4th ed.). Dover Publications. ISBN 0486206300.

Barnes, Jonathan, (1982), *The Presocratic Philosophers*, rev. ed., London and New York: Routledge.

Berdoulay, V. (1978), "The Vidal-Durkheim Debate", in Ley, D. and M. Samuels (Eds.), *Humanistic Geography: Prospects and Problems*, Maroufa Press, Chicago.

Berry, B.J.L. (1964), "Approaches to Regional Analysis: A synthesis", *Annals of Association of American Geographers*, Vol. 54, No. 1, pp. 2-11.

Berry, B.J.L. (1964), "Cities as Systems within Systems of Cities", *Pap. Regional Science Association*, 13, 147-63.

Bertanlanffy, L. von. (1968), *General System Theory: Foundation, Developments, Applications*, George Braziller, New York.

Bhargava, M.L. (1964), *The Geography of Rigvedic India*, The Upper India Publishing House, Lucknow.

Bleicher, Josef (1982), *The Hermeneutic Imagination: Outline of a Positive Critique of Scientism and Sociology*. London: Routledge and Kegan Paul.

Blij, Harm J. de (1971), *Geography, Regions and Concepts*, John Wiley and Sons, Inc., New York.

Block, Ned. (1980a), "Introduction: What Is Functionalism?" in *Readings in Philosophy of Psychology*. Cambridge, MA: Harvard University Press.

_______ (1996), "What is functionalism?" in *The Encyclopedia of Philosophy Supplement*, Macmillan.

Bodde, Derk (1991), *Chinese Thought, Society, and Science*. Honolulu: University of Hawaii Press.

Bogdan, Robert, and Steven J. Taylor (1975), *Introduction to Qualitative Research Methods: A Phenomenological Approach to the Social Sciences*. New York: Wiley.

Bornes, Johnathan (1978), *Early Greek Philosophy*, New York: Penguin Books.

Bose, Sugata and Ayesha Jalal (2003), *Modern South Asia: History, Culture, Political Economy*, Parts 2-3, Routledge.

Bourbaki, Nicolas (1998), *Elements of the History of Mathematics*, Berlin, Heidelberg, and New York: Springer-Verlag, ISBN 3540647678.

Bowman, John S. (2000), *Columbia Chronologies of Asian History and Culture*. New York: Columbia University Press.

Boyer, Carl Benjamin (1949), *The History of the Calculus and its Conceptual Development*. Hafner. Dover edition 1959, ISBN 0-486-60509-4.

_______ (1991), "The Mathematics of the Hindus". *A History of Mathematics* (Second Ed.). John Wiley and Sons, Inc. ISBN 0471543977.

Bradley, Patrick J. (2010), "Irony and the Narrator in Xenophon's Anabasis", in *Xenophon*. Ed. Vivienne J. Gray. Oxford University Press.

Braithwaite, R.B. (1960), *Scientific Explanations*, Harper Torchbooks, New York.

Breasted, J.H. (1916), *Ancient times, a history of the early world: an introduction to the study of ancient history and the career of early man*, Boston: Ginn and Company.

Britannica Online Encyclopedia.

Broek, J.O.M. (1965), *Compass of Geography: Social Science Perspective*, Columbus, Ohio, USA.

Brukfield, H.C. (1969), *Progress in Geography*, Vol. I, pp. 51-80.

Bruyn, Severn T. (1966), *The Human Perspective in Sociology: The Methodology of Participant Observation*. Englewood Cliffs, N.J.: Prentice-Hall.

Bunge, W. (1966), *Theoretical Geography*, Lund Studies in Geography. C No. 1 (Second Edition).

Burton, I. (1963), "The Quantitative Revolution and Theoretical Geography", *Canadian Geographer*, 7, 4, pp. 151-182.

_______ (1972), "Quantitative Revolution and Theoretical Geography" in W.K.D. Davies, *The Conceptual Revolution in Geography*, Rowman and Littlefield, New Jersey, pp. 140-156.

Buttimer, A. (1976), "Grasping the Dynamism of Lifeworld", *Annals of Association of American Geographers*, Vol. 66, pp. 277-292.

_______ (1978), "Charisma and Context: The Challenge of *La Geography Humaine*", in Ley, D. and Samuels, M.S. (Eds.), *Humanistic Geography: Prospects and Problems*, Maroufa Press, Chicago.

_______ (1974), *Values in Geography*, Commission on Geography, Resource Paper 24, Association of American Geographers, Washington D.C.

Calinger, Ronald (1999), *A Contextual History of Mathematics*. Prentice-Hall. p. 150. ISBN 0-02-318285-7.

Catholic Encyclopedia (Internet).

Cerchai, Luca (2004), *The Greek Cities of Magna Gracia and Sicily*, J. Paul Getty Trust Publications, USA.

Cerbone, David R. (2006), *Understanding Phenomenology*. Chesham: Acumen.

Chan, Alan Kam-leung and Gregory K. Clancey, Hui-Chieh Loy (2002), *Historical Perspectives on East Asian Science, Technology and Medicine*. Singapore: Singapore University Press. ISBN 9971-69-259-7.

Chemla, Karine; Cohen, Robert S.; Renn, Jürgen *et al.* (2004), *History of Science, History of Text (Boston Series in the Philosophy of Science)*, Dordrecht: Springer, Netherlands.

Chopra, Pran Nath (1982), *Religions and Communities of India*, Vision Books,

Chorley, R.J. (1962), *Geomorphology and General Systems Theory*, U.S. Govt. Print. Off., Washington.

_______ (1964), "Geography and Analogue Theory", *Annals of Association of American Geographers*, Vol. 54, 127-37.

Chorley, R.J. and P. Hagget (1967), *Integrated Models in Geography*, Part IV, Methuen and Co. Ltd., London.

Christopher, R.W. (1994), *Geosystems: An Introduction of Physical Geography*, Macmillan College Pub. Co., Englewood Cliffs, New Jersey.

Clagett, M. (1955), *Greek Science in Antiquity*, Dover.

Claeys, G. (2000), "The "survival of the fittest" and the origins of social Darwinism". *Journal of the History of Ideas* 61(2).

Clifford, J. (1988), *The predicament of Culture, Nineteenth Century Ethnography*, Literature and Art, Harvard University Press, Cambridge.

Cole, L. (1958), "The Ecosphere", *Scientific American*, Vol. 1. 1984, pp. 83-92.

Collins, J. (1952), *The Existentialists: A Critical Study*, Chicago: Henry Regnery Company.

Cook, R.U. and J.H. Johnson (1969), *Trends in Geography*, Pergamon Press, London.

Copenhaver, Brian and Charles Schmitt (1992), *Renaissance Philosophy*, Oxford University Press.

Copleston, F., (1950), *A History of Philosophy*, Volume II: From Augustine to Scotus (Burns and Oates).

_______ (1953), *A History of Philosophy, Volume III: From Ockham to Suarez* (The Newman Press).

Coppock, J.T. and J. H. Johnson (1962), "Measurement in Human Geography", *Economic Geography*, Vol. 38, No. 2, pp. 130-137.

Costelloe, Timothy M. (1996), "Between the Subject and Sociology: Alfred Schutzof's Phenomenology of the Life-World.", *Human Studies*, 19:247-266.

Crawford, O.G.S. (1927), *Antiquity*. [Gloucester, Eng.]

Craig, J. Calhoun (2002), *Classical Sociological Theory*. Wiley-Blackwell.

Crespigny, Rafe de. (2007), *A Biographical Dictionary of Later Han to the Three Kingdoms* (23-220 CE). Leiden: Koninklijke Brill.

Crystal, D. (2002), *The New Penguin Encyclopedia*, London.

Dacey, M.F. (1965), "The Geometry of Central Place Theory", *Geographical Annlr Series*, B, 47, pp. 111-124.

Daintith, J.; Mitchell, Sarah, *et al.* (1994), *Biographical Encyclopedia of Scientists*, Second Edition, Institute of Physics Publishing, Bristol.

Daniel Dennett. (1998), *Postmodernism and Truth*.

Darroch, Vivian, and Ronald J. Silvers (eds.) (1982), *Interpretive Human Studies: An Introduction to Phenomenological Research*. Washington, D.C.: University Press.

Dasgupta, Surendranath (1992), *A History of Indian Philosophy*, Volume 1. Motilal Banarsidass Publ.

Davenport, Harold (1999), *The Higher Arithmetic: An Introduction to the Theory of Numbers* (7th ed.), Cambridge University Press, Cambridge, U.K.

David McLellan (1973), *Karl Marx: His life and Thought*. New York: Harper and Row.

David Stewart and Algis Mickunas (1990), *Exploring Phenomenology: A Guide to the Field and its Literature*, Athens: Ohio University Press.

Davies, J.L. (1961), "Aim and Method in Zoogeography", *Geographical Review*, 51, pp. 412-17.

Davies, W.K.D. (1972), *Conceptual Revolution in Geography*, Rowman and Littlefield, New Jersey.

Dear, M. (1988), "The Postmodern Challenge: Restructuring Human Geography", *Transactions, Institute of British Geographers*, NS 13, pp. 262-274.

De Blij (1974), *Man Shapes the Earth*, Hamilton Publishing Co., California, pp. 147-241.

_______ (1973), *Systematic Political Geography*, II Ed., John Wiley and Sons, New York.

_______ (1996), *Human Geography: Culture, Society and Space*, Fifth Edition, John Wiley and Sons Inc., New York, pp. 228-31.

De Blij, H.J. and P.O. Muller (1971), *Geography: Regions and Concepts*, John Wiley and Sons Inc., New York, pp. 1-18.

Devlin, Keith (1996), *Mathematics: The Science of Patterns: The Search for Order in Life, Mind and the Universe* (Scientific American Paperback Library).

Dickinson, R.E. (1969), *The Makers of Modern Geography*, Praeger Publishers, New York.

_______ (1970), *Regional Ecology: The Study of Man's Environment*, John Wiley, New York.

_______ (1976), *Regional Concept: The Anglo American Leaders*, Routledge and Kegan Paul, London.

Dikshit, R.D. (1994), *The Art and Science of Geography: Integrated Readings*, Prentice Hall of India.

_______ (1997), *Geographical Thought: A Contextual History of Ideas*, Prentice Hall of India.

Downs, R.M. and D. Stea (Eds.) (1973), *Image and Environment*, Edward Arnold, London.

Durant, Will (1939), *The Life of Greece*. New York: Simon and Schuster.

Ebrey, Patricia Buckley, Anne Walthall, and James B. Palais (2006), *East Asia: A Cultural, Social, and Political History*. Boston: Houghton Mifflin Company. ISBN 0-618-13384-4.

Edwards, Paul (1967), *Encyclopedia of Philosophy*, New York: MacMillan Pub. Co.

Encyclopedia.com (Internet).

Encyclopedia Britannica.

Encyclopedia of Science and Religion.

Encyclopedia of the Orient (Internet).

Encyclopedia-Yahoo Education (Internet).

English, P.W. and R.C. Mayfield (1972), *Man, Space and Environment*, Oxford University Press, London.

Entrikin, J.N. (1976), "Contemporary Humanism in Geography", *Annals of the Association of American Geographers*, Vol. 66, pp. 615-632.

Eves, Howard, (1969) *An Introduction to the History of Mathematics*, Holt, Rinehart and Winston.

Eyre, S.R. (1964), "Determinism and Ecological Approach to Geography", *Geography*, Vol. I, No. 49, pp. 369-376.

Farrington, B. (1961), *Greek Science: Its Meaning for us*, Penguin Books.

Fielding, G.J. (1974), *Geography as a Social Science*, Oxford University Press, Harper and Row Publishers, New York.

Fien, J. (1983), "Humanistic Geography", in Huckle, J. (Ed.) *Geographical Education: Reflection and Action*, Oxford: Oxford University Press, pp. 43-55.

Flynn, T. (2006), *Existentialism: A Very Short Introduction*, Oxford: Oxford University Press.

Fodor, J. (1968), *Psychological Explanation*, New York: Random House.

Fordyce, J. (1888), *The New Social Order*. London: Kegan Paul, Trench and Co.

Foster, J. (2003), "Between Economics and Ecology: Some historical and philosophical considerations for modelers of natural capital" . *Environmental Monitoring and Assessment* **86**.

Fowler, Jeaneane (2005), *An Introduction to the Philosophy and Religion of Taoism: Pathways to Immortality*. Brighton: Sussex Academic Press. ISBN 1-84519-085-8.

Fuson, R.H. (1969), *Geography of Geography*, W.M.C. Brown Company Publishers, Dubuque.

Gender, Place and Culture — *A Journal of Feminist Geography*. Routledge.

Giddens, Anthony (1974), *Positivism and Sociology*. Heinemann. London.

Gilbert, E.W. (1960), "The Idea of the Region", *Geography*, Vol. 45, 1960, pp. 157-175.

Goodman, R. (ed) (1995), *Pragmatism*, London: Routledge.

_______ (ed) (2005), *Pragmatism: Critical Concepts in Philosophy* (four volumes covering: volume one: Pragmatism's first decade; volume two: metaphysics and epistemology; volume three: moral and political issues; volume four: neopragmatism and aesthetics). London: Routledge.

Gold, J.R. (1980), *An Introduction to Behavioural Geography*, Oxford University Press.

Golledge, R.G., L.A. Brown and F. Williamson (1972), "Behavioural Approaches in Geography: An Overview", *The Australian Geographer*, Vol. 12, pp. 59-79.

Gould, P.R. (1966), "On Mental Maps", *Michigan Inter-University Community of Mathematical Geographers*, Discussion Paper 9, reprinted in Downs, R.M. and D. Stea (Eds.) (1973), pp. 182-220.

Gould, P.R. and R. White (1974), *Mental Maps*, Hammondsworth: Penguin Books.

Gracia, Jorge J.E. in Nicholas Bunnin and E.P. Tsui-James (eds.) (2002), *The Blackwell Companion to Philosophy*, 2nd ed. Blackwell.

Gregor, H.F. (1963), *Environment and Economic Life*, Princeton, New Jersey, pp. 367-430.

Gregory, D. (1978), *Ideology, Science and Human Geography*, Hutchinson, London.

_______ (1981), "Human Agency and Human Geography", *Transactions*, Institute of British Geographers, NS-6, pp. 1-6.

Grigg, D. (1967), "Regions, Models and Classes" in Chorley, R.J. and P. Hagget: *Integrated Models in Geography, op. cit.*, pp. 461- 501.

Grosvenor, Edwin Augustus (1899), *Contemporary history of the world*, T.Y. Crowell & Co. New York and Boston.

Guelke, L.C. (1971), "Problems of Scientific Explanation in Geography", *Canadian Geographer*, Vol. 15, pp. 38-53.

Guignon, C., and D. Pereboom (eds.), *Existentialism: Basic Writings*, Indianapolis: Hackett.

Gurjar, R.D. (2005), "Cognitive System and Paradigms in behavioural Geography" in H.S. Sharma, *Trends in Geographical Research*, Department of Geography, Rajasthan, Jaipur, pp. 1-19.

Haack, S. (ed) (2006), *Pragmatism, Old and New*, Amherst NY: Prometheus.

Haggett, P. (1965), *Locational Analysis in Geography*, London.

Haggett, P., A.D. Cliff and A. Frey (1977), *Locational Models*, Edward Armold (Publishers) Ltd., London.

Hall, A.D. and R.E. Fagen (1956), "Definition of System", *General Systems Yearbook*, Vol. I, Ann Arbour, Mich., pp. 18-28 (mimeographed).

Hanfling, Oswald (1981), *Logical Positivism*. Oxford: B. Blackwell.

Harris, C.D. (1964), "Economic Regionalization", *Geographia Politica*, Vol. 4. pp. 59-86.

Harris, R.C. (1971), "Theory and Synthesis in Historical Geography", *The Canadian Geographer*, Vol. 15, pp. 157-172.

Harrison, C.M. (1969), "The Ecosystem and The Community in Biogeography". in Cooke, R.U. and J.H. Johnson, *Trends in Geography: An Introductory Survey*, Pergamon Press, pp. 36-44.

Hartshorne, R. (1939), *The Nature of Geography: A Critical Survey of Current Thought in the Light of the Past*, Association of American Geographers, Lancaster.

_______ (1959), *Perspective on the Nature of Geography*, Rand McNally and Co., Chicago.
Harvey, D. (1969), *Explanation in Geography*, Arnold Publishers, New Delhi.
Harvey, M. E. and B.P. Holly (1989), *Themes in Geographic Thought*, Rawat Publications, Jaipur and New Delhi.
Hayashi, Takao (2003), "Indian Mathematics", in Grattan-Guinness, Ivor, *Companion Encyclopedia of the History and Philosophy of the Mathematical Sciences*, 1, pp. 118-130, Baltimore, MD: The Johns Hopkins University Press, 976 pages, ISBN 0801873967 .
_______ (2005), "Indian Mathematics", in Flood, Gavin, *The Blackwell Companion to Hinduism*, Oxford: Basil Blackwell, 616 pages, ISBN 9781405132510 .
Heintzelman, O.H. and R.M. Highsmith (1973), *World Regional Geography*, Engelwood Cliffs, New Jersey, Prentice Hall.
Helle, Horst J., ed. (1991), *Verstehen and Pragmatism: Essays in Interpretative Sociology*. Frankfurt, Federal Republic Germany: Peter Lang.
Henry Kitchell (2004), *Early European History*, Hutton Webster.
Herbert, Spiegelberg (1983), *The Phenomenological Movement: A Historical Introduction*, 3rd ed. The Hague: Martinus Nijhoff.
Herbertson, A.J. (1905), "The Major Natural Regions: An Essay in Systematic Geography", *Geography*, pp. 300-310.
Hitzler, Ronald, and Reiner Keller (1989), "On Sociological and Commonsense Verstehen", *Current Sociology*, 37, pp. 91-101.
Holt-Jenson, A. (1980), *Geography: Its History and Concepts*, Harper and Row, London.
Hoosan, D.J.M. (1968), "The Development of Geography in Pre-Soviet Russia", *Annals of Association of American Geographers*, Vol. 58, June, pp. 250-272.
Howard Eves (1990), *An Introduction to the History of Mathematics* (6 Ed.). Saunders College Publishing House, New York. p. 237.
Hsu, Mei-ling (1988), "Chinese Marine Cartography: Sea Charts of Pre-Modern China," *Imago Mundi*, Volume 40, pp. 96-112.
_______ (1993), The Qin Maps: A Clue to Later Chinese Cartographic Development," *Imago Mundi*, Volume 45, pp. 90-100.
Huang, Ray (1997), *China: A Macro History*. New York: An East Gate Book, M. E. SHARPE Inc.
Huckle, J.(ed.) (1983), *Geographical Education: Reflection and Action*, Oxford University Press, Oxford, pp. 43-45.
Huffman, Carl A. (1993), *Philolaus of Croton: Pythagorean and Presocratic*, Cambridge University Press.
Husain, M. (1984), *Evolution of Geographical Thought*, Rawat Publications, Jaipur.
_______ (1988), *Evolution of Geographical Thought*, Second Revised Edition, Rawat Publications, Jaipur.
_______ (1993), Perspectives in History and Nature of Geography, Vol. 2, 3 and 5, Anmol Publications, New Delhi.
_______ (1995), *Evolution of Geographical Thought*, Third Revised Edition, Rawat Publications, Jaipur.
Husserl, Edmund. (1999) [1907], *The Idea of Phenomenology*. Trans. Lee Hardy. Dordrecht: Kluwer.
Ifrah, G. (2000), *The Universal History of Numbers*. John Wiley and Sons, New York.
Internet Encyclopaedia of Philosophy.
James, P.E. and C.F. Jones (1954), *American Geography: Inventory and Prospects*, Syracuse University Press, Syracuse.
James, P.E. and G.J. Martin (1972, 1981), *All Possible Worlds: A History of Geographical Ideas*, Third Edition, John Wiley and Sons, New York.

James, W. (1907), *A New Name for Some Old Ways of Thinking: Popular Lectures on Philosophy*, Longmans; Harvard University Press, 1975.

Jenny Teichmann and Katherine, C. Evans (1999), *Philosophy: A Beginner's Guide*, Blackwell Publishing.

Jewish Encycopedia.

Jon Elster, (1986), *An Introduction to Karl Marx*. Cambridge, England.

Johnston, R.J. (1979), *Geography and Geographers: Anglo-American Human Geography Since 1945*, 4th Edition, Edward Arnold, London.

_______ (1983), *Philosophy and Human Geography*, Edward Arnold, London.

_______ (2004), *Geography and Geographers: Anglo-American Human Geography Since 1945*, 6th Edition, Edward Arnold, London.

Johnston, R.J., D. Gregory, and D.M. Smith, (1994), *The Dictionary of Human Geography*, (3rd edition), USA: Blackwell.

Johnston, R.J., J. Hauer and G.A. Hoekveld (1990), *Regional Geography: Current Developments and Future Prospects*, Routledge and Kegan Paul, London.

Kant, Immanuel. (1963), *Critique of Pure Reason*. Trans. N.K. Smith. London: MacMillan and Co.

Katz, Victor J. (1995), "Ideas of Calculus in Islam and India", *Mathematics Magazine (Math. Assoc. Amer.)* 68 (3): 163-174.

Kim, Yung Sik. (2000), *The Natural Philosophy of Chu Hsi (1130-1200)*. DIANE Publishing.

Kirk, W.K. (1951), "Historical Geography and the Concept of Behavioural Environment", *Indian Geographical Journal*, Vol. 25, pp. 125-160.

Kirk, G., J. Raven, M. Schofield (2003), *The Presocratic Philosophers*. Cambridge University Press.

Kohn, Livia (2000), *Daoism Handbook: Handbook of Oriental Studies—Part 4: China, 14*. Boston: Brill Academic Publishers.

Kuhn, S. Thomas (1962), *The Structure of Scientific Revolutions*, 1st. ed., Chicago: University of Chicago Pr.

Kurt A. Raaflaub and Richard J.A. Talbert (2009), *Geography and Ethnography: Perceptions of the World in Pre-Modern Societies*, John Wileyand Sons.

Langsdorf, Lenore (1995), "Treating Method and Form as Phenomena: An Appreciation of Garfinkel's Phenomenology of Social Action.", *Human Studies*, 18:177-188.

Levin, Janet. (2004), "Functionalism", *The Stanford Encyclopedia of Philosophy* (Fall 2004 Edition), E. Zalta (ed.).

Lewis, D. (1966), "An Argument for the Identity Theory". *Journal of Philosophy*, 63: 17-25.

Ley, D. and M.S. Samuels (Eds.) *Humanistic Geography: Prospects and Problems*, Chicago: Maroufa Press.

Linton, C.M. (2004), *From Eudoxus to Einstein: A History of Mathematical Astronomy*. Cambridge University Press. p. 52. ISBN 0521827507.

Livingstone, David, (1992), *The Geographical Tradition: Episodes in the History of a Contested Enterprise*, Blackwell Publishers.

Loewe, Michael. (1968), *Everyday Life in Early Imperial China during the Han Period 202 BCE–CE 220*. London: B.T. Batsford Ltd.; New York: G.P. Putnam's Sons.

_______ (1988), "The Oracles of the Clouds and the Winds", *Bulletin of the School of Oriental and African Studies*, University of London, Volume 51, Number 3, pp. 500-520.

Long, A.A. and D.N. Sadley (1987), *Hellenistic Philosophers*, Vol. 1: 'Translations of the Principal Sources, with Philosophical Commentary', Cambridge.

Lowenthal, D. (1961), "Geography, Experience and Imaginations: Towards a Geographical Epistemology", *Annals of the Association of American Geographers*, Vol. 51, pp. 241-260.

Luper, Steven (ed.) (2000), *An Introduction to Existential Thought*. Mountain View, California: Mayfield. ISBN 0-7674-0587-0.

MacIntyre, A. (1967), "Existentialism," in *The Encyclopedia of Philosophy*, Vol. III. Ed. Paul Edwards. New York: Macmillan Publishing Company.

Mackinder, H.J. (1930), "The Content of Philosophical Geography", *Proceedings of the International Geographical Union*, Cambridge [1928], pp. 305-312.

MacMillan Encyclopedia of Sex and Gender.

Macmillan, C. (1956), "The Status of Plant Ecology and Plant Geography", *Ecology*, 37, pp. 600-602.

Macmillan, W. (Ed.) (1989), *Remodelling Geography*, Basil Blackwell, Oxford.

McDowell, Linda (1993), "Space, place and gender relations" in *Progress in Human Geography*, 17(2).

Majumdar, R.C. (1960), *The History and Culture of the Indian People*, Vol. II. (The Age of Imperial Unit).

Manley, Gordon (1958), "The Revival of Climatic Determinism", *Geographical Review*, Vol. 48, pp. 98-105.

Marcel, G. (1968), *The Philosophy of Existentialism*, New York: Citadel Press.

Marino, Gordon (ed.) (2004), *Basic Writings of Existentialism*. New York: Modern Library. ISBN 0-375-75989-1.

Marr, D. (1982), *Vision: A Computational Approach*. San Francisco: Freeman and Co.

Martin, Geoffrey J. (1994), "Richard Hartshorne, 1899-1992", *Annals of the Association of American Geographers*, Volume 84, Issue 3, pp. 480-492.

McBride, W. (ed.) (1997), *The Development and Meaning of Twentieth Century Existentialism*, New York: Garland. Publishers.

McLellan, David (2007), *Marxism After Marx*. Basingstoke: Palgrave Macmillan.

Melnyk, Andrew. (2003), *A Physicalist Manifesto: Thoroughly Modern Materialism*. Cambridge: Cambridge University Press.

_______ (2003), "Some Evidence for Physicalism." In *Physicalism and Mental Causation: The Metaphysics of Mind and Action*, eds. Sven Walter and Heinz-Dieter Heckmann, 155-172.

Menand, L. (ed) (1998), *Pragmatism*, New York: Random House.

Michael Hammond, Jane Howarth, and Russell Kent (1995), *Understanding Phenomenology*, Oxford: Blackwell.

Mill, John Stuart. *August Comte and Positivism*. web-books.com.

Minshull, R. (1971), *Regional Geography*, Hutchinson University Library, London.

Misak, C.J. (ed) (1999), *Pragmatism*, Calgary: University of Calgary Press.

Mises, Richard von. (1951), *Positivism: a study in human understanding*. Harvard University Press. Cambridge, Mass.

Misra, H.N. (1989), "Traditional and Contemporary Paradigms of Urban Geography", *Annals of the NAGI*, Vol. 9, No. 1, June, pp. 49-58.

Mohn, Peter (2003), *Magnetism in the Solid State: An Introduction*. New York: Springer-Verlag Inc. ISBN 3-540-43183-7.

Money, D.C. (1967), *Introduction to Human Geography*, University Tutorial Press, London.

Moran, D. (2000), *Introduction to Phenomenology*, Oxford: Routledge.

Mourelatos, A.P.D. (1971), *The Route of Parmenides*, New Haven: Yale University Press.

Muir, R. (1981), *Modern Political Geography*, Macmillan, London.

Muley, G. (2000), "When Exactly Does the 20th Century Begin?" *Employment News*, 22-28 January, p. 5.

Myron Orleans, *Encyclopedia of Sociology*.

Needham, Joseph (1986), *Science and Civilization in China: Volume 1, Introductory Orientations.* Taipei: Caves Books Ltd.

_______ (1986), *Science and Civilization in China: Volume 3, Mathematics and the Sciences of the Heavens and the Earth.* Taipei: Caves Books Ltd.

_______ (1986), *Science and Civilization in China: Volume 4, Physics and Physical Technology, Part 2: Mechanical Engineering.* Taipei: Caves Books Ltd.

_______ (1986), *Science and Civilization in China: Volume 4, Physics and Physical Technology, Part 3, Civil Engineering and Nautics.* Taipei: Caves Books Ltd.

_______ (1986), *Science and Civilization in China: Volume 5, Chemistry and Chemical Technology, Part 1: Paper and Printing.* Taipei: Caves Books Ltd.

Nelson, Howard, (1974), "Chinese Maps: An Exhibition at the British Library," *The China Quarterly* (Number 58): 357-362.

Neugebauer, O. and A. Sachs. (1945), *Mathematical Cuneiform Texts,* New Haven, CT, Yale University Press.

New World Encyclopedia.

Odum, H.W. and H.F. Moore (1938), *American Regionalism: A Cultural Historical Approach to National Integration,* H. Holt and Company, New York.

Oxford Dictionary of Geography.

Park, R.E. (1934), "Human Ecology", *American Journal of Sociology,* 43, pp. 1-15.

Park, R.E. (1936), "Succession and Ecological Concept", *American Sociological Review,* 1, pp. 171-79.

Park, R.E. and E.W. Burgess (1921), *Introduction to Science and Sociology,* Chicago.

Pattison, W. (1964), "Four Traditions in Geography", *Journal of Geography,* Vol. 63, pp. 211-216.

Paul Oskar Kristeller (1990), *Humanism,* pp. 113-4, in Charles B. Schmitt, Quentin Skinner (editors), *The Cambridge History of Renaissance Philosophy.*

Pearcy, E. (1957), *World Political Geography,* Crowell Company, New York.

Peet, R. (1998), *Modern Geographical Thought,* Blackwell.

Peet, R. and N. Thrift (Eds.) (1989), *New Models in Geography,* Unwin Hyman, London.

Pickering, Mary. (1993), *Auguste Comte: An intellectual biography.* Cambridge University Press, Cambridge.

Platt, Robert S. (1947-48), "Environmentalism versus Geography", *American Journal of Sociology,* Vol. 53, pp. 351-58.

Pocock, D.C.D. and Hudson, R. (1978), *Images of the Urban Environment,* MacMillan, London.

Pounds, N.J.G. (1972), *Political Geography,* Ed. 2, McGraw Hill, New York.

Pred, A.R. (1967 and 1969), *Behaviour and Location: Foundations for a Geographic and Dynamic Location Theory,* Parts 1 and 2. Lund Studies in Geography, B, 27 and 28, Lund: Gleerup.

Premwati (2006), "Development of Women Through Empowerment in India", *Geo-Vision: Annual Geographical Magazine,* Vol. I, Feb., Swami Shradhanand College, Alipur, Delhi, pp. 19-22.

Putnam, H. (1960), "Minds and Machines", in S. Hook (ed.), *Dimensions of Mind,* New York: University of New York Press, pp. 148-80.

_______ (1967a), "The Mental Life of some Machines", in Hector-Neri Castañeda (ed.), *Intentionality, Minds and Perception,* Detroit: Wayne State University Press, pp. 177-200.

_______ (1967b), "The Nature of Mental States", in Captain, W.H. and Merrill, D.D. (eds.), *Art, Mind and Religion,* Pittsburgh: University of Pittsburgh Press, pp. 37-48.

_______ (1994), *Pragmatism,* Oxford: Blackwell.

Radhakrishnan, S.; Moore, C.A. (1957), *A Source Book in Indian Philosophy.* Princeton, New Jersey: Princeton University Press. ISBN 0-691-01958-4.

Rajagopal, C., Venkataraman A. (1949), "The sine and cosine power series in Hindu mathematics". *Journal of the Royal Asiatic Society of Bengal (Science)* **15**: 1-13.

Rana, L. (2008), *Geographical Thought: A Systematic Record of Evolution*, Concept Publishing Company, New Delhi.

_______ (2011), *Ancient Tradition in Geography: The Western and Oriental Perspective*, Project Report (unpublished), ICSSR, New Delhi.

Rawson, Elizabeth (1975), *Cicero, A portrait*, Allen Lane, London.

Relph, E. (1970), "An Inquiry into Relations between Phenomenology and Geography", *The Canadian Geographer*, Vol. 14, pp. 193-201.

_______ (1976), The Phenomenological Foundation of Geography", *Discussion Paper No. 21*, Department of Geography, University of Toronto, Toronto.

Reynolds, J. (2006), *Understanding Existentialism*. London: Acumen.

Richard M. Zaner (1970), *The Way of Phenomenology*, Indianapolis: Pegasus.

Robert Sokolowski (2000), *Introduction to Phenomenology*, Cambridge: Cambridge University Press.

Robinson, G.W.S. (1953), "The Geographical Region: Form and Function", *The Scottish Geographical Magazine*, Vol. 69, No. 2, September 1953, pp. 49-58.

Roepke, H.G. (1967), *Readings in Economic Geography*, Wiley, New York, pp. 17-37.

Roman K. Kovalev, Alexis C. Kaelin (2007), "Circulation of Arab Silver in Medieval Afro-Eurasia: Preliminary Observations", *History Compass* 5 (2), pp. 560-80.

Rowe, J.S. (1961), "The Level-of-Integration Concept and Ecology", *Ecology*, 42, pp. 420-427.

Rose, Gillian (1993), *Feminism and Geography: The Limits of Geographical Knowledge*, University of Minnesota Press.

Ruggles, D. Fairchild (2003), "Botany and the Agricultural Revolution", *Gardens, landscape, and vision in the palaces of Islamic Spain*, Penn State University Press, pp. 15-34 [31],

Samuels, M.S. (1978), "Existentialism in Human Geography", in Ley, D. and Samuels, M.S. (Eds.), *Humanistic Geography: Prospects and Problems*, Maroufa Press, Chicago.

Sartre, Jean-Paul (1945), *Existentialism and Humanism*.

Schaefer, F.K. (1953), "Exceptionalism in Geography: A Methodological Examination", *Annals of Association of American Geographers*, Vol. 43, 226-249.

Schmitt, Charles and Quentin Skinner (eds.) (1988), *The Cambridge History of Renaissance Philosophy*, Cambridge University Press.

Seidenberg, (1978), "The origin of mathematics", *Archive for the history of Exact Sciences*, Vol 18.

Siddall, W.R. (1959), *Idiographic and Nomothetic Geography, The Application of Some Ideas in the Philosophy and History and Science to Geographic Methodology*, Ph.D. Thesis, University of Washington.

Simon, H.A. (1957), *Models of Man*, Wiley, NewYork.

Singh, R.L. and P.K. Dutt (1979), *Elements of Practical Geography*, Students' Friends, Allahabad, India.

Sivin, Nathan (1995), *Science in Ancient China: Researches and Reflections*. Brookfield, Vermont: VARIORUM, Ashgate Publishing.

Smith, D.M. (1977), *Human Geography: A Welfare Approach*, London: Edward Arnold.

Soja, E.W. (1997), *Postmodern Geographies: The Reassertion of Space in Critical Social Theory*, Rawat Publishers, Jaipur and New Delhi.

Solomon, R. (ed.), (1974), *Existentialism*, New York: Random House.

Sonnenfeld, J. (1972), "Geography, Perception and the Behavioural Environment", in *Man, Space and Environment*, Oxford University Press, London, pp. 244-250.

Spate, O.H.K. (1952), "Toyanbee and Huntington: A Study in Determinism", *Geographical Review*, Vol. 68, pp. 406-28.

Sprigge, T.L.S. (1998), "Idealism." In E. Craig, ed. *Routledge Encyclopedia of Philosophy*. London: Routledge.

Staal, Frits (1999), "Greek and Vedic Geometry", *Journal of Indian Philosophy*, **27** (1-2): 105-127.

Stamp, D. (1957), "Major Natural Regions: Herbertson after Fifty years", *Geography*, Vol. 42, pp. 201-216.

Stanford Encyclopedia of Philosophy.

Stauffer, R.C. (1957), "Haeckel, Darwin and ecology." *The Quarterly Review of Biology* **32** (2).

Stoddart, D.R. (1965), "Geography and the Ecological Approach", *Geography*, 50, 252-251.

_______ (1966), "Organism and Ecosystem as Geographical Models" in Chorley and Hagget, *Integrated Models in Geography, op. cit.*, pp. 511-539.

_______ (1972), "Darwin's Impact on Geography" and "Geography and the Ecological Approach" in W.K.D. Davies, *The Conceptual Revolution in Geography*, Rowman and Littlefield, New Jersey. pp. 52-68 and 301-311.

_______ (1986), *On Geography and its History*, Basil Blackwell, Oxford.

Taaffe, E.J. (1974), "The Spatial View in Context" in *Annals of Association of American Geographers*, Vol. 64, No. 1, pp. 1-16.

Tansley, A.G. (1946), *Introduction to Plant Ecology*, Allen and Unwin, London.

Taylor, G. (1951), *Geography in the Twentieth Century*, Butlar and Tanner Ltd., London.

Taylor, P.J. (1976), "An Interpretation of the Quantification Debate in British Geography", *Transactions of the Institute of British Geographers*, N.S. 1, pp. 129-142.

Tewari, Preeti (2010), *Division of Work and Gender Space in Delhi*, Ph.D. research work submitted at Department of Geography, University of Delhi.

The Columbia Electronics Encyclopedia.

The Dictionary of Philosophy, Progress Publishers.

"The Early History of World Maps", *www.maps-gps-info.com.*

The Internet Encyclopedia of philosophy.

Thayer, H.S. (ed) (1982), *Pragmatism: The Classic Writings*, Hackett.

The Catholic Encyclopedia.

Tom Rockmore (2002), *Marx after Marxism: The Philosophy of Karl Marx*. John Wiley and Sons.

Trevelyan, G.M. (1961), *British History in the 19th Century and After*, Macmillan, London.

Truzzi, Marcello (1974), *Verstehen: Subjective Understanding in the Social Sciences*. Reading, Mass.: Addison-Wesley.

Tuan, Yi-Fu (1971), "Geography, Phenomenology, and the Study of Human Nature", *The Canadian Geographer*, Vol. 15, pp. 181-192.

_______ (1974), *Trophilia: A Study of Environmental Perception, Attitudes and Values*, Englewood Cliffs, Prentice Hall, NJ.

_______ (1976), "Humanistic Geography", *Annals of the Association of American Geographers*, Vol. 66, pp. 266-276.

Turing, A.M. (1950), "Computing machinery and intelligence". *Mind*, 59, 433-460.

Unstead, J.F. *et al.* (1933), "Classification of Regions of the World", *Geography*, Vol. 22, pp. 253-282.

Vlastos, Gregory (1975), *Plato's Universe*. Seattle: University of Washington Press.

Walmsley, D.J. and Lewis, G.J. (1982), *Human Geography: Behavioural Approaches*, Longman, London.

Walmsley, D.J. and G.J. Lewis, (1993), *People and Environment: Behavioural Approaches in Geography*, Longman, London.

Wanklyn, H. (1961), *Friedrich Ratzel: A Biographical Memoir and Bibliography*, University Press, Cambridge.

Watson, Andrew M. (1974), "The Arab Agricultural Revolution and Its Diffusion, 700-1100", *The Journal of Economic History* 34 (1).

Watson, J.W. (1975), "Perception and Place", *Geographical Journal*, Volume 141, pp. 271-274.

Wheen, Francis (2002), *Karl Marx: A Life*. New York: Norton.

Wikipedia: *The Free Encyclopedia*.

Winters, C. (1991), *International Dictionary of Anthropologists*, Garland Publishing, New York.

Wright, John K. (1947), "Terrae Incognitae: The Place of Imagination in Geography", *Annals of the Association of American Geographers*, 37: 1-15.

Wu, Kuang Ch'ing. (1943), "Ming Printing and Printers," *Harvard Journal of Asiatic Studies* (February): 203-260.

Yan, Hong-sen. (2007), *Reconstruction Designs of Lost Ancient Chinese Machinery*. Dordrecht: Springer.

Yao, Xinzhong. (2003), *Routledge Curzon Encyclopedia of Confucianism: Volume 2, O-Z*. New York: Routledge.

Young, G.L. (1974), "Human ecology as an interdisciplinary concept: A critical inquiry". *Advances in Ecological Research* 8.

Index